Construction Management

Construction Management: Theory and Practice is a comprehensive textbook for budding construction managers. The range of coverage makes the book essential reading for students studying management courses in all construction related disciplines and ideal reading for those with non-cognate degrees studying construction management masters courses, giving them a broad base of understanding about the industry.

Part I outlines the main industry players and their roles in relation to the Construction Manager.

Part II covers management theory, leadership and team working strategies.

Part III details financial aspects including: sources of finance, appraisal and estimating, construction economics, whole life costing and life cycle analysis, bidding and tendering as well as procurement methods, types of contract and project costing.

Part IV covers construction operations management and issues such as supply chain management, health and safety, waste, quality and environmental management.

Part V covers issues such as marketing, strategy, HRM, health, stress and well-being.

Part VI concludes the book with reflections on the future of the industry in relation to the environment and sustainability and the role of the industry and its managers.

The book keeps the discussion of current hot topics such as building information modelling (BIM), sustainability, and health and well-being included throughout and is packed with useful figures, tables and case studies from industry.

Chris March worked for John Laing Construction and later for John Laing Concrete where he became Factory Manager. On entering higher education he worked in both the UK and Hong Kong before joining the University of Salford, UK, becoming Senior Lecturer and then the Dean of the Faculty of the Environment. He is a former winner of the Council for Higher Education Construction Industry Partnership Award for Innovation.

Construction Management

Theory and practice

Chris March

Routledge
Taylor & Francis Group
LONDON AND NEW YORK

First published 2017
by Routledge
2 Park Square, Milton Park, Abingdon, Oxon OX14 4RN

and by Routledge
711 Third Avenue, New York, NY 10017

Routledge is an imprint of the Taylor & Francis Group, an informa business

British Library Cataloguing-in-Publication Data
A catalogue record for this book is available from the British Library

Library of Congress Cataloging in Publication Data
Names: March, Chris, author.
Title: Construction management : an introduction to theory and practice / Chris March.
Description: Abingdon, Oxon : Routledge, 2017. |
Includes bibliographical references and index.
Identifiers: LCCN 2016035350| ISBN 9781138694453 (hardback : alk. paper) | ISBN 9781138694477 (pbk. : alk. paper) | ISBN 9781315528175 (ebook)
Subjects: LCSH: Building—Superintendence. | Construction industry—Management.
Classification: LCC TH438 .M30685 2017 | DDC 624.068—dc23
LC record available at https://lccn.loc.gov/2016035350

ISBN: 978-1-138-69445-3 (hbk)
ISBN: 978-1-138-69447-7 (pbk)
ISBN: 978-1-315-52817-5 (ebk)

Typeset in Sabon
by Florence Production Limited, Stoodleigh, Devon, UK

Printed and bound in Great Britain by
TJ International Ltd, Padstow, Cornwall

Contents

Introduction

This single book replaces a trilogy of construction management books published in 2009. Inevitably, besides bringing the text up to date and adding new thoughts, a difficult decision had to be made as to what should be condensed or removed. The main omissions are work-study, an analysis of the clauses in ISO 9001 and some of the discussions on communications.

The original intention of the trilogy was to write it with construction management undergraduate students in mind, but as the project developed it became clear that much of the subject matter was appropriate for all the construction disciplines. In more recent times the industry, being undersupplied with good construction graduates, has turned to recruiting non-cognate degrees holders and many of these will be studying Master's courses in construction management. This text is also appropriate for them.

Rather than having a large number of references and bibliographies at the end of each chapter, the author has limited these to a few well-established texts, some referenced in more than one chapter, so that the reader is directed to only a few, if wishing to read further and in more depth on the subject. The chapters vary in length considerably, depending upon the amount of information the author believes is relevant at this level.

The book is divided into six parts as outlined below. The reader needs to be aware that some subjects such as risk assessment, welfare, sustainability and environment are relevant to more than one topic and will be found in more than one chapter.

Part I

There are many different professions in which people are involved in a major construction process, besides the clients, the designers and the contractors. This part looks at the key players and gives a brief description of their roles.

Part II

The author believes that to develop as a manager it is essential to have an understanding of the development of management thought. This is why this is placed at the beginning. The next requirement is to understand what makes a leader, the types there are and what makes good teams function. It is argued that only when the reader has a basis of this knowledge can one begin to consider how to manage. In summary: what makes people tick, how to lead them and make groups work effectively.

Part III

This part takes the reader through the cost implications from when a client first contemplates a new project, through the design and construction processes in accordance with the Royal Institute of Chartered Surveyors (RICS) New Rules of Measurement 2 (NRM2).

Part IV

This part is concerned with the on site activities looking at the thought processes required for setting up the site; planning the work; looking after the health, safety and well-being of all those either working on, visiting or passing by the site; waste management; managing the construction materials from when they arrive on site to being incorporated into the building; managing the supply chain; and quality assurance

Part V

Supporting the site activity is the running of the business as a whole. This part looks at the way companies are organised, the way they strategically plan for the future; how to market their services and mange their human resources. Risk is a theme that runs throughout the text and has been discussed where appropriate in previous sections such as in health and safety. A chapter has been devoted to the subject in this section in particular to summarise the risks in all activities of the company's business.

Part VI

Environment and sustainability are implicit in much of the text such as in life cycle costing and waste management. However, the author believes that the reader should stand back from the day-to-day issues and consider where they and the industry fits into what is an extremely complicated, yet crucial, global problem. This part attempts to define sustainability, introduces the reader to the relationship between technological and social economic solutions and gives guidance on how to break down the issues affecting construction into manageable and appropriate lots.

I would like to acknowledge the contribution made by Mark Shepherd, Manchester University, for Chapter 6; Professor David Eaton for his advice on all matters pertaining to finance; Peter Roberts, Director of Peroman, on operation management topics; Professor Steve Curwell for his help on Chapter 31; and Duncan Cartlidge for the work he carried out in bringing Chapters 12 and 13 up to date to include NRM2.

Finally, my wife Margaret, who had to suffer many hours on her own while I locked myself away in the study but never ceased to give her support and encouragement throughout.

Part I

The players

1 The main participators

Before considering the development, design and construction process, it is important to realise the wide range of different people who can become in*f*volved. Not all those described will be involved as it depends upon the type and scale of the development. In certain instances some of those identified are described in more detail than others, especially if their role extends into the construction phase as well as the development. There can also be overlap in the roles, whereby a developer could employ several specialists, or for smaller development projects, go to a more generalist service provider. A large developer might have some of the personnel employed in its own organisation supplemented by outside specialists as required.

1.1 Landowners

All developments need either virgin, previously used or reclaimed land, all of which will be owned by an individual, group of people, corporate body or public authority. Without their permission, unless compulsorily purchased, a development cannot take place. Landowners may either initiate the development or be approached by others to release land for the development to take place.

1.1.1 Traditional landowners

These include:

- the Church
- the landed aristocracy
- the landed gentry
- the Crown Estate.

Jointly they own a significant amount of the national land bank. For example the Duchy of Cornwall comprises some 57,000 hectares, the late Duke of Westminster owns an even greater amount of land including 120 hectares of Mayfair and Belgravia in London. The Church owns 5000 hectares of agricultural land, some 16,000 churches, of which 12,000 are listed as being of special architectural or historic interest; some of these are coming onto the market as 'redundant churches'. The Crown Estate is part of the hereditary possessions of the Sovereign from which the profit is paid into the Exchequer. It incorporates an urban estate that includes significant London holdings in Regent Street, Regent's Park and St James's, as well as just under 5000 hectares of agricultural land and extensive UK marine assets.

Generally, in these cases, the motives for ownership are more than just capital return and include social, political and ideological reasons often based on tradition and heritage.

1.1.2 Industrial and commercial landowners

These include:

- farmers
- manufacturers
- industrialists
- extractive industries
- retailers and a variety of other service providers such as banks
- public authorities such as central, local and nationalised industries that also own land.

The value of the land is usually incidental to their main purpose. Primarily, farmers use the land to produce crops or rear livestock. While the initial purchase price of the land is a contribution to the overheads the farmer incurs, the fact that the land may be an appreciating asset, doesn't affect the price of the produce as this is determined by others and only comes into play if the farm is sold. The only other occasion when the value of the land is likely to be taken into account is if the farmer wishes to borrow money against it. The same logic can similarly apply to the others listed above.

While the land is fundamental to their business, it is not perceived as an asset in the same way as those who purchase land with a view to capitalise upon it. This means that the economic reasons for releasing land are not always obvious to them. As a result, they may be reluctant to cooperate with a potential developer unless a good case can be made to them. An example of how the land value could be of importance is, if it is of high value it may be profitable to sell the land and with the monies released relocate elsewhere and build up-to-date, purpose made, energy efficient premises, thereby reducing the overheads and becoming more profitable.

1.1.3 Financial landowners

These people and groups purchase land, with or without property built on it, as an investment and therefore are more likely to cooperate with a developer and include:

- builders
- property developers
- pension funds
- insurance companies.

The latter two have invested from between 5 and 15 per cent of their funds in property investment over the last 20 years and have a considerable portfolio of property.

Those purchasing land are playing a long-term game anticipating healthy profit in the future, knowing that the land, even if not appreciating in the short term, will almost certainly become saleable. Land banks are crucial for the long-term viability of residential developers if they wish to maintain a flow of completed properties. This can include green belt land, held in hope the Secretary of State will release it for building as a result of pressure in the housing market. It can be derelict areas in inner cities where it can be seen there is a potential for future development, and many areas in inner cities that are being regenerated due to the demand for having accommodation closer to one's place of work. In essence the purchaser is looking for potential opportunity.

1.2 Private developers

Private sector developers come in a variety of sizes, from one person to multinationals that may be publicly quoted on the Stock Exchange. Their purpose is to make a direct financial profit from the process of development. They operate as either traders who sell the property they develop or as investors who make their money from renting the property and from an appreciating asset. They may also develop the property for their own occupation and use. Examples of the latter include the banks and building societies that have a considerable property portfolio.

Traditionally those involved in residential development were usually traders who build to sell. However, in recent years housing associations have had a significant impact on the housing stock and they build with a view to let. The other major change in development has been with the advent of the Public Finance Initiatives (PFI). These fall into two main categories. The first is where a new building with a specific purpose in mind such as a hospital or school is built and the developer then rents the property back to the user and contracts to maintain the building over a period of years, usually 25 to 35 years. This form of maintenance contract is usually referred to as Facilities Management. The second is where an existing building is to be refurbished, and the whole or part is to be handed back to the client supported with a facilities management contract. The Treasury Building in London is an example of being able to use half of the floor area for development other than Treasury business. In essence the contract was to find alternative accommodation for the Treasury employees during the refurbishment, redesign the interior in a more effective way so that on returning they could function in half the footprint of the building and then the developer could use the other half for other purposes that were acceptable to the client. The developer could then make a profit from this development as well as from the facilities management provided for the client.

In the total development process it is the developers and those that provide the finance for the project who take the greatest risk. Post-war history is littered with examples of those who have made great profits when the rents and property values have risen, and failures when the bottom has dropped out of the market. It is extremely difficult to predict market trends with certainty as fluctuations in the world economy can occur as a result of unpredictable events. The high increase in oil prices in 1973, the miners strike during the Thatcher era, the dismantling of the Berlin Wall and the unification of Germany, the Exchange Rate Mechanism crisis, September the 11th, and a deterioration or improvement in major economies such as the American, Chinese and Japanese economies, are examples of such events.

1.3 Public sector and government agencies

There was a time when central government carried out or supported the local authorities in doing a considerable amount of direct development; Socialist administrations more than Tory, but in recent years, the policies of both parties have converged to one of carrying out little direct development. Increasingly, if Government requires work to be done, it uses the PFI and other variants (see 11.7) of this, using finance other than monies direct from the Treasury.

Local government can become involved in local developments within their boundaries, but it depends much on the community's interests. If the area is derelict or run down, developments that improve the local amenities are likely to be welcome, whereas if the area is highly sought after, the community may not wish development to occur as they may consider this to be a loss of amenity.

Some interesting work has been done in partnership with the private sector. An example of this is the Barnsley Metropolitan Borough Council and Costain Construction partnership where the local authority enlisted the help of the construction company to plan and improve the area as well as seeking financial support and investment, for which they obtained a fee. While the company was also able to tender for construction work they were not always successful as systems were put in place to ensure the local authority got best value for money (see 11.6).

The more active authorities can act as a catalyst by supplying the land, which they lease to the property developer or by investing in the infrastructure. Salford Quays was an example of the latter. The authority with the aid of central funding carried out the refurbishing of the harbour structures themselves and installed good quality paved roads to attract developers and at the same time set a standard of quality that would encourage the developers to follow. For example, the materials used for the car park areas provided by the developer are generally of similar quality to those used by the local authority.

In attempting to foster development, especially in inner cities, government have produced urban regeneration initiatives administered through various government agencies such as:

- Urban Development Corporations
- Homes and Community Agency
- Scottish Enterprise.

These groups see themselves as assisting developers with land assembly, site reclamation, provision of infrastructure and financial grants.

Other initiatives aimed at attracting occupiers with financial incentives include Enterprise Zones and Regional Selective Assistance. There are also Regeneration Zones specified by the EC, which also give financial incentive for development.

1.4 Planners

The most significant change in the planning system in the UK occurred with the enactment of the Town and Country Planning Act 1947, from which all subsequent legislation has been developed.

Both politicians and professional planners control planning. The former in both local and central government are responsible for approving or refusing development plans in accordance with the policy laid down by them. They will be advised and guided by the professional planners who also administer the system on behalf of the politicians.

This normally means that unless the development is controversial, the politicians will accept the planner's recommendations. The basis for determining permission is laid down in statute and a variety of central government policy guidance notes. Local authorities are obliged to work to these, but will determine their own local policy normally using the local development plans that will have been approved by central government.

There are two main reasons for planning. The first is to prevent development that is undesirable in so far as it is, for example, out of character with its surroundings aesthetically, is for a use not compatible with other users in the area, such as positioning industrial premises in the middle of a residential area, or places excessive demands on the existing infrastructure. The second is to encourage development where it is appropriate. For example, if an area is declining it is important to encourage investment to breathe new life into the area either by improving the overall amenities or by attracting new industry and commerce.

In practice there are a lot of opportunities for interpretation of the local plans and developers may employ their own planning consultants to assist in negotiating with the planners. There is also the issue of 'planning gain'. Developers may be expected to, or offer to, provide something extra as part of the agreement. This may be an improvement in the local roads, landscaping or other amenity for the local community.

There is no national standard as to what development is acceptable. This depends upon the local area and whether or not there is a need to encourage development. Clearly if the authority wishes to attract investment, it is likely to require lower standards, and impose higher standards if it wishes to slow down or deter development.

There has been an increasing trend by developers to use the planning appeal procedures as a result of conflict between the developers and the planners and sometimes because of intervention by the politicians who have ignored their planners' advice.

Planning for some of the larger government-driven contracts infrastructure projects, such as Crossrail in London and the High Speed 2 (HS2) railway, is complex as ways to speed up the planning and development processes have been introduced.

1.5 Financial institutions

Financial institutions usually refers to pension funds, insurance companies, clearing and merchant banks (both UK and foreign) and the World Bank. Building societies also provide finance although most of their funding is for the private residential market.

The types of financial institutions and how they provide finance is discussed in more detail in Chapter 5 Sources of Finance.

1.6 Agents

Commercial or estate agents (in the case of residential) may see a potential opportunity for development and bring together the key players in the process. Increasingly, they are more likely to be commissioned by the developer to find suitable sites or properties for development, redevelopment or refurbishment.

The knowledge these players bring to the process at the early stages is an awareness of the current market – for example, what the current demand is, and what current rents and prices are – essential to assess the project viability. Linked to this is advice on potential occupiers, the mix of tenants and their likely requirements in terms of design, layout and space. However, it should be noted this is a specialist area and if the agent does specialise the developer would commission market research independently for more detailed information. They will also be aware of the planning implications of the project in the location being considered and be able to advise. Depending upon the experience of the client they will assist or advise on how to obtain finance for the project.

They are commonly employed as the selling or letting agents. Here their role is to advertise, negotiate with potential future occupants or owners, and in the case of tenants, manage the tenant agreements on behalf of the developer. This entails the collection and processing of rents, ensuring the property is being maintained as agreed and administrating any dilapidations schedules at the end of the tenancy.

They can also be employed by the landowner to protect their interests when approached by the developer. The landowner may not be aware of the development potential and could receive a lower price than it is worth. If on the other hand the landowner wishes to sell land or property they may not know its likely value and will employ an agent to advise. A simple example of this is when a homeowner wishes to sell their property and employs an estate agent to value the property and act as agent in the selling process.

1.7 Building contractors

Traditionally, building contractors were brought in to construct the building after the project had been financed and designed. However, in recent years many different forms of procurement and contract forms have been developed and used. Typical examples are design and build, management contracting, construction management, PFI and its derivatives (see 11.7). Developers now often engage a contractor, or project manager, to manage the whole process including financing, the design and building of the project, and the facilities management. For example, the last two developers competing for the Treasury Building PFI contract employed MACE and Bovis Lend Lease as the project managers during the bidding process.

Developers, especially retailers such as Marks & Spencer and Tesco, sometimes develop partnerships with contractors to build their outlets. There are advantages in doing this for both parties. The developer can negotiate competitive rates with the contractor and use their expertise to improve the performance of the design and specification. The contractor has guaranteed work over a period of time and can designate certain key personnel to these projects because of the continuity of work, which in turn benefits the developer. The contractor can develop partnerships with subcontractors and suppliers that also can produce similar benefits and again the developer can benefit as a result. In essence it enhances the possibility of good supply chain management (see Chapter 24).

Contractors can also be the developers, especially in the provision of residential accommodation. Some have built up a significant land bank over many decades that they release for their own speculative development. Such organisations will also purchase run-down premises, notably in inner cities, and either upgrade them or convert them into private residential developments.

1.8 Project design and development team

Once the decision is made to seriously consider putting a development project into practice a team of experts needs to be put together. Most developers do not have the skills or expertise in-house to carry out major development so they employ professionals/specialists/experts to advise them at the various stages of the process. These may include all or some of the following:

1.8.1 The architect

Anybody can call themselves a builder, surveyor or engineer, unless prefaced by the term Chartered, but in terms of designing buildings the title 'architect' is protected by law. They have to be qualified and registered with the Architects Registration Board.

Traditionally, the architect was responsible for the design of the building, obtaining planning permission and supervising the construction operation on behalf of the client. However, while this still occurs, on most major projects management of the whole process is carried out by others, the role of the architect being confined to the design of the building only. On some projects specialist space planners may support their work (1.8.2).

It is a popular misconception to believe the architect's function is solely to produce an aesthetically pleasing building or a 'work of art'. While this is in part their role, the talent they bring to the process is considerably more than this. There are instances where it is important to make a dramatic statement such as The Lowry, the millennium landmark project for the arts at Salford Quays, Canary Wharf in London and The Guggenheim Museum, Bilbao. Normally the design of the building has to blend in with adjacent buildings or those in the locality using appropriate building materials, style, form and colour.

However, hidden behind the building's façade is where the architect's other key contribution can be found: making the building work for the owners and occupiers. The first stage of this process is the designing of the footprint of the building and its orientation. Fundamental strategic decisions are made at this stage, especially those concerned with environment, in particular the amount of energy required to run the building. If as much natural daylight and ventilation are to be used as possible, this limits the depths of rooms measured from the outer wall. The orientation of the building can affect such issues as solar gain or noise pollution. The way in which the architect creates working/living and circulation spaces within this footprint is of crucial importance if the building is to function properly and efficiently. They are also involved in the specification and selection of components and materials used for the construction of the building.

1.8.2 Space planner

Space planning is one of the roles architects provide as part of the overall design service. However, specialist planners are often employed, especially on large projects, to organise spaces, such as open-plan offices, to accommodate the client's needs. This will often involve establishing the client's requirements either through discussion or to an already prescribed brief. They may also, in the process of their work, demonstrate the space available is unsatisfactory for the client's work practices.

1.8.3 Structural engineer

The prime role of the structural engineer is to ensure the building will not collapse or deform significantly when in use. Secondary roles are to achieve this as economically as possible, to provide a solution that satisfies the architect's design needs, and to accommodate the building services.

The engineer has to take into account the final use of the building and the resultant loads that will be placed on the structure through to the foundations. The architect will have produced floor plans and elevations as well as information on suggested floor-to-floor and floor-to-ceiling dimensions. The engineer has then to design solutions that satisfy these requirements if possible. If not, then the architect may well have to reconsider their design and agree a compromise. The materials for the structural frame may have also been determined for a variety of reasons such as speed of construction, the amount of repetition and client's specifications.

A considerable amount of money is spent on the foundations. However, if the project is being fast tracked, much of the information may not be available at the time of the design, resulting in the engineer being cautious and producing a more expensive solution than is finally necessary.

During the construction phase, the structural engineer will inspect and monitor the project to ensure that the structural work complies with the specifications stipulated. This will be done with close liaison with the contractor's engineering staff who will also discuss any structural design issues that occur as the project progresses.

1.8.4 Building services engineer

The costs of the building services as a percentage of the final cost of the building work can vary from 25 per cent to in excess of 60 per cent depending upon the nature of the building's use. Mostly hidden from view it comprises a labyrinth of pipes, ducts and wires travelling both vertically and horizontally throughout the building delivering services that include water, power, lighting, communication, heating and ventilation. Much of these will require plant of some significant size and weight, which has to be supported structurally as well as given a place in the building. Added to these could be means of transporting people and goods such as escalators, travellators and lifts.

Clearly, the input of the building services engineer into an integrated design process is crucial if a satisfactory solution for the functioning of the completed building, its maintenance and its buildability during construction is to occur. They often are able to contribute to the environmental debate especially in terms of energy consumption by proffering ideas and suggestions on how strategic and detailed design decisions can be modified, reducing the building's reliance on certain services. This could result in cost savings on the size and capacity of plant not to mention reducing the weight on the structure and possibly the ceiling-to-floor dimensions in which ducts have to be accommodated.

The selection of plant and equipment to run the building services is also important, as they will have to be replaced and maintained during the life of the building. The cheapest may not be the most economical, looking at the long-term financial implications as it may break down more often, require more frequent servicing and have to be replaced in a shorter time than a more expensive equivalent (see Chapter 9).

1.8.5 Quantity surveyor

The quantity surveyor (QS) is concerned with the cost of the construction of the project and these days also needs to consider the future life and cost of maintaining the building. The developer needs an indication very early on what the building on completion will cost. This allows for a decision to be made as whether or not to progress with the project, either in its current form or in some modified way. In the latter case, this may require producing a less ambitious solution or, more likely, cutting back on the standard of finishing and site works to make the necessary savings. The amount of information available will determine the technique used for producing the cost plan. Using these methods permits the surveyor to produce costs of alternative design solutions relatively rapidly.

A key role for the QS is to draw up the contract documents that will be used for the selection of a contractor and subsequent cost control of the project during construction. Traditionally this was the bill of quantities and one of the standard forms of contract such as JCT Standard Building Contract. However, fewer and fewer contracts now use the bill of quantities and other less cumbersome forms of control have been developed. This has enabled the time from initial design to contractor selection to be reduced considerably.

Finally in liaison with the contractor's QS, the amount paid by the client to the contractor for work done each month is agreed, any variations from the original design and assumptions are validated and the final account settled at the end of the contract. While not part of the design and development team, the contractor's quantity surveyor is discussed here because they represent the contractor's close link with the process during construction.

The contractor's QS is there to look after the contractor's financial interest as the contract progresses and is completed. As indicated before, each month they agree the measure of work that has been done and the amount of materials stored on site for use in the contract so the contractor can obtain payment for the work done each month. This is important, as it is from this money the contractor is able to pay their subcontractors and suppliers. During the contract they record and notify any variations to the work that has been tendered and any legitimate delays beyond the contractor's control, and do similarly with the subcontractors and suppliers who themselves may have claims.

They will also advise the contractor's management about any contractual implications in the tender documents during the estimating process. Finally they work to ensure that at the end of the contract the contractor is paid for everything they are entitled to.

1.8.6 Environmental consultant

Any development, whether it is a refurbishment, a new building or infrastructure, will have an impact on the environment both during the construction and during the life of the building. Some contracts such as Mass Transit Railway in Hong Kong and the proposed HS2 rail link, not only take a long time to complete but also extend through large tracks of urban communities. The impact on the community therefore can be considerable, not just during the construction process, but long afterwards as in, for example, the levels of noise generated through a 24-hour operating day.

The environmental consultant carries out environmental impact assessments and advises on measures that can be taken to alleviate the problems.

Environmental consultants need to be brought in at the earliest stage of the design process to impact on the strategic design decisions such as the footprint and orientation of the building, otherwise their contribution to the environmental debate will be minimal. As the design process moves to the detail, the consultant advises on the selection of materials. Issues for consideration are the embodied energy of the product or material, pollution implications during manufacture and eventual disposal or demolition, whether it is sourced from sustainable sources, its long-term durability and the environmental implications of maintenance such as cleaning – all of these in the context of ensuring the material or component satisfies the technical criteria of the design for its application and is within sensible cost parameters (see Chapter 9).

1.8.7 Building surveyor

Traditionally, building surveyors were concerned with the state of the existing building, analysis of the causes of the defects and producing solutions to rectify the situation. This has placed them in an ideal position to be involved in development of existing property, be it maintenance or refurbishment. When a property is being considered for refurbishment, which may also mean a change in use, such as Victorian warehouses being converted into residential units, the property has to be initially surveyed to establish its condition but also to see whether it is practical to convert it to its new use. Design solutions can then be drawn up for consideration and the process then continued in a similar manner to any new-build development. Resulting from their expertise, building surveyors are sometimes employed as the project manager for the whole project.

1.8.8 Lawyers

Lawyers are involved at various stages of the project, and on major developments this may involve using specialists from different practices. This starts with the acquisition of the land and or existing property that may require negotiation with several interested parties if the development is to cover several landowners, some of whom may only wish to lease, whereas others may wish to sell.

If there is a need to go through the planning appeal process, solicitors and barristers may be used to present the developer's case at the enquiry. Often there has to be put in place legal agreements with those funding the development and contracts drawn up between the developer and the professionals concerned with the design and management of the project. These contracts may be significantly different from the standard contracts normally in use such as the JCT form. It also may be necessary, for example in the case of Government work, to ensure that all those who are given access to certain material sign the Official Secrets Act.

Finally contracts have to be drawn up between the developer and those occupiers who are either leasing or purchasing part or the entirety of the development.

1.8.9 Accountants

Governments change the tax regimes and VAT regulations, usually in the annual Budget and subsequent announcements, on a regular basis. Development funding instigated by Government and the European Commission also change from time to time. Many of these changes are very complex and need specialist accountants to understand the ramifications. They also can be used to advise on financing agreements, methods of raising finance, as well as advising on the structure of partnerships.

1.8.10 Valuation surveyors

Valuation surveyors are usually brought in very early in the process to assess the viability of a project on the proposed land. They may be used to seek out different potential sites for the development and conduct valuation surveys to establish the most suitable for the developer's needs. They will provide such information as the cost of the land and the likely returns the developer could be expected to make by leasing or selling. They would be expected to provide knowledge on taxation. Much of what they do can be provided in more specialist forms by some of the participants previously described, but they provide an extremely useful service in their own right. They can be employed during the formative design process to value different solutions and uses for the proposed development.

1.8.11 Facilities management consultant

There are many large projects, especially PFI type, that require the property to be maintained after it has been built or refurbished, for a period up to 25 or 35 years as part of the overall contract. The provision can include all types of maintenance such as a cleaning service, repairs and planned maintenance. The latter involves both the maintenance of existing plant and the fabric of the building as well as replacing components at the end of their useful life. It would cover day-to-day repairs such as damage and failure from light bulbs to lift breakdowns. Checking of fire alarms and systems would be included and in certain circumstances security, car parking and the postal services within and to and from the building. The cost implications of providing this all-inclusive service are immense and it is essential that the correct advice be given at the design stage as incorrect decisions made then have a major knock-on effect in the future (see Chapter 9).

1.8.12 Project management

Finally someone has to manage the process of the contributors to this section. This may be the developer themselves or they may delegate this to one of the major players. Traditionally this would have been the architect, but in recent years it is more likely that a management contractor or project management organisation would be employed. For a refurbishment project a building surveyor could be employed.

1.9 Other stakeholders

Objection to development has almost become part of our national life. They can be issues of relatively minor significance, except to the objectors, to issues of regional and national importance. Opposition can be costly to the developer, because of delays incurred, the cost of planning enquiries, increases in standards of materials and components imposed and, at worst, the project having to be abandoned. However, with foresight much of this can be anticipated and contained.

There are several categories of objectors:

- self-interested neighbours of the proposed development often referred to as NIMBYS (not in my back yard). They can become well organised and significantly obstruct the progress of the development proposals and have been known to successfully cause delay, as well as cause the project to be abandoned;
- New Age protestors or eco-warriors who come from outside the local area. These can be very persistent in their protest and take illegal action by occupying the site, putting themselves at danger by tunnelling, tree dwelling and moving to anywhere where it is difficult and dangerous for them to be evicted. They are, of course, seeking high levels of publicity and aim to draw attention to the environmental issues that concern them. Note it is usually the contractor who is only the artisan in the process who usually gets the bad publicity rather than the developer or planner. It is an interesting discussion point to formulate ways of dealing with these situations if they arise;
- the professional, permanent bodies at local and national level such as the Victorian and Georgian societies, Friends of the Earth and Greenpeace. There are also the official quangos such as Natural England, Scottish Natural Heritage, Natural Resource Wales and Historic England. All of these organisations are well organised, with a high level of devoted supporters and the capability, because of their knowledge and experience of the planning and development processes, to cause problems especially to an inadequately thought-out project that has neglected the issues these kinds of objectors are concerned with.

While not objectors as meant in this section, it should not be forgotten that politicians can overturn a development project if they believe it not to be in the public interest or could affect their likelihood of re-election, and at national level the Secretary of State has powers to overrule either the development proposal or those who are objecting to it. The planners employed by local government can recommend a development should not take place although they can be overruled by the politicians.

1.10 Occupiers

Often unknown at the beginning of the development process, their needs are not always fully researched, but should be. Development is often tuned to suit the needs of the financial institutions rather than the occupier. There is a growing realisation this has to change and for example in shopping malls the development may well not commence until some of the bigger retailers are signed up. If the occupier is known at the beginning of the design process then they can become involved in some of the key decisions so their requirements can be met.

Whereas retailers often see the space around that which they occupy as part of the shopping experience, commercial occupiers are more likely to regard the building as an overhead incidental to their business and do not always appreciate the importance of the building's design, layout and provision of services in contributing to the successful running of their operation, especially in terms of accommodating change in the way they work.

Occupiers are demanding much more flexible lease arrangements to enable them to react to changing needs. The financial institutions would prefer long-term lease agreements for obvious reasons, but are having to come to terms with the changing market place.

Part II

So you want to be a leader?

Management theory, leadership and team building

2 Pioneers of management theory

2.1 Definitions

There are terms regularly used in management 'speak' so the author has given his own definitions based upon his own readings and experiences.

Management: each situation is almost invariably unique. Management should assess all the relevant factors available at that moment in time and make a decision concerning the problem in question.

Administration: administrators work primarily to a set of rules and decisions are made based on these guidelines. There can be some flexibility in 'bending the rules' but usually administrators are concerned that a new precedent might be set if the rules are not adhered to. It should be noted that in most organisations, especially medium and large, both management and administration functions exist side by side. For example, wages and salaries would be seen as an administrative function.

Accountability: an employee is accountable for the actions they take and to their immediate supervisor.

Responsibility: an employee is responsible for the actions they take, but is also responsible for the actions of those reporting directly to them. Responsibility registers highly in employee job satisfaction.

Delegation: this is probably the hardest act to carry out by a manager. In essence it means delegating part of one's responsibility to a subordinate and allowing that person to have complete responsibility for the work delegated. Many managers fall down on this by either delegating roles and jobs they do not wish to carry out, or worse still, continue to take actions and decisions relating to the tasks delegated. This later failure is often due to their lack of confidence in others. It can be extremely frustrating for the subordinate!

2.2 Introduction

There are far too many contributors to management theory to cover them all, so those selected are people who have made, in the author's opinion, significant contributions and to whom he can closely relate, relative to his own experiences. This is important for the reader to understand, as there are many others who have contributed equally and may well be more in tune with the experiences of the reader. Of those referred to, the author has also 'cherry picked' the topics that have influenced him and in no way represent a complete coverage of all the respective writers' work. Hence, when browsing through the many management texts on offer, the reader should widen their knowledge beyond the content of this book.

2.3 Development of management theory

Serious management thinking has been developing for just over 100 years. It can be broadly classified into four categories: classical, human relations, systems and contingency, which have occurred sequentially as approximately demonstrated in Figure 2.1. It is difficult to indicate precisely when one finishes and another starts. Pre-classical management was based primarily on trial and error, but the age of the classical school saw the need to apply scientific approaches to management, at the exclusion of the human factor. This exclusion was addressed at the next phase when the human element was considered; after the Second World War the thought was redirected to looking at systems and processes. More recently there was awareness that all schools had something to offer and a more holistic approach has been taken.

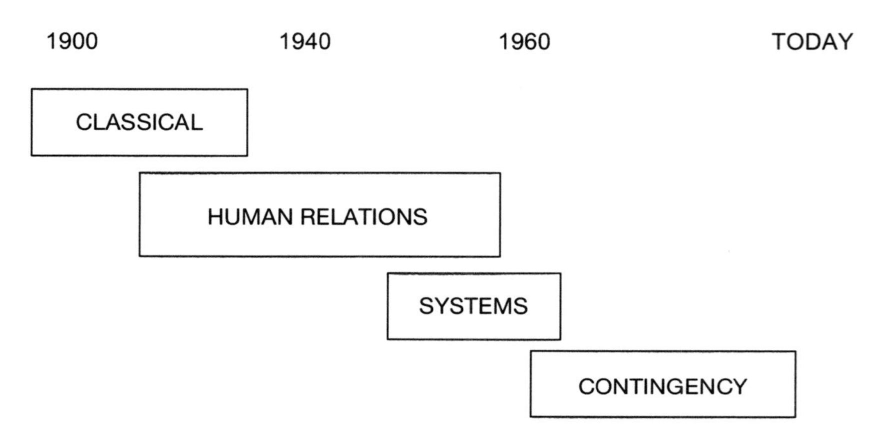

Figure 2.1 Timeline of management schools of theory

2.4 The classical school

It is necessary to place this development of management thought into its historical context. At the end of the nineteenth century labour was cheap, plentiful and considered primarily as a resource, rather than a 'human resource'. If the steel furnaces were not being filled with coal fast enough, you just added extra manpower. Little or no thought was placed on the efficiency of labour. There are three key theories in this school: scientific, administrative and bureaucratic developed primarily by Taylor, Fayol and Weber respectively.

2.4.1 Scientific management – F.W. Taylor (1856–1915)

Taylor joined the Midvale Steel Works, USA, as a machine shop labourer in 1878 and within six years had risen through the ranks to become chief engineer after which he moved to the Bethlehem Steel Works. He observed workers were not working at their full potential and came to the conclusion it was because they were concerned that if they worked flat out, others would lose their jobs. The wage systems in place discouraged higher productivity, as they were often structured so that if productivity increased above the set standard, incentive payments would be cut. Many of the working methods had been handed down from generation to generation of employees without any consideration of how to improve productivity.

He was interested in applying scientific techniques to management, believing this was the way to deal with low productivity, and developed his Four Principles of Scientific Management:

1 Evolve a science to study each element of a person's work to develop the best method disregarding traditional rule of thumb methods.
2 Scientifically select then train, teach and develop the workers using these methods (in the past the workers had chosen their own work and trained themselves).
3 Obtain cooperation with the employees to ensure all the work was carried out to best possible method.
4 Divide work and responsibility so management takes the responsibility for planning work methods, using the scientific principles, and the workers are responsible for carrying out the work to these methods. In the past workers had chosen how to carry out the tasks.

Taylor was interested in measuring the output of operatives to establish the best way to carry out tasks which he called time study, subsequently referred to as time and motion study which became the basis of work-study as known today. He also developed wage incentive schemes to increase productivity.

One of his famous studies was the use of shovels in the Bethlehem steel works. He noted operatives used the same shovel irrespective of the type and weight of the material they were loading. Using scientific measurement he established what mattered was the total weight of both the shovel and the materials, which he determined to be 21 pounds. He was able to increase shovelling productivity almost fourfold as a result.

2.4.2 Administrative management – H. Fayol (1841–1925)

While Taylor was developing his ideas in America, Henry Fayol was revolutionising management thinking in France. He was brought up in an upper-middle-class family and trained as a mining engineer, working his way up to become managing director of Commentry-Fourchampboult-Decazeville, a coal mining and foundry combine. Whereas much of Taylor's work concentrated on the production part of the process, Fayol was interested in management principles from the chief executive's point of view. He was concerned with designing an efficient organisation structure based on administration.

He developed a definition of management based on five functions (note some texts refer to six, separating forecasting and planning) that in essence are still valid. These are:

- forecast and plan
- organise
- command
- coordinate
- control.

Any business needs to examine and forecast what is likely to happen in the future, be it the market place in which one is to sell one's products or services, i.e. the demand side, or on the supply side to establish the availability of resources, notably labour and materials – all of this taking account of economic trends. The company has then to produce a corporate plan that brings together the needs of all sections of the business to meet the corporate objectives, followed by operational plans to assist in executing these needs in practice. These should be flexible enough to take account of changing circumstances and assist in predicting courses of action.

The management needs to provide the material and human resource and build up an organisational structure so the work can be executed to meet the plan. This requires management to stand back and decide upon a structure that permits this to happen. In other words, how the business should be divided into groups of activities or functions, who should be responsible to whom, the number of people one should be responsible for and so on.

Logically following from the above two is the need to have the authority to command so the organisation, now structured, can be made to work to enable optimum return from all employees in meeting the corporative objectives. Command is the relationship between a manager and their immediate subordinates. The quality of it results from the manager's ability to demonstrate knowledge of the business, of their subordinates, the quality of contact with them, and to inspire confidence.

However, this in itself will be insufficient if the interface between the different parts of the organisation is not explored and coordinated. Fayol wrote that coordination is necessary in 'binding together, unifying and harmonizing all activity and effort'. Too often departments work in isolation without understanding or consideration of how their actions impact on another. It is interesting this issue is addressed in ISO9001: 2004 Quality Management Systems Requirements written a century later (see 25.6).

The final element is to control and monitor to check all the previous elements are functioning in line with the plans, commands and instructions.

He further expounded 14 principles as follows:

1 *Division of work*. By performing only one part of the job, a worker can produce more and better work for the same effort. Specialisation is the most efficient way to use human effort. This view was the basis of much of the thinking in mass production, such as cars, where each worker was given a clearly defined task.
2 *Authority and responsibility*. Authority is the right to give orders and obtain obedience, and responsibility is the corollary of authority. This ignores the human element that today might question the competence of the superior, requiring that authority should also be linked to the need to obtain respect. However, at the time the position automatically commanded respect and hence authority.
3 *Discipline*. Obedience to organisational rules is necessary. The best way to have good superiors and clear and fair rules and agreements is to apply sanctions and penalties judiciously. Nobody would disagree with this in principle, but the emphasis is on punishment rather than on congratulations and positive motivation. There is a danger that by taking this too literally there becomes a conflict with 14 below.
4 *Unity of command*. There should be one and only one superior for each individual employee. This on the face of it is correct especially in the context of executive line management, but as can be seen in functional management organisations (see

26.4.2) There are legitimate situations where a person is responsible to two people, one for discipline and performance and the other for the quality of their expertise.

5 *Unity of direction*. All units in the organisation should be moving towards the same objectives through coordinated and focused effort. This would be a valid statement today.
6 *Subordination of individual interest to general interest*. The interests of the organisation should take priority over the interests of an individual employee. This should be at all levels of the organisation. While the concept is valid, there can be conflicts with this and the ambitions of employees who, to get on, can often usurp this position by providing their superior with the views and answers they think they want so as to impress and be noticed, whereas a healthy debate and disagreement may serve the organisation more fruitfully.
7 *Remuneration of employees*. The overall pay and compensation for employees should be fair to both the employees and the organisation. Any organisation needs to balance what it can afford, with an equitable distribution to its employees. There is no perfect system even today as the endless debates over 'fat cats', gender discrimination and pay compatibility demonstrate.
8 *Centralisation*. There should be a balance between subordinates' involvement through decentralisation and managers' retention of final authority through centralisation. Centralisation is always present to a greater or lesser extent depending upon the size of the company and quality of its managers. The larger the organisation, the more likely the lines of communication between the board and the lower parts of the organisation are to become stretched. This makes control more difficult as the further away from the centre they are, the more autonomous the parts can become. How much is permitted should be a function of the capabilities of the personnel.
9 *Scalar chain*. Organisations should have a train of authority and communication that runs from the top to the bottom and should be followed by managers and subordinates. A follow on from centralisation, the longer the chain of the command is, the more difficult it is for the top management to know what is going on at the bottom and they become cut off. On the other hand, if these lines of communication are reduced, the top manager has many more people to control which in turn can cause difficulties. Lateral communication is encouraged, providing the vertical management chain is kept informed.
10 *Order*. People and materials must be in suitable places at the appropriate time for maximum efficiency; that is, a place for everything and everything in its place. No change here – indeed this might be used as the starting point for just-in-time deliveries (see 23.6).
11 *Equity*. Good sense and experience are needed to ensure fairness to all employees who should be treated as equally as possible. This is a law of management that most strive to achieve.
12 *Stability of personnel*. Employee turnover should be minimised to maintain organisational efficiency. There is no argument to be found with this.
13 *Initiative*. Workers should be encouraged to develop and carry out plans for improvement. This is a pre-runner of McGregor's Y theory (see 2.5.5).
14 *Esprit de corps*. Management should promote a team spirit of unity and harmony among employees. To create this unity and harmony requires a deep understanding of motivation theory and the cooperation of all levels of management.

2.4.3 Bureaucratic management – M. Weber (1864–1920)

The German sociologist, Max Weber, believed organisations should operate on a rational basis and not on the arbitrary decisions made by the owners, which were usually based upon nepotism. He believed this to be not only unfair, but also a waste of talent. He advanced the concept of bureaucracy. Today the term bureaucracy has connotations of red tape and excessive numbers of rules and regulations, but Weber used it in the context of the way authority was exercised with organisations. He distinguished power as being the ability to make people obey regardless of their resistance, and authority as commands being obeyed voluntarily. He recognised three pure types of legitimate authority:

1 *Charismatic.* The authority based upon leadership qualities or strength of personality. There are many examples of these throughout history in politics, the military and religion, but especially since the industrial revolution, in industry as well. The problem arises when considering succession, as the children, or the 'groomed' replacement may not have the same charismatic personality. Indeed, in such an event history is littered with examples of rivalry between potential successors.
2 *Traditional.* The authority based upon custom, tradition and precedence. An example of this is heredit when the son and heir takes over, with everybody expecting and assuming this will happen. Another form is patronage, the giving of position for example by the monarch or prime minister.
3 *Rational/legal.* It is called rational because the means are designed to achieve specific goals. Personnel are selected because of their ability and they have a clear understanding of their part in the overall process, each part working as if in a well-oiled machine. The legal authority is based upon their position in the hierarchy at that time using established rules and procedures. The individual only has the authority while in post and ceases to have that specific authority if moved elsewhere in the organisation. Weber believed that a bureaucratic organisation as he defined it was the best.

His organisational structure has the following characteristics:

- tasks are divided into very specialised jobs;
- a rigorous set of rules must be followed to ensure predictability and eliminate uncertainty in task performance;
- there are clear authority–responsibility relationships that must be maintained;
- superiors take an impersonal attitude in dealing with subordinates;
- employment and promotions are based on merit;
- lifelong employment is an accepted fact.

These types of organisations still exist, but tend to be rigid, inflexible and heavily reliant on red tape and it is often difficult to identify clearly the more capable employees.

2.4.4 Others of interest

The reader should also be aware of the work of Henry Gantt (1861–1919) noted for the 'Gantt Chart', these days referred to as the 'Bar Chart', still in use today, and the development of incentive schemes for supervisors that were linked to the output and incentives paid to their workers. Frank Gilbreth (1868–1924) was interested in time and motion studies and efficiency of human movement. He did a considerable amount of work on the bricklaying process and by designing appropriate scaffolding and stipulating the consistency of mortar was able to almost treble output. His wife Lillian (1878–1972) worked with him and was also interested in the human aspects of work such as the selection, placement and training of personnel.

2.5 Human relations school (behavioural sciences)

This school of thought emerged as workers became more organised and managers sought to achieve increased productivity and man-management. It was in part a reaction to the rigidity of the classical school and an awareness of manpower being a more complex resource than had previously been considered. The key theories are group work, democratic decisions, motivation and participation.

There is often considerable overlap between these so it is not proposed to discuss them individually, but to look at the contributions that various pioneers made and to allow the reader to distinguish accordingly.

2.5.1 Elton Mayo (1880–1949)

Elton Mayo was an Australian academic who spent most of his working life at Harvard University and is credited by most as being the founder of both the human relations movement and industrial sociology. The behavioural approach is generally considered to have started with the experiments carried out at the Hawthorne plant of Western Electric Company (1924, 1927–1932). These are referred to as the 'Hawthorne Studies'. In the first studies, called the 'Illumination studies', the control group worked with constant light, whereas the experimental group had the lighting steadily reduced until such time as the workers in the group complained there was inadequate light to work. At this point output began to decline with the experimental group, but surprisingly up until then production had risen almost identically in both groups as demonstrated in Figure 2.2. It would be expected the control group's output would remain constant and the experimental group's would decrease, but in fact both groups' productivity increased. The researchers deduced that something other than the levels of lighting was affecting performance.

This unexpected result needed further resolution and between 1927 and 1932 Mayo led a team of researchers to explore the matter further in what was to become known as the Relay Assembly Test Room experiment. A group of six women employed to assemble telephone relays were separated from the rest of the workforce so their output and morale could be observed when changes were made to the working arrangements. The normal working week at the start of the experiment was a 48-hour week with no rest breaks and included working Saturdays. Over a period of five years the following changes were made:

- A special group incentive scheme was introduced. Prior to this they had been grouped together with a hundred other operatives for incentive purposes.
- Two five-minute rest periods were introduced, one in the morning the other in the afternoon.

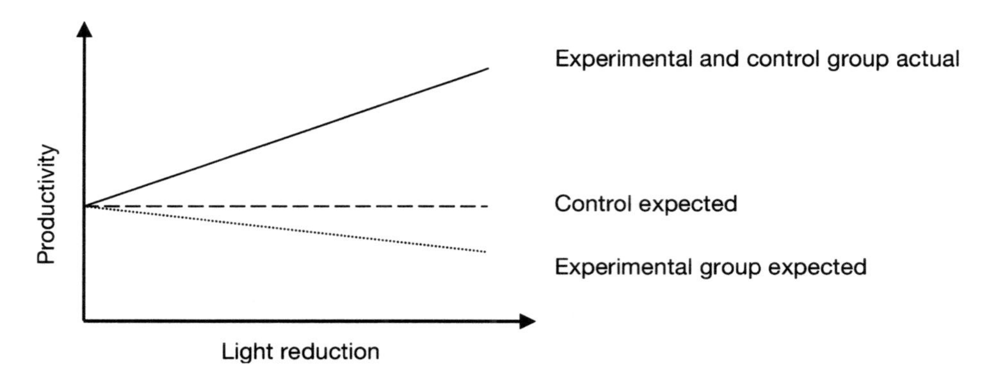

Figure 2.2 Hawthorne illumination studies

- These rest periods were increased to ten minutes.
- Six five-minute rest periods were introduced, but the women complained their work patterns were broken up by so many breaks.
- The two ten-minute breaks were introduced and during the first break the company provided hot food at no charge.
- They were allowed to finish work at 4.30 instead of 5.00 pm.
- This was then brought forward to 4.00 pm.

In all cases output increased except when the six five-minute breaks were introduced, when output fell slightly, and the 4.00 pm finish when output remained the same. It would not be unreasonable to expect this to have happened as by and large work conditions were being enhanced. Then the researchers removed all the improvements and the women reverted to working a 48-hour week, no breaks or free meals and Saturday working. Output increased further.

At the time this increase was regarded as a mystery and it was only in hindsight it was realised there can be informal organisations within the formal organisation. Further, these informal groups exercise a strong influence over the behaviour of the workers especially if they are given the freedom to take over responsibility for the way they can work within the group, which is what happened at Hawthorne. Here they were being observed by the researchers rather than being supervised as those not in the selected group. In other words, workers should not be seen in isolation but as members of a group, in which case supervisors should realise this and see that the group will, if allowed, have common purposes and objectives.

2.5.2 Mary Parker Follett (1868–1933)

Follett was born in Boston, educated at Harvard and Cambridge and studied political science, history and philosophy, becoming a social worker with particular interests in the workplace. She promoted many ideas, many of which were largely ignored at the time, perhaps because they came from a woman at a time when business was very much a male domain. Her profound thoughts and writing are now accepted as very relevant today and it is perhaps significant the Japanese hold her in the very highest esteem. She believed organisation would only work well as a whole if all the parts worked together to meet the company objectives. Her views on conflict are interesting and underpin much of her writings. She believed conflict could not be avoided, but should be used to work for us. Indeed differences could be turned to become a positive asset in an organisation. She advocated that one should not ask who or what is right, but rather to assume both sides are right, viewing the issues from different standpoints having asked different questions. An industrial dispute is an obvious example of this, each side placing a different emphasis or spin on the issue. A dismissed employee will often believe they have been unfairly dismissed whereas the employer perceives the 'facts' differently. The resolution of conflict is not to submit to, or have a victory over the other, nor to produce a compromise, but to find an 'integration' of interests or joint problem solving. Her belief was that working with someone was better then working over or under someone.

A summary of these thoughts is given in the following quotation from her writings: 'One person should not give orders to another person, but both should agree to take their orders from the situation. If orders are simply part of the situation, the question of someone giving and receiving does not come up.'

She also postulated responsible people must be in direct contact irrespective of their position in the organisation. Horizontal communication is equally important as vertical chains of command, a significant change from the classical school. She went on further to say that people concerned should be involved in policy or decisions when they are being formed and not brought in later. By doing this, morale and motivation will be increased. It was some 60 years later before the construction industry belatedly adopted this philosophy by involving the contractor in design decisions.

2.5.3 Chester Barnard (1886–1961)

President of the New Jersey Bell Telephone Company, Chester Barnard emphasised communication as an important means of achieving goals. He also introduced a new acceptance theory of authority that argues subordinates will only accept orders if they understand them, see they are consistent with the aims of the organisation, are in line with their needs and those of their colleagues, and they are willing and able to comply with them. So, for example, if a system of quality assurance is imposed on staff and they are not advised as to the reasons and its purpose, or they consider it to serve no useful purpose, they will either not use it or alternatively will sign each stage off without giving it the appropriate attention.

He also said that it was no good producing organisation charts if the personnel needed to fill the roles were neither available nor able enough and that if necessary the organisation should be amended to take account of the human resource available.

2.5.4 Abraham Maslow (1908–1970)

A psychologist by background, Maslow eventually became chairman of the psychology department at Brandeis University. Without doubt one of the most commonly referred to theories on motivation is his 'Hierarchy of Needs'. The basis of his theory is threefold:

- that human beings have needs that are never fully satisfied;
- that there is a hierarchy of needs from basic to higher level needs;

- that once one level of need is satisfied the next level needs to be satisfied. However it should be noted the line between the different levels is blurred. You don't wake up one morning and decide you have been promoted to the next level of need.

He also defined five different levels of need:

Physiological and biological needs. These are the basic needs for sustaining human life such as food, water, clothing, shelter, sleep and sexual satisfaction. Examples of this are regularly seen in famine struck areas where those fit enough will fight and scramble to obtain any food available. Once this need has been satisfied they will wait in orderly queues for food to be distributed. This raises them to the second level.

Safety and security needs. These are the needs that occur once the physiological needs are satisfied, referred to as security, order and stability. People generally like order and predictability in their lives, which is probably why most do not become entrepreneurs. They feel safe when, as a child, their parents provide a safe and secure environment in which to live and, as adults, they are secure in the knowledge that their income is sufficient to pay the bills and provide a reasonable standard of living. This level of need might also explain why religion or other philosophy such as communism, acting as a familiar, well-structured and organised environment tends to be more prominent in the poorer communities of the world.

Social and belonging needs. This third level need is sometimes referred to as the love needs but should not be confused with sex needs, which are at first level. It is about being accepted by others, feeling part of a group, enjoying and seeking friendship. The need to belong perhaps best sums up this need. In a developed society this is seen very clearly in the development of children when they reach an age when they move from wishing to be solely dependent upon their family to developing friendships with other children and even at an early age demonstrating group tendencies in fashion and team games. When no longer permitted in the group as a result of bullying or exclusion, the impact on the child's happiness is clear for all to see.

Esteem needs. Once people feel accepted they tend to want to be held in esteem by themselves and others. Maslow stressed this should include esteem for others also. Satisfaction of this need generates such feelings as power, prestige, status and self-confidence. If this need is thwarted, the person enters a feeling of inferiority, lack of confidence and inadequacy and therefore a major lack of motivation. This is important for managers to realise when dealing with their subordinates.

Self-fulfilment and self-actualisation needs. This is regarded as the highest level as, when all the previous needs have been met, people are self-fulfilled and have reached their full potential. At this stage it is likely that people wish to put something back into society also. That is not to say that those at the lower levels would not.

2.5.5 Douglas McGregor (1906–1964)

McGregor spent most of his working life as professor of industrial management at the Massachusetts Institute of Technology. His most famous concepts are known as Theory X and Theory Y. His view was that managers' assumptions on how workers thought and reacted were at one extreme X or at the other Y and they treated their workers accordingly. In practice, managers' assumptions are not always as black and white as X and Y and fall some way between the two. It is important to stress it is what the managers assume and believe rather than what workers actually are.

Theory X

The average human being has an inherent dislike of work and will avoid it if possible. Because of this human characteristic, most people must be coerced, controlled, directed, or threatened with punishment to get them to put forth adequate effort in order to achieve organisational objectives. The average human being prefers to be directed, wishes to avoid responsibility, has relatively little ambition, and wants security above all.

Theory Y

The expenditure of physical and mental effort in work is as natural as play or rest, and the average human being, under proper conditions learns not only to accept but also to seek responsibility. People will exercise self-direction and self-control to achieve objectives to which they are committed. The capacity to exercise a relatively high level of imagination, ingenuity and creativity in the solution of organisational problems is widely, not narrowly, distributed in the population, and the intellectual potentialities of the average human being are only partially utilised under the conditions of modern industrial life.

The more enlightened manager favours the Theory Y approach, which, if they take full advantage of it, can capitalise on the considerable skill and motivation in the workforce.

2.5.6 Frederick Hertzberg (1923–2000)

Hertzberg was the professor of management in Cape Western University where he established the Department of Industrial Mental Health. During this period he was a consultant to many multinational companies such as British Petroleum and General Motors. His main study, the motivation-hygiene theory, was recorded in the *Motivation to Work*, its objective to test the hypothesis that man has two sets of needs, firstly to avoid pain and secondly to grow psychologically.

The motivation-hygiene theory

Two hundred engineers and accountants, who represented a cross-section of Pittsburgh industry, were interviewed. Each was asked about occasions and the type of experience they had experienced at work that had either resulted in a noticeable improvement or reduction in their job satisfaction. They were also asked how strongly they felt about each of these experiences, be they good or bad, and the duration they continued to have feelings about them. The areas covered that generally were satisfiers were achievement, recognition, the job itself, responsibility, whereas company policy and administration, supervision, salary, interpersonal relationships and work conditions were measurements of dissatisfaction.

Examples of the results include that when people are given responsibility they feel very good about it for a reasonable length of time and while not feeling quite as great about achievement, this has a very long-lasting effect. On the other hand, all those administrative rules and regulations cause irritation for some considerable time and lack of recognition gives a strong feeling of dissatisfaction, but over a shorter period of time.

2.5.7 Rensis Likert (1903–1981)

Rensis Likert was an American social psychologist. He was the first Director of the Institute for Social Research at the University of Michigan and is acknowledged as one of the leaders of research into human behaviour in industrial organisations. He wrote on many subjects, but of particular interest are his findings on the way the style of supervision can affect productivity, and his four definitions of management style.

He investigated how employee-centred and job-centred supervisors have an impact on performance. To measure this he used several criteria to evaluate their administrative effectiveness. They included productivity per man-hour, job satisfaction, turnover and absenteeism, costs, scrap loss, and employee and management motivation. He clearly established that supervisors who concentrate on the job itself by breaking down the job into simple clearly defined tasks, determining the method of work, employing and training personnel to do the specific task and checking they are performing to the standards set generally are found to have a lower output rate than those supervisors who are interested in the personnel. In this case they are more interested in their problems and building effective working teams. This means giving the operatives the opportunity to make decisions for themselves, delegating responsibility and leaving them to get on with the job rather than continually monitoring performance.

He identified four management systems that can be found in organisations:

Exploitative–Authoritative. This is when top management is very autocratic, makes all the decisions, motivates using threats and discourages any input from below. Senior management takes most of the responsibility, the lower levels having very little. What communications there are is top down, resulting in little teamwork. Subordinates do not feel free to discuss anything about their work with their manager. This is a management style based on the premise stated in McGregor's X Theory (see 2.5.5).

Benevolent–Authoritative. As with the exploitive style, senior management makes most of the decisions, although in this case some lesser decisions may be made at lower levels. A condescending attitude is usually displayed in communicating with subordinates, resulting in a subservient attitude towards superiors. There may be some minimal flow of ideas from subordinates to managers, but generally there is very little communication between the two. Many of the old family firms adopted this benevolent attitude where members of the family would be referred to by their Christian name always prefixed by Mr. They often felt they had a responsibility for their employees.

Consultative. Although senior management still reserves the right to make decisions and control the business, ideas are sought from below. Usually they only have a certain level of trust in their subordinates and are therefore reluctant to fully delegate to them. As a result there is some two-way communication and some teamwork. While subordinates feel relatively free to discuss things about their job they can become frustrated as often their opinion is ignored if it does not coincide with that of the manager. In many ways this is the worst management style to be a subordinate to.

Participative. This is by far the most satisfactory work environment. In this case senior management has complete confidence in its subordinates and is prepared to fully delegate responsibility to them. This results in higher levels of satisfaction, although sometimes more sleepless nights as a result of the responsibility. Subordinates feel completely free to discuss issues with their managers and equally managers ask for advice and opinion. This develops a strong group and teamwork spirit. In the ultimate case the manager will accept a strong majority view even though it is not the same as his own. In other words the group itself makes the decision. There will of course be occasions when the manager feels so strongly they will overrule the group, but experience indicates the rest of the group will accept this, knowing in the end the manager has to take the responsibility. Generally Likert found that managers adopting this style were the most successful.

2.5.8 Robert Blake and Jane Mouton

Both American psychologists, Blake and Mouton were president and vice-president, respectively, of Scientific Methods Incorporated, providing behavioural consultancy services to industry. They developed their Managerial Grid, which, similar to Likert, combines two fundamental aspects of managerial behaviour, that of concern for people and concern for production. The term production is meant to cover all tasks, be it manufacturing, construction, volume of sales, number of accounts processed and so on. The grid demonstrates there are a variety of combinations for the degree of concern for people and production.

Country club management has a high degree of concern for people, but little concern for production. On the positive side, as with participative style of management (see 2.5.7), personnel are encouraged and supported by management, mistakes made are overlooked on the grounds that the person making them is aware of this, and there is an overall camaraderie that helps the production process. On the downside there is the danger people avoid conflict, and problems are watered down rather than being properly addressed. Any new ideas someone has that might be controversial tend not to be brought forward for discussion as this might cause upset.

Impoverished management, where there is neither concern for people nor production, is a ticket to organisational failure. The supervisor avoids responsibility and blames his superior for any difficult or unpopular directive or tells the superior it is the personnel below him who have made the mistake. On the surface it seems hard to believe this kind of management could exist, but people that have been passed over for promotion can show this tendency either for a short or prolonged period. This is often referred to as 'taking one's bat home'.

Task management focuses primarily on production and has little concern for people; it was defined by Likert as job-centred. This type of manager expects people to do as they are told without question and for programmes to be met on time. In the event of something going wrong they will look for someone to blame. They will not accept any disagreement and it will be perceived as insubordination. It is suggested this style of management was reflected in the struggles between the trade unions and certain companies especially in the 1960s and 1970s, which resulted in strengthening unions and greater industrial unrest.

Team management is focused on both concern for people and production and believes these two requirements are compatible. They believe everybody in the team can contribute in some way to achieving the ultimate goal of higher achievable levels of productivity and in doing so satisfy their own needs. The manager's role is to ensure work is planned or organised, using the expertise of the others in the team, resulting in their full engagement in the process and goals. When conflict does occur, and it almost certainly will, then it should be confronted head on and openly and not seen as a battle of sides, but rather as a mutual problem that can be satisfactorily resolved (see 2.5.2 Parker-Follett). Generally this will be the most effective style.

Dampened pendulum represents what often happens in practice: that managers swing between usually 'country club' and 'task management' styles. They start by being much focused on production and when the levels of discontent and arguments arise, move to concentrating on the concern of people at the expense of production and as this improves move back towards production focus and so on. This produces a satisfactory level of production and generally keeps morale at an acceptable level, but does not reach the full potential available.

2.5.9 Victor H. Vroom (1932–)

Victor Vroom was born in Montreal, Canada. He is Professor of Psychology at Yale School of Management and is a leading authority on the psychological analysis of behaviour in organisations. Among many significant contributions to management thought, he is particularly noted for his Expectancy Theory of Motivation in which he examines why people choose to pursue a particular course of action. He argues there are three main issues people consider before expending sufficient effort to complete the task at the appropriate standard. Figure 2.3 demonstrates the basic components and their interrelationship.

Effort–Performance (E→P) expectancy is a self-assessment of the probability that the efforts put in will realise the required performance level. The assessment of probability will depend upon the person's own belief in their ability to accomplish the task and the availability of resources of any kind it is believed are necessary to carry out the work. As an example, a newly promoted project manager may feel ill-equipped to run a type of project they have had no experience of whereas if the project is similar to others they have worked on before promotion they will feel much more confidence in completing the task satisfactorily. In the first case the E→P expectancy is low and in the latter high. This is only the first stage of the process.

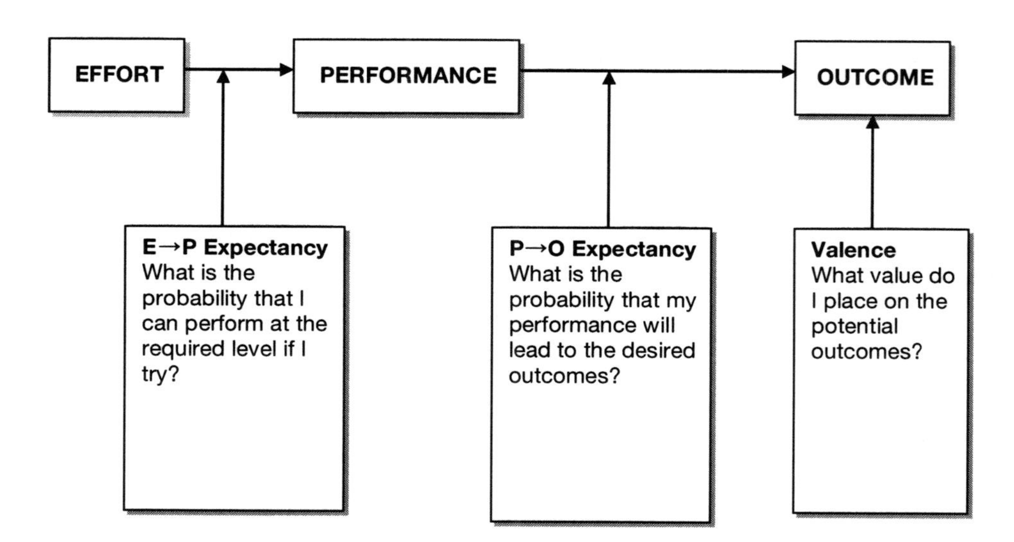

Figure 2.3 Basic components of expectancy theory

Source: Adapted from Bartol and Martin (1994), © McGraw-Hill.

The second stage is known as Performance–Outcome (P→O) expectancy. In this case, it is an assessment of the probability that successful performance will lead to certain outcomes. On the positive side these could include, enhanced prospects of promotion and extra remuneration, known as extrinsic rewards, and job satisfaction, recognition and self-development, known as intrinsic rewards. Against this there is the consideration on the negative side, of perhaps having to work longer hours affecting family life and increasing pressure and stress. The likelihood of these desired outcomes is then assessed against the experience previously had within the organisation such as if hard work is acknowledged and rewarded. If this always happens the P→O will be high, if never, then low, and if sometimes, then the assessment may only be 50/50 or some other proportion.

The third part of the process is called Valence. This is an assessment of the worth or value of the anticipated outcomes. As indicated before there can be positive outcomes such as extra money and negative such as loss of time at home. Depending upon the strength of feeling, these will determine whether the valance is high or low. If it is very positive then motivation will be high. This is a personal judgement. A struggling student will more likely be motivated to take a mundane job in their spare time to make ends meet than if born with a silver spoon in their mouth; whereas a rich person may carry out volunteer work because of the satisfaction obtained as a result of helping others.

The hypothesis is that people will only put effort into a task after considering these three elements: E→P Expectancy, P→O Expectancy and Valence. This is called the expectancy theory:

$$(E \rightarrow P) \times (P \rightarrow O) \times \text{Valence} = \text{Motivation}$$

It can be seen from this formula that if any of the three parts of the process is zero or approaching it, then the motivation will also be zero or close to it. Equally, if they are all high then motivation will be high. The outcome of the formula is not the same for each person and management should be aware of this. Each individual has different perceptions of their own ability and the performance outcomes will vary dependent upon personal circumstances and aspirations. This means that personal knowledge of the employee will help considerably, especially when evolving their personal development plans as well as their motivation.

2.6 Systems theory

Up until the 1950s there had been relative stability in society in spite of the Second World War, but this was to dramatically alter in the 1960s. There was a major revolution in every walk of life as the developed world came out of restrictions placed on it by rationing and so on. In the Arts came The Beatles, The Rolling Stones, Elvis Presley, *That Was the Week That Was*, *Beyond the Fringe*, Mary Quant, David Bailey, Harold Pinter and so on; the list is endless. Students demonstrated by sitting in at British Universities, they took to the streets in Paris and the Campaign for Nuclear Disarmament marches to Aldermaston started. Trade Unions showed their muscle, notably in the car industry, the press and docklands, and management was often unable to react positively and effectively, with the resulting industrial unrest.

A new management approach was needed, now called the systems approach. The main developers of the systems approach were Richard Johnson, Fremont Kast and

James Rosenzweig. They defined the definition of a system as 'an organised or complex whole: and assemblage or combination of things, or parts forming a complex or unitary whole'. Whereas previous management thinkers had looked at individuals, groups or components of the organisation, the systems analyst looks at the organisation as a whole and views it as a system with the different parts interacting with each other. An analogy is with a healthy human body in which each organ is related to, interacts with, and is dependent on the proper functioning of all the other organs. If any of these organs malfunctions then there can be a knock-on effect resulting in extreme situations, total body failure.

A system is made up of four components as shown in Figure 2.4. The 'inputs' are from various sources such as human, financial, materials, equipment and information. The 'transformation process' is the operations and processes where the inputs are converted into and are a function of the management and technical abilities of the organisation. The 'outputs' are products and services, but also include the profit or losses achieved and the levels of satisfaction employees obtain. 'Feedback' is an assessment of the results achieved, the organisation's reputation in society and reaction from society and the working environment, both externally and internally. The external environment includes customers, shareholders, trade unions, governments and the general public.

There are two types of system: closed and open. A closed system is one that operates within itself and has little or no contact with the environment around it, resulting in little or no feedback. The only truly closed system is the universe. The nearest to a closed system in practice is such as a domestic hot water heating system in operation, but even this has some interaction with outside such as when radiators need to be bled or maintained. It would be very difficult to imagine a closed system for an organisation as most, if not all, must have some communication with the outside; otherwise what is the point of their existence?

An open system usually has a dynamic and continual interaction with both the external environment and between the component parts within the organisation. While the inputs and outputs are very important, the crucial element is the transformation

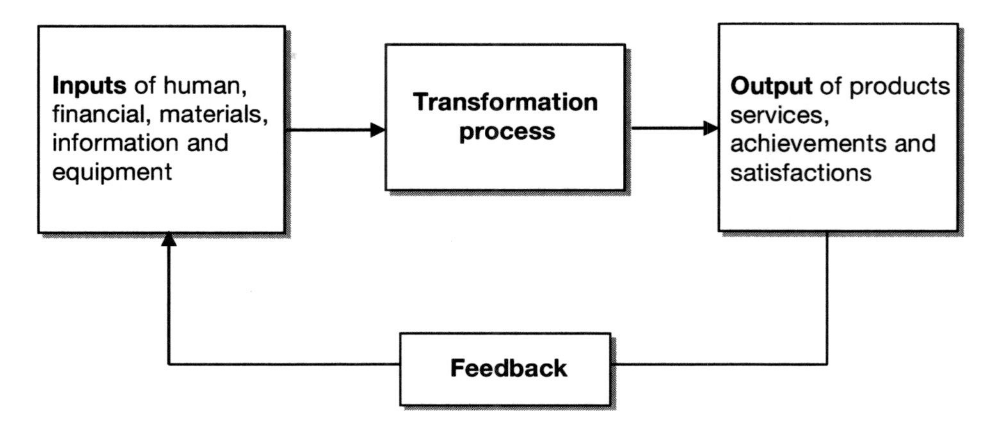

Figure 2.4 The systems approach

Source: Adapted from Bartol and Martin (1994) *Management*, 2nd edn, © McGraw-Hill.

process where there are large numbers of subsystems to be considered and analysed. The system must be adaptable to meet the demands of the external environment by continuously amending its internal systems. This may mean modifying the product or service, reacting to competition, market demand and price. These in turn may affect the inputs such as quality of materials, equipment used and amount of labour. This may require training in new techniques and staff development in general, adapting total quality assurance systems (see 25.3) and dealing with the management of change (see 4.7).

There are three main objectives of the systems approach to management. The first is to define the relationships between the various parts of the organisation with each other, and with the outside environment. The second is to establish how these relationships work and the third is establishing the purpose of these relationships.

Kast and Rosenzweig (1985) postulated there were three main characteristics of open systems, which they called negative entropy, differentiation and synergy. Entropy is the rate of decay measured over time, so negative energy, in an open system is the ability to bring new energy into the process so as to slow down or stop the rate of decline. Differentiation is when the system becomes more complex as a result of adding new units to the organisation to cope with the changes and new challenges, for example to satisfy government requirements, employment policy, quality assurance, safety and environmental issues. Note there is a danger that these can increase to such an extent that the administration becomes top heavy so they must always be kept under review. Synergy is the concept that the whole works better than the parts, so if the parts are working and interacting properly the business will perform better.

2.7 Contingency theory

Systems approaches, however, are not the total answer either, which is not surprising with the rapid changes in the world. Modern management thought is moving to the contingency approach; combining the best aspects of all the other theories that have gone before, including those that preceded the Classical approach, that of trial and error: in other words, the need to be adaptable and flexible and to continue to seek new ways to manage the organisation.

The overriding principle is that the managerial action taken is determined by the particular circumstances of the situation rather than by using universal principles that apply to every situation. It is worthwhile noting that Parker-Follett had suggested this years before but had been largely ignored.

2.8 Further thoughts

To bring this section up to date it is important to mention the concepts of lean thinking, benchmarking, best practice and theory Z. Much work and research in all of these areas has been done. It is not the intention to develop these ideas further in this section other than to say that lean thinking (see 10.7) is primarily to do with the banishing of waste of all types, not just material waste. Benchmarking (see 25.8) is concerned with comparisons between organisations and establishing key performance indicators (not just necessarily in one sector), and best practice, linked to benchmarking, is concerned with establishing the best way to carry out or organise and control work.

William Ouchi has suggested a way to combine the Japanese way of running organisations with that of the American to get the best of both worlds. He called this Theory Z. This includes:

- *Mutual trust.* Trust, integrity and openness are essential for an effective organisation as this results in a reduction of conflict and all work to achieve the company's objectives.
- *Strong bond between organisation and employees.* This is the offer of lifetime employment; shareholders accepting none or reduced dividends during recession; the opportunity to move sideways if promotion is limited; and all employees are given career plans with the aim to use their talents to the full.
- *Employee involvement.* Employees should be involved in decisions, especially those that affect them.
- *Integrated organisation.* Job rotation is encouraged so that staff appreciate the interdependence of tasks.
- *Human resource development.* Managers should look for the talents of those they are responsible for so that they can be developed through career training.

While the easer may find these ideas sensible from a western perspective, a word of caution: multicultural groups may have difficulty in accepting this because of their traditions and culture, for example the position of women in many countries and the Indian caste system.

In the future the author suggests the impact of sustainability issues will become a further stage in the development of management thinking. While nothing written before will become obsolete, personal and society values will change and there will be a move towards much more self-reliance and regional sustainability and this could affect the decision-making process.

Bibliography

Barnard, C.I. (1938) *The Function of the Executive*. Cambridge, MA: Harvard University Press.

Bartol, K.B. and Martin, D.C. (1994) *Management*, 2nd edn. New York: McGraw-Hill.

Bartol, K.B. and Martin, D.C. (1998) *Management*, 3rd edn. New York: McGraw-Hill.

Blake, R.R. and Mouton, J.S. (1985) *The Management Grid III: The Key to Leadership Excellence*. Houston, TX: Gulf Publishing.

Fayol, H. (1949) *General Industrial Management*. London: Pitman.

Graham, P. (ed.) (1995) *Mary Parker Follet: Prophet of Management: A Celebration of Writings from the 1920s*. Boston, MA: Harvard Business School Press.

Hertzberg, F., Mausner, B. and Snyderman, B.B. (1959) *The Motivation to Work*, 2nd edn. New York: John Wiley & Sons.

Johnson, R.A., Kast, F.E. and Rosenzweig, J.E. (1963) *The Theory and Management of Systems*. New York: McGraw-Hill.

Kast, F. and Rosenzweig, J.E. (1985) *Organisation and Management: A Systems and Contingency Approach*. New York: McGraw-Hill.

Likert, R. (1961) *New Patterns of Management*. New York: John Wiley & Sons.

Luthans, F. (1976) *Introduction to Management: A Contingency Approach*. New York: McGraw-Hill.

McGregor, D. (1960) *The Human Side of Enterprise*. New York: McGraw-Hill.

Maslow, A. (1954) *Motivation and Personality*. New York: Harper & Row.

Mayo, E. (1933) *The Human Problems of an Industrial Civilisation*. New York: Macmillan.
Megginson, L.C, Mosley, D.C. and Peitri, P.H. (1989) *Management: Concepts and Applications*, 3rd edn. New York: Harper & Row.
Ouchi, W.G. (1981) *Theory Z*. New York: Avon Books.
Ouchi, W.G. and Jaeger, A.M. (1987) 'Theory Z Organisations: Stability in the Midst of Mobility'. *Academy of Management Review*, 3, 308–311.
Peters, T.J. and Waterman, R.H. Jr. (1982) *In Search of Excellence: Lessons from America's Best-Run Companies*. New York: Harper & Row.
Taylor, F.W. (1947) *Scientific Management*. New York: Harper & Row.
Vroom, V.H. (1995) *Work and Motivation*, revised edn. San Francisco, CA: Jossey-Bass.
Weber, M. (1947) *The Theory of Social and Economic Organisations*. New York: Free Press.
Womack, J.P., Jones, D.T. and Roos, D. (1997) *The Machine that Changed the World*. New York: Macmillan.

3 Leadership

3.1 Definitions of leadership

Various definitions have been written over the years, but in essence the five given below perhaps summarise the essence of the subject.

1 Leadership is the process of influencing individual and group activities towards achieving the objective.
2 The organisation, the leader and the group may have different objectives and interests. The leader's role is to find the right balance between these.
3 A leader is one who succeeds in making others in the group follow their lead.
4 A good leader enthuses others in the group to want to achieve the goal.
5 Leaders are the ones who create the environment that others are motivated to work in.

It should be noted it is a misconception to think managers are the leaders. They may well be, but not necessarily. A manager has responsibility and therefore occupies a formal position in the organisation but does not always have the ability to lead others as defined above. It is not uncommon to find a person below the manager is the one who leads and actually gets things done.

3.2 The power of leadership

Leaders require power to be able to lead, but power comes from a variety of sources. These include:

- legitimate power that results from the position one holds within the hierarchy of the organisation and the authority that goes with that position;
- reward power that goes with the ability to be able to reward people for their work. This can include financial reward such as pay rises, bonuses and promotion, but also non-financial rewards such as recognition, staff development programmes, interesting work and high profile work;
- coercive power which is the ability to punish subordinates in a variety of ways. These include reprimands, criticisms, suspensions, warning letters, demotions and termination of employment. This should not be confused with bullying;
- information power, which is when a manager has access and control over information that subordinates do not have. This can also work in the opposite

direction when a subordinate has knowledge and access the manager does not have, such as IT skills. This overlaps with the next item;

- expert power which is expertise accumulated and valued by others, because the subordinate's expectation is to learn about them and assist in their future promotion and success. This, as with information, can work in reverse as, for example, on site many of the support functions such as planning and quantity surveying are specialist. If the manager comes from a different background they can be 'subservient' to those with this expertise;
- personality power which is when a manager is respected and admired, but not necessarily liked. Some of our great leaders also have charisma, are difficult to quantify, but they tend to 'fill the room' when they are present.

3.3 Leadership style and types

There are various ways leadership style is defined, some more complex then others. Those given below are only a summary of some of these and are based on the behaviour of the leader rather than a measurement of the skills of leadership.

3.3.1 Autocratic

They believe that decisions and the authority to make them must remain with the leader. This is often because they believe the subordinates are incompetent and lazy and reflects McGregor's Theory X (see 2.5.5). When they give orders the subordinates are expected to follow.

The advantage of this style is the tasks are completed efficiently as there is no time for two-way communication, providing the leader is competent. The problem is subordinates are told what to do and not why. They follow instructions even knowing they are wrong. This type of leadership normally leads to poor morale and low productivity.

3.3.2 Democratic or participative

These leaders delegate authority to subordinates and allow them to make some decisions depending upon their perceived competence and interest in dealing with the task. The leader involves the subordinates in discussing the objectives of the task, developing strategies to accomplish it and determining who carries out what.

It is argued the advantage of this style of leadership improves productivity because the subordinates are engaged in the decision making and feel useful and hence have increased job satisfaction as a result. Because they are involved in the decision-making process, decisions made are of a higher quality because of the extra thought given, and if it is necessary to change the methods of work or control mechanisms they are less resistant to it because they have an understanding of why change is required.

However, this assumes the subordinates wish to be involved in participation, which is not always the case. Whether or not this is due to the way management conducts themselves can be a factor. However, it is clear that if management is just consulting rather than involving them in full participation, the subordinates will soon lose interest and stop cooperating in the process. Often subordinates have been conditioned to do as they are told and not be consulted. To be confronted with participation is a culture shock they may not wish to become involved in.

3.3.3 Laissez-faire

Here the leader abdicates the leadership position. They handle the group loosely, allowing them to do more or less what they want, usually handing over the leadership to someone else in the group. The usual reasons for this stem from lack of confidence, fear of failure or not wanting to be part of the group, especially if having been promoted from within it. It can work if the group is very experienced and highly motivated, but it is more likely to fail.

A further way to classify leaders is, as Likert suggested, in the manner in which they get the task done. He suggested there were two types, namely task or production oriented on the one hand and on the other, people- or employee-centred leaders (see 2.5.7).

3.4 Theories of leadership

There are several theories of leadership, which have developed over the years, the earlier ones based upon observation and assumption rather than scientific study.

3.4.1 The great man approach

The adjective great should not be confused with moral. Indeed, many great leaders in history have been immoral. The great man approach is based upon the assumption that men and women of great vision, personality and ability rise to positions of prominence and affect and change the course of history. There is an underlying assumption that great leaders are born and not made. Well-known historical leaders who are used to demonstrate the great man approach are Alexander the Great, Genghis Khan, Julius Caesar, Joan of Arc, Elizabeth I, Napoleon Bonaparte, Isambard Kingdom Brunel, Pope John XXIII, Winston Churchill, Mahatma Gandhi, Joseph Stalin, Mao Tse Tung, Henry Ford, Martin Luther King, Nelson Mandela and Bill Gates. (The author's apologies to any reader who feels they have been left out.)

Advocates of this approach also believe that very successful or great men and women can be found in certain families. In modern times there are examples of dynasties in certain walks of life such as in politics with the Kennedy and Bush families, and in the theatre the Redgrave and Fox families.

The problem with this approach is that if it accepted that leadership is inherited, favouritism in promotions is inevitable. If it is believed leaders are born and not made, then all leaders in the organisation have to be staffed with born leaders. This will inevitably cause a problem because it is clear from the examples of great leaders that there will be insufficient to go around. Further, if leaders are born, then there is no point in providing staff development for others who were not born to lead. Finally, since leaders are born, they will define their own parameters for the job and the concept of producing job roles and specifications becomes redundant. It can be seen from this argument that while it is true to say there are occasionally 'great' born leaders, it can be concluded this approach is severely limited.

3.4.2 The trait approach

Similar to the great man theory, this case assumes a leader's personal attributes are the key to leadership. Traits are distinctive qualities or characteristics an individual possesses. These can be physical qualities, personality, skills and abilities, and personable skills such as the ability to socially interact with others.

The difference with great man theory is that the trait theorists do not necessarily assume leaders are born with these traits, but only need some of them. Examples of suggested traits have included height and stature, vitality and energy, drive, enthusiasm, decisiveness, good looks, knowledge and intelligence, imagination, sociability and friendliness, courage, self-confidence, honesty and integrity, fluency of expression, and self-control.

Researchers have found that several traits seem to be related to effective leadership. Ralph Stogdill (1974) found that certain characteristics seem to be important to effective leadership. The traits include:

- being able to adapt to the situation;
- being alert and aware of the social environment in which one is working;
- being achievement orientated;
- being assertive, cooperative, decisive, persistent, self-confident and dependable;
- having the desire to influence people;
- having high levels of energy and a tolerance to stress;
- being prepared to take responsibility.

The skills include:

- being intelligent and having common sense;
- being creative and conceptually skilled;
- being tactful and diplomatic when required;
- being persuasive and socially skilled;
- having organisational and administrative ability;
- having good control of language;
- have full knowledge of the group task and required outcomes.

Edwin Ghiselli (1971) found the main characteristics for effective leadership were:

- supervisory ability or performing basic functions of management, especially leading and controlling the work of others;
- need for occupational achievement including seeking responsibility and desiring success;
- intelligence such as the quality of judgement, reasoning and reactive thinking;
- decisiveness in having the ability to make decisions and solve problems capably and well;
- self-assurance in believing oneself capable of coping with problems;
- initiative in having the ability to act independently, develop courses of action not readily apparent to others and to find new or innovatory ways to do things.

While on the surface this approach seems to make sense and there is much validity, there are certain limitations to the trait theories. Just because one has some of these traits, does not necessarily make a person a good leader. There are many tall people who fail and historically many small people who have made 'great' leaders such as Napoleon, Hitler and Ghandi. Many academics have intelligence and are very knowledgeable, but by no means all make good leaders.

The definition of the traits can be misunderstood and different interpretations made. For example in the previous paragraph the word intelligence is used. Many will interpret this as being 'brainy' and educated, but there are many intelligent people who have not had the benefit of a good education, but who have a high level of common sense and who are equally if not more intelligent than the former. Others would consider a great athlete to be 'intelligent' because of the manner the brain works in, for example, coordinating the limbs, reaction and vision. How do you define dependability? Is it being trustworthy or always being there when needed?

It is often assumed a leader has good values, which in turn gains respect. Research in prisons has demonstrated that often the most effective leaders within the prison populations are homosexual, neurotic and psychopathic. Further trait theories imply success is determined by inherent qualities and ignore the environment in which people have been brought up.

How can traits be measured? Height can be physically measured as can intelligence to a certain extent, but how are looks, which are very subjective and cultural? Some Eastern societies believe that a tanned skin is bad as it portrays workers in the fields whereas many in the West strive to obtain one. Even if it can be measured, what scale is used and can one trait be compared against another using the same scale? Some people in a given job will demonstrate traits they believe to be required even though it goes against their natural tendency. For example, they may be naturally gregarious, but try to remain aloof and distant from subordinates, or they make quick decisions rather than give them the thought they would normally. Finally, most people in society want a leader, so anybody who decides to become one may well be followed irrespective of their ability. However, in spite of the points made above it does not mean trait theory is totally invalid.

3.4.3 The behavioural approach

The concept behind this is to concentrate on the behaviour of leaders, which can be measured. Behaviourists assume leaders are not born, but can be developed through a learning process. Hence the focus is on what leaders do rather than what they are.

In the 1930s Kurt Lewin (1890–1947) carried out research on small-group behaviour that had a great influence on further researchers. He divided a bunch of boys into small groups to which each was allocated an adult leader. One of these was autocratic, another democratic and the other laissez-faire (3.3.1, 2, 3). In the group led by the autocratic leader there was no long-range planning, a considerable amount of aggression occurred and, while present, productivity was similar to the group led by the democratic leader, although the quality was lower. When the leader left, productivity stopped altogether. Two of the boys left the group because they believed they had been made scapegoats for the failure of the group. In the group led by the democratic leader, work continued even when he left. Productivity was at its lowest in the group led by the laissez-faire leader; the boys did as they pleased and they all became frustrated. In conclusion it was seen the democratically led group was the most successful. To complete the experiment, the leaders were moved round to take charge of the other groups. In each case, after an initial adjustment in the boys getting used to the new style of leadership, the groups performed the same as in the first allocation of leaders.

This was followed by two major research projects at the Ohio State University and the University of Michigan carried out in the 1940s. Ohio's aims were to identify the behaviours exhibited by leaders, to determine the effect these behaviours had on both employee satisfaction and performance, and finally to identify the best leadership style. By questioning both civil and military personnel about the behaviour of their supervisors, they found subordinates generally saw their leadership behaviour in two categories they called the 'Initiating Structure' and 'Consideration'. Initiating structure is the extent to which a leader defines the activities of subordinates in attaining their organisational goals, sometimes referred to as being production- or task-orientated. Consideration is the extent to which leaders are concerned with developing mutual trust between them and their subordinates as well as respect for their feelings and opinions, sometimes referred to as being employee- or human-relations orientated. The Michigan studies' conclusions were very similar.

Further views on leadership are the managerial grid produced by Blake and Mouton, McGregor's Theory X and Y and Likert's systems of management' as described in Chapter 2.

3.4.4 Contingency theories of leadership

Both the great man and trait theories had been of limited use, and the behavioural approach did not produce the answers. This was mainly because on investigation of different successful leaders, while many did, not all fell into the categories as defined above, so the contingency or situational approach was adopted. In other words there is no one leadership style appropriate to every manager to cover all circumstances.

The Fiedler model

Fred Fiedler developed one of the first leadership contingency theories. The basic premise behind it is that the effectiveness of a group or an organisation depends upon the interaction between the leader's attitude and personality, and the situation. The situation is defined as the extent the leader has the power, control and influence over the situation, and the amount of uncertainty the situation causes the leader. This clearly differs depending upon the type of situation. A soldier in battle has a different expectation from the leader to that of a secretary in the office. Fiedler's contingency model is used to identify which type of leader will probably do best. His least preferred co-worker scale (LPC) as shown in Table 3.1, reproduced with permission of Professor F. Fiedler, identifies 16 personality traits that the leader is asked to consider that best describe the person(s) they are responsible for. Generally if the leader describes the others in the team in more negative terms, i.e. a low score, they tend to be task orientated and if the tendency is to describe them in more positive terms, i.e. a high score, they are more likely to be people orientated.

After the person's basic leadership style has been assessed using the LPC scale, the situation is evaluated to match the leader's style with the specific situation. Fiedler argued there are three situational variables that interact with the leader's style that determine the effectiveness of a leader. These are:

Table 3.1 Used by Fiedler to establish their least preferred co-worker scale (LPC)

Least preferred co-worker scale (LPC)									
Pleasant	8	7	6	5	4	3	2	1	Unpleasant
Friendly	8	7	6	5	4	3	2	1	Unfriendly
Rejecting	8	7	6	5	4	3	2	1	Accepting
Helpful	8	7	6	5	4	3	2	1	Frustrating
Unenthusiastic	8	7	6	5	4	3	2	1	Enthusiastic
Tense	8	7	6	5	4	3	2	1	Relaxed
Distant	8	7	6	5	4	3	2	1	Close
Cold	8	7	6	5	4	3	2	1	Warm
Cooperative	8	7	6	5	4	3	2	1	Uncooperative
Supportive	8	7	6	5	4	3	2	1	Hostile
Boring	8	7	6	5	4	3	2	1	Interesting
Quarrelsome	8	7	6	5	4	3	2	1	Harmonious
Self-assured	8	7	6	5	4	3	2	1	Hesitant
Efficient	8	7	6	5	4	3	2	1	Inefficient
Gloomy	8	7	6	5	4	3	2	1	Cheerful
Open	8	7	6	5	4	3	2	1	Guarded

1 *Leader–member relations.* This is the extent to which the subordinates support the leader. To the leader this is a measure of how loyal and trustworthy the subordinates are when given an instruction. Shown as good or poor in the model demonstrated in Figure 3.1.
2 *Leader position power.* This is the amount of power the leader has been given by the organisation to carry out the tasks. On the one hand they may have the authority to dole out rewards or punishment, but on the other hand if this has not been given, they rely upon other methods to influence their subordinates in obeying instructions. Shown as strong or weak in the model.
3 *Task structure.* This is about the clarity with which tasks are defined. How clear are the directions given for each stage of the task? If the task is highly structured, with detailed directions, it is easier for the leader to monitor performance and influence the way the task is carried out by the subordinate. If the task is unstructured, the leader has not decided, or does not know the best way to carry out the task; they cannot monitor the performance of the subordinate. Shown as high or low in the model.

Fiedler concluded task-orientated leaders generally perform better in situations that were favourable to them (categories 1–3 and 7–8) and less well in situations that were unfavourable. On the other hand relationship-orientated leaders perform better in moderately favourable conditions (categories 4–6) as demonstrated in Figure 3.1. He believed the leadership style was fixed so the only way to improve the effectiveness of the leader was to change the situation. An analogy would be in cricket, where a bowler can only bowl to their capabilities, as the batsman to theirs. Therefore the captain will change the bowler depending upon the style of the batsman and the state of the pitch and the weather.

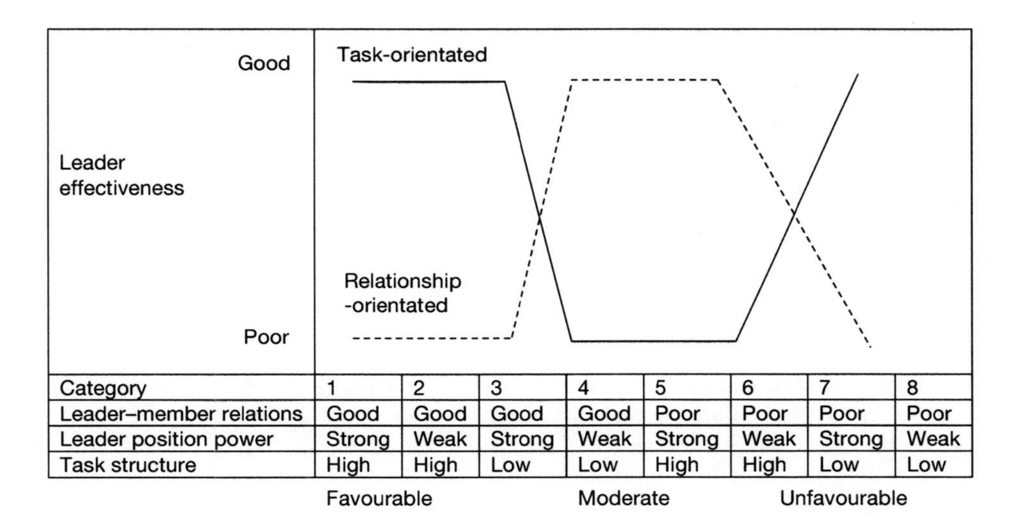

Category	1	2	3	4	5	6	7	8
Leader–member relations	Good	Good	Good	Good	Poor	Poor	Poor	Poor
Leader position power	Strong	Weak	Strong	Weak	Strong	Weak	Strong	Weak
Task structure	High	High	Low	Low	High	High	Low	Low

Figure 3.1 Findings of the Fiedler model
Source: Fiedler (1967). Reproduced with permission.

Leader-participation model or the normative leadership model

Developed by Victor Vroom and Philip Yetton in 1973 the leader-participation model relates leadership behaviour and participation to the decision-making process. It is based upon the quality and acceptance of the leader's decision. Unlike the trait and great man theories, it is assumed leadership style can be adapted to suit the situation and is not necessarily inherent in the leader.

They defined a range of leadership styles all of which they argued can be effective, depending upon their answers to a set of questions with quality, information, structure, commitment, goals and conflict. The answers to these questions direct the leader to the style that will be the most effective. These are as adapted from Vroom and Yetton's (1973) *Leadership and Decision Making*:

1 You solve the problem or make the decision yourself, using information available to you at the time.
2 You obtain the necessary information from your subordinates, and then decide the solution to the problem yourself. You may or may not tell the subordinates what the problem is in getting the information from them. The role played by your subordinates in making the decision is clearly one of providing the necessary information to you, rather than generating or evaluating alternative solutions.
3 You share the problem with the relevant subordinates individually, getting their ideas and suggestions without bringing them together as a group. Then you make the decision, which may or may not reflect your subordinates' influence.
4 You share the problem with your subordinates as a group, obtaining their collective ideas and suggestions. Then you make the decision, which may or may not reflect your subordinates' influence.

5 You share the problem with your subordinates as a group. Together you generate and evaluate alternatives and attempt to reach agreement on a solution. Your role is more like that of a chairman. You do not try to influence the group to adopt 'your' solution, and you are willing to accept and implement any solution that has the support of the entire group.

Path–goal theory

Although several people have looked at this theory, Robert House is credited for the development of this contingency model of leadership. The term path–goal is derived from the concept that leaders influence the manner subordinates perceive work goals and the possible paths to attain both work and personal goals. The theory is a development from the expectancy theory of motivation (2.5.9), and uses it to determine ways the leader could make the achievement of work goals easier and more desirable. This involves making clear what the goals are, reducing problems that get in the way of achievement, and acting in a manner that increases the subordinates' job satisfaction – in other words how leader behaviour can influence the motivation and job satisfaction in a beneficial way.

House identified four leadership behaviours that can be used to affect subordinates' performance:

- Directive leadership is about letting subordinates know what is expected of them. This involves giving them specific instructions and guidance. The subordinates are then expected to follow the instructions to the letter. This is similar to the Ohio studies' initiating structure or task-oriented type.
- Supportive leadership behaviour is about showing concern for the needs and welfare of the subordinates and being friendly and approachable. This is similar to the Ohio studies' consideration or relationship-orientated type.
- Participative leadership behaviour is about consulting subordinates, encouraging them to participate and then evaluating their views and suggestions before making a decision.

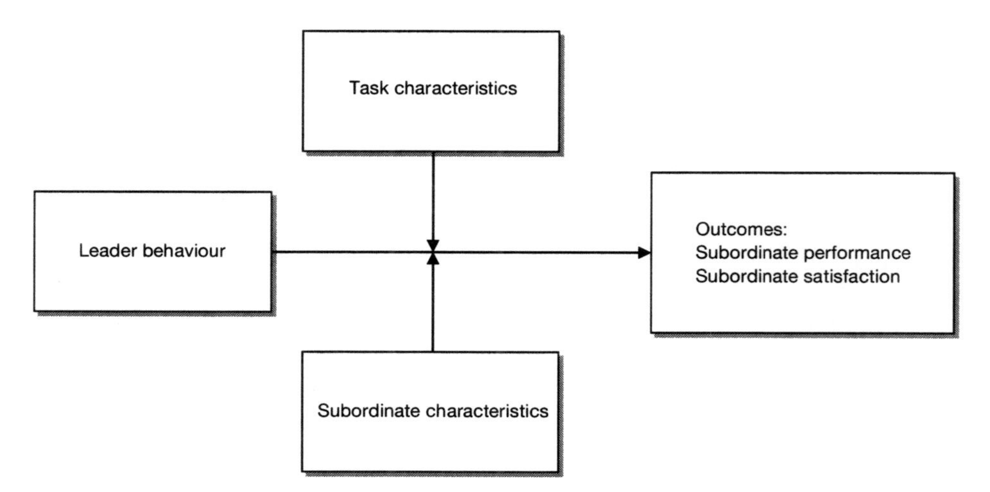

Figure 3.2 Path–goal theory model

- Achievement-oriented leadership behaviour is about setting challenging goals for subordinates. The aim of doing this is to improve their performance and demonstrate confidence in their ability to complete the tasks well.

The path–goal theory argues the behaviour of the leader in one situation should not necessarily be the same as in another. It depends upon the situation and the leader should adapt their behaviour to suit the situation. This is demonstrated in Figure 3.2, which shows the behaviour of the leader as defined above, being influenced by the demands of the task and the characteristics of the subordinates such as their experience and perceived ability, resulting in the outcomes of performance and satisfaction of the subordinates.

Three-circle model of leadership

A further approach to leadership is that the leader of the team is related to the task to be carried out and the time available to complete it. It also considers the needs of the group of people making up the team and their individual needs, as is shown Figure 3.3. The team leader has to understand all these three aspects in making decisions and achieving the objective of the task.

Adair (1983) developed the three-circle model of leadership by proposing a number of questions the leader might ask as shown in Figure 3.4.

The leader has to be clear about the objectives of the task and what has to be achieved. However, the rest of the team may not be clear and equally may disagree with the way this objective is to be met. The leader has to communicate clearly with all members of the team and engage them in the process, so they all pull together.

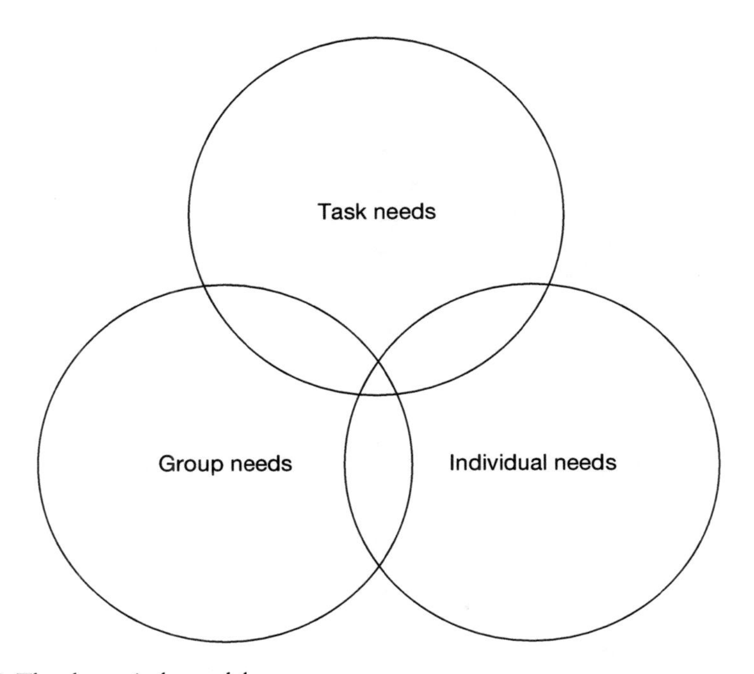

Figure 3.3 The three-circle model

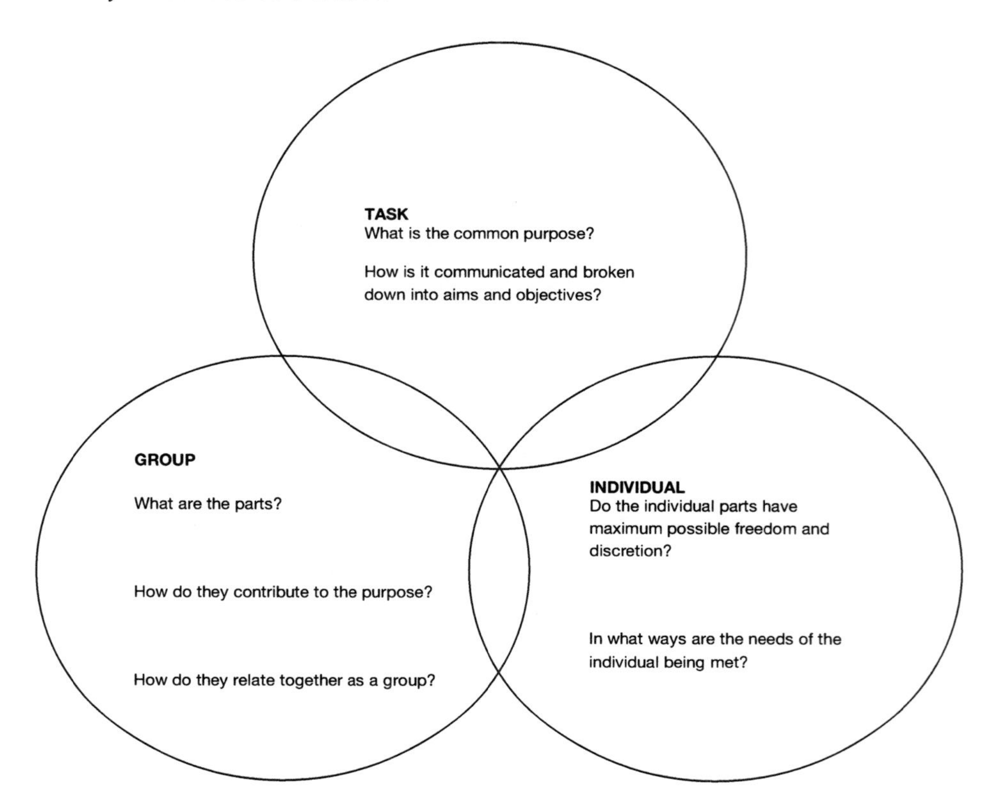

Figure 3.4 Development of the three-circle model of leadership
Source: Adair (1983).

The team leader has to understand the group composition and the various talents within so work can be allocated to these strengths, and at the same time understand how each of the individuals within the group relates to the others so conflicts between individuals can be anticipated and managed. On an individual basis the leader has to decide how much latitude is given to people to make decisions for themselves with the best interests of the team at the centre. In other words how much delegation of responsibility is given to meet the aspirations and ambitions of the individual.

3.5 Female and male leaders

While perhaps controversial, an introductory chapter on leadership would not be complete without raising the issue of whether women and male leaders carry out their function differently. Historically, leaders have been generally men, with notable exceptions in England such as Boudicca (Boadicea) and Elizabeth I, but until recently they have been relatively few. In modern times the world has had several eminent female politicians, such as Golda Meir (Israel), Sirimavo Bandaranaika (Sri Lanka), Indira Gandhi (India), Margaret Thatcher (UK), Gro Harlem Brundtland (Norway) and Angela Merkel (Germany). Increasingly, women are taking up senior roles in industry and commerce, in what is still a male dominated workplace, and it is anticipated this trend will continue, so the question posed at the onset is a valid and relevant one.

A word of caution as pointed out in Robbins (1994): if the styles of women and men are different, could this mean one or the other is inferior and if they are, there is a danger that gender leadership style can be stereotyped. Research appears to indicate the two sexes do use different styles. Women tend to use a more democratic and participative style by encouraging participation and sharing power and information. Men tend to be more directive orientated by commanding and controlling, using rewards for good performance and punishment for poor performance, although it should be noted not all do. However, it has also been found that when women are in male-dominated jobs they allow their natural styles to be over-ridden and they act more like men do in such positions.

All the theory tends to lean towards the needs in a modern society for teamwork, information sharing and trust, and the ability to listen and support subordinates as well as provide motivation. Generally, women do this better than men and have the ability to negotiate and hence compromise rather than try to win and make the other lose. It is interesting to refer back to Mary Parker Follet (see 2.5.2) and read she was advocating this approach nearly one hundred years ago.

3.6 Further thoughts

Observations by the author have noted that the better leaders generally show the following characteristics:

- Probably the most important asset they have is the ability to communicate both verbally and in print. When walking the site they will stop and talk to operatives and craftsmen as well as the foreman and section managers irrespective of who is employing them. They conjure up a feeling of belonging and common purpose.
- They stand back, sometimes aloof, from the action and direct the talent in their control to resolve the problem, rather than mucking in, except when the extra pair hands will clearly be of use. This means that they can see what is happening, which they would not see if they were actively engaged.
- They always seem to have time. Their diaries are rarely full, giving them time to think, and appointments to see them can almost invariably be arranged quickly. Those managers whose diaries are full for the next month or more generally do not make good leaders.
- They always appear to know what questions to ask that get straight to the heart of the problem. These challenge those present to be positive and imaginative. The question 'why not?' is a typical example and has floored many an expert who has given reasons why something cannot be done.
- They have the ability to join a meeting and know who really is in charge and has the knowledge and ideas, rather than those theoretically in control.
- Finally they are firm but fair.

Bibliography

Adair, J. (1983) *Effective Leadership*. London: Pan Books.

Bartol, K.M. and Martin, D.C. (1998) *Management*, 3rd edn. New York: McGraw-Hill.

Campbell, A. (2015) *Winners*. London: Hutchinson.
Fiedler, F.E. (1967) *A Theory of Leadership Effectiveness*. New York: McGraw-Hill.
Ghiselli, E. (1971) *Explorations in Management Talent*. Pacific Palisades, CA: Goodyear.
Gray, J.L. and Starke, F.A. (1988) *Organizational Behaviour: Concepts and Applications*, 4th edn. Columbus, OH: Merrill Publishing.
Handy, C. (1985) *Understanding Organisations*, 3rd edn. Harmondsworth: Penguin Books.
Hersey, P. and Blanchard, K. (2000) *Management of Organisational Behaviour: Utilising Human Resources*, 8th edn. Englewood Cliffs, NJ: Prentice Hall.
Luthans, F. (2005) *Organisational Behaviour*, 10th edn. New York: McGraw-Hill.
Megginson, L.C., Mosley, D.C. and Peitri, P.H. (1989) *Management Concepts and Applications*, 3rd edn. New York: Harper & Row.
Robbins, S.P. (1994) *Management*, 4th edn. Englewood Cliffs, NJ: Prentice Hall.
Robbins, S.P. (2005) *Management*, 10th edn. Englewood Cliffs, NJ: Prentice Hall.
Stogdill, R.M. (1974) *Handbook of Leadership: A Survey of Theory and Research*. New York: Free Press.
Vroom, V.H. and Yetton, P.H. (1973) *Leadership and Decision Making*. Pittsburgh, PA: University of Pittsburgh Press.
Wright, P.L.(1996) *Managerial Leadership*. London: Routledge.
Yukl, G.A. (2003) *Leadership in Organisations*, 8th edn. Englewood Cliffs, NJ: Prentice Hall.

4 Team or group working

4.1 Introduction

Unlike most other industries, there is less opportunity for the majority of construction personnel to develop long-term working relationships with other employees. This is because by the very nature of the business, a project manager is selected for a project and while they may choose some of the personnel, many of the rest are provided for the project as required or when they can be freed up from another project. Add to this that most contracts will have a different client and design team members; it can be seen that team building and the dynamics of the group are very important to ensure a successful conclusion to the project. Since construction projects are on average 12 to 24 months in duration, rapid team building is essential. It would also be difficult to think of situations in the construction process where an individual could operate other than in a group scenario. The project manager relies heavily on the support and cooperation of those in the team and, with the industry moving away from confrontational contracts and towards more partnership types, the team now embraces many more members. Some of these are based on site, whereas others are in the regional, head or design partner's offices. In conclusion, one cannot emphasise enough the importance of group working as a core trait of both the design and production processes.

While the author has an affinity with the word 'team' as a concept commonly used in construction, many of the texts use the term 'group'. To start with, 'group' will be used in discussing the manner in which people can be organised, but in the latter part the emphasis is on how the group becomes a team and what constitutes an effective team.

4.2 What is a group?

A group is two or more individuals who have a common task objective, interact with each other, are aware of the others in the group and consider they are part of the group. There are two types of groups, namely formal and informal as shown in Figure 4.1.

The formal groups are defined as:

- *command groups*: these are the groups as shown on the company organisation chart. They have been positioned in the chart to shown their relationships with others in the organisation demonstrating the chain of command;

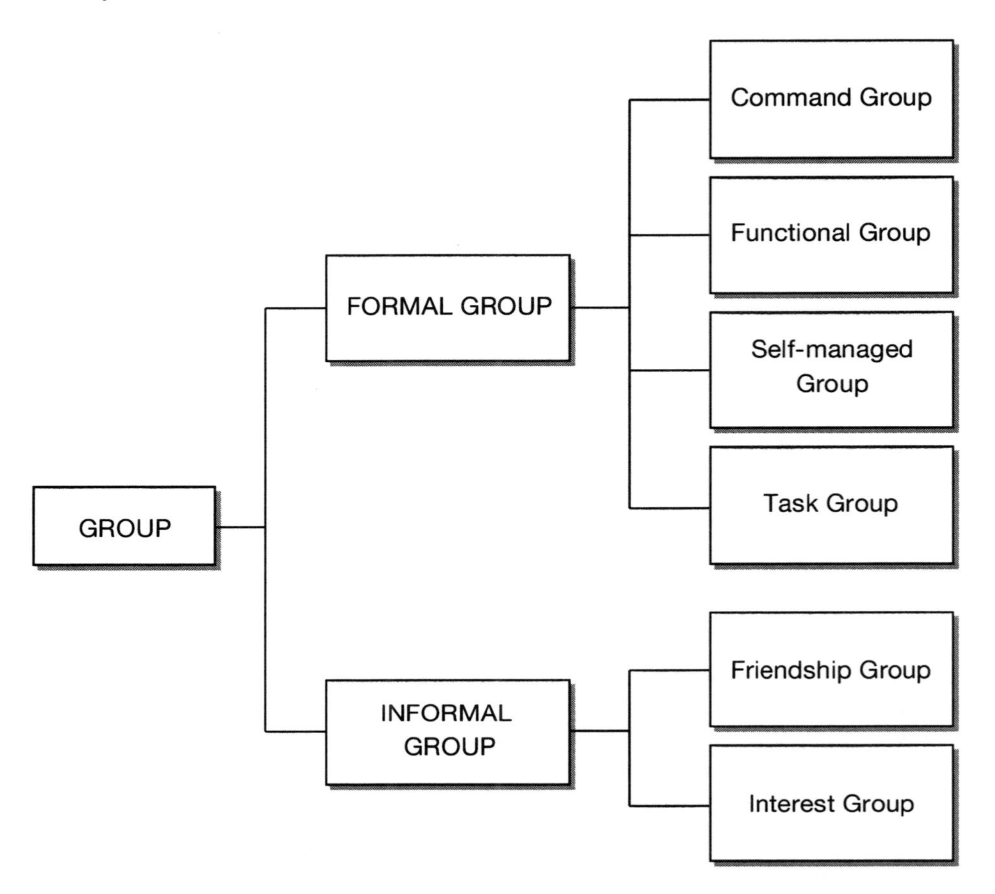

Figure 4.1 Types of work group

- *functional groups*: these are groups where personnel from different disciplines are brought together to resolve problems. Sometimes members of these groups have been trained to carry out other group members' jobs. The construction site uses functional groups to carry out many of its tasks, as the site organisation comprises personnel from different disciplines working together to construct the project;
- *self-managed groups*: personnel in this group besides carrying out their normal tasks, have the authority to carry out their own planning, evaluation and discipline and the group is therefore a self-contained unit. That does not mean it is not accountable to others for its actions. A construction site team can have this character. Another example is a ship's company when at sea;
- *task groups*: this group is set up specifically to solve a problem or complete a task. When completed, the group is then disbanded. Examples of this include committees, working parties and royal commissions.

Informal groups occur in a variety of ways within an organisation and are determined by the employees rather than by management. It is interesting to go into an organisation and ask employees to say whom they are responsible for and whom they are accountable to. It is not unusual to find their answers do not tally with those drawn on the organisation chart. The effect of this is the informal group can cut across the formal structure. Understanding these informal groups can assist in using the organisation more effectively. For example, the author when working in Hong Kong discovered informal groups based upon status were very strong, so all the chief administrators in each department regularly did, or could, meet for lunch. By speaking to each other they could readily find out what was happening in another department and report back, by-passing the formal lines of communication that were often slow and sometimes obstructive. This was an example of an interest group. Another could be the company sports club or horticultural society where personnel of different status participate.

Friendships develop between personnel, but over time promotion may separate their status within the organisation. This does not mean the friendship wanes and they will cease communicating. One of the most important methods of informal communication is called networking where friendships and good working relationships enable parties to cut through normal lines of communication and permit assistance. This can occur within the organisation and between different organisations. While this is not an informal group as such, it does indicate the strength of informal links.

4.3 Why are groups formed?

Groups are formed either formally or informally for a variety of reasons. Formal groups are put together by the organisation for good practical reasons. These include the following:

- The task requires more than one person to complete it. Bricklayers cannot lay bricks efficiently by themselves. They need someone to mix the mortar if they are to lay bricks all day and generally speaking a labourer can keep two bricklayers provided with materials: hence the gang size.
- The task requires a variety of skills, experience and knowledge. A construction site is a good example of this.
- Management hopes that by putting people together in a group it will encourage an exchange of ideas so the work process might change for the better, although much will depend upon the environment the supervisor creates.
- It is hoped that group peer pressure will keep in line anyone whose behaviour is perceived as unacceptable.

The members of the group can also derive benefit, which can include:

- a feeling of belonging. There is a feeling of security by belonging to a group rather than being isolated as an individual. Most people find the companionship a positive;
- if the group is successful, a high feeling of satisfaction and status;
- if the group works well as a group and individuals' contribution is recognised, a feeling of self-esteem;
- protection by the group of a member of the group against outside pressures and threats.

4.4 Group cohesion

Generally the greater the cohesion in the group the more effective the group will be, but there are situations where the group is very cohesive and rebels against the organisation's set objectives. Part of the leader's function is to ensure the group is both cohesive and directed towards attaining the goal. There are several factors that can affect the amount of group cohesion.

The composition of the group. The larger the size of the group the more difficult it is to control and maintain cohesion. Communication between members becomes harder and smaller factions might form, which can be counterproductive. There is also the risk that some members may take a free ride. There is no magic number that determines the optimum size, as many variables come into play, but generally once the group exceeds 12, difficulties are likely to emerge. The compatibility of members or otherwise can play an important role. Personalities, the amount of common interest, the level of competition and conflict and compatibility of talent are issues that determine the success of the group. Another key issue in construction is the permanence or otherwise of team members. It takes time for people to get to know each other and engender team sprit and good working relationships. The frequency of turnover of members can also affect the cohesion.

The workplace. The type of work undertaken can mean either the group is close together or separated. In the latter case communication becomes more difficult, but in spite of this if communication is good, often due its frequency, cohesion can still be high. This is important in construction, as many of the team players are not located on the construction site, such as the design team and departments such as purchasing and estimating. The physical environment can also have a part to play, although much depends upon the expectations and experiences of the group members. Working on a construction site is not a clean, quiet, comfortable environment, but these conditions are normal and expected. Workers in a normal office would be upset if they had to accept anything like those on the site. In an office environment the debate over the pros and cons of open-plan offices as against personal offices is another example of how the working environment can have a part to play. The open plan certainly enhances communication, but many prefer their private space.

Management. The type and style of leadership from management will affect the relationship of the group to the needs of the organisation. If the group is successful in achieving the goals set or the goals they have set, the group will be more cohesive. Cohesiveness can also be achieved when external threats occur and group members rely on each other for survival. Both of these can clearly be seen on outside activities courses designed to develop management and leadership skills. Delegates confronted with a strange and seemingly tough environment can be seen to bond together against, in this case, the environment as they are outside their own comfort zone, and when achieving the goal are generally highly elated.

Group development. Groups develop over a period of time as relationships mature, but not without pain. The most notable explanation of this identifies four stages referred to as forming, storming, norming and performing. Forming is when group members, in this new environment, somewhat anxiously test each other out to get to know each other. As they know each other better they are more prepared to be open and express views and opinions, known as storming. Conflicts and disagreements can occur as a result, so the next stage, norming, is when members start to accept there has to be compromise and cooperation for the task to be properly executed. At this point the group can commence to perform as a cohesive group. A more detailed discourse on these is given in 4.5.

4.5 Managing conflict within and between groups

It is rare that conflict will not occur within a group. Indeed, it can be argued there is something wrong if there is not some conflict, as this will mean there is apathy among members of the group, or they are so similar in nature and views alternative opinions are not aired and discussed. There are various factors that can cause conflict. These include:

- *task dependency*: this is when one member of the group cannot commence or complete their task until another has finished theirs or when they are mutually dependent and cannot proceed unless both accomplish their respective tasks;
- *scarcity of resources*: limited resource availability can be a cause of conflict. These resources include equipment, support staff, office space, funding, pay increases and staff development. If two people require access to the same piece of equipment or cannot get a pay rise they think they merit, they become upset and annoyed;
- *target conflicts*: the organisation can give different, conflicting target objectives to different parts of the organisation. For example the purchasing department is charged with obtaining materials and subcontractors at the lowest cost. The site then can be confronted with a selected subcontractor that is under performing, but is still expected to bring the contract on time and within budget;
- *personnel characteristics*: a major source of conflict occurs as a result of the individuals within the group. Loss of patience by an experienced member because a less experienced member is slow on the uptake or makes mistakes is a typical example. A member may be reluctant to own up to a mistake, whereas another believes they should. There can also be personality clashes, some members finding it difficult to work with another;
- *communication breakdown*: it has been a common theme throughout the book, to stress the importance of good communications. Lack of communication or imprecise or incorrect communications can often lead to conflicts between parties;
- *pay and bonuses*: however much management wishes to keep remuneration to staff confidential between them and the individual, it will eventually almost certainly be revealed from one source or another. There is little more disruptive when members of a group (or other groups or organisations) are paid disproportionately to each other.

There are several methods managers can use to try to resolve and minimise conflict. The first stage is to anticipate likely points of conflict and take action before conflict takes place. When conflict has occurred the manager needs to evaluate the situation and establish the cause of the conflict and having found the cause take action changing the factors causing the problem. However, it may not be possible to change these circumstances either because it will be beyond the current availability of resources and too expensive, or it is not practical to do at that point in time. It is sometimes possible to persuade the group about the bigger picture than the objectives initially set for the group. For example, if the contract is successful and the client satisfied, there is very likely to be more work for the company and hence those employed by it, although care has to be taken in adopting this strategy, because, if work is not forthcoming in spite of a successful conclusion, then trust will be lost for the future.

Conflict is not always a bad thing as it can stimulate discussion and ideas. In certain circumstances it should be encouraged, providing it is monitored and controlled. An athlete in competition is in conflict with the competition or opposition, but the successful ones are usually in control. Lack of conflict within a group usually means that unless it is an exceptionally harmonious group, then there is a state of apathy and lethargy, resulting in low levels of performance.

4.6 Team roles

There are classifications of roles from other sources with a similar theme. However this section relies heavily on the work and research Meredith Belbin and researchers developed primarily at the Administrative Staff College, Henley and the Industrial Training Research Unit at Cambridge.

A team comprises a group of individuals all of whom have different personalities and talents. The skill of management is not just to motivate and obtain the highest collective output from them, but, if the opportunity arises, put together a team that will have a range of talents and abilities that collectively will add value to the group as a whole. To do this it is necessary to establish what the talents and abilities of each member or perspective member are, and what the mix of these characteristics should be to obtain optimum performance. As will be seen, individuals can be broadly classified into various types, each characteristic of which, if identified and understood, can be utilised for the betterment of the whole. It can be used to identify a person's strengths and their weaknesses; the latter not being used to exploit but to protect, by ensuring that person is not expected to perform using their weakness, that another in the group who has strengths in this area is utilised instead. A team is after all only as good as the weakest link. Belbin initially defined eight different characteristics. Later this was modified to nine by adding the role of Specialist as shown in Table 4.1.

Note, originally the Chairperson was called the Chairman and in some texts is now referred to as the Coordinator and the Company Worker called the Implementer. These were changed in the relentless pursuit of political correctness.

As can be seen from these descriptions in Table 4.1, there are potentially a wide range of abilities and talents to choose from if each of these types of role was represented in a team. However, in reality it is rare for this to occur, not least of all because there is a tendency for appointers of staff to recruit and attract people similar to themselves. For example, the author used the Belbin analysis for several years on groups of students all sponsored by contractors on a construction management course. It was found nearly all the students fell into two or three of the categories described, with just a few falling into the others.

Putting together a team that will function well together is difficult, as for example, Completers, with their intense concern for detail, are very useful in a team, but tend to annoy those who fall into such categories as Plant and Resource Investigators, hence the need for a coordinator to manage these conflicting personalities. A team full of

Table 4.1 The nine team roles

Role		*Contribution to the team*	*Allowable weaknesses*
Plant	PL	Creative, imaginative and unorthodox. Able to solve difficult problems.	Ignore details and are often ineffective communicators as they are preoccupied with the problem.
Resource Investigator	RI	Extrovert, enthusiastic and communicative. They explore opportunities and are good at developing contacts.	Over-optimistic and soon lose interest once the initial enthusiasm has passed.
Chairperson	CH	Mature and confident. Make good chairpersons. They clarify goals, promote decision making and are good at delegating.	Can be seen as manipulative. They delegate their personal work.
Shaper	SH	Challenging, dynamic individuals who thrive on pressure. Have the drive and courage to overcome obstacles.	Can provoke others and hurt others' feelings.
Monitor Evaluator	ME	Sober, strategic and discerning. They see all the options and judge things accurately.	Lack drive and the ability to inspire others. They are overly critical.
Team Worker	TW	Co-operative, mild, perceptive and diplomatic. They listen, build, avert frictions and calm the waters.	Indecisive in crunch situations and can be easily influenced.
Company Worker	CW	Disciplined, reliable, conservative. They turn ideas into practical actions.	Somewhat inflexible and are slow to respond to new possibilities.
Completer/ Finisher	CF	Painstaking, conscientious and anxious. They search out errors and omissions. They deliver on time.	They are inclined to worry unduly and are reluctant to delegate. Can be 'nit-pickers'.
Specialist	SP	Single-minded, self-starting and dedicated. They provide knowledge and skills that are in short supply.	Contribute on only a narrow front. They dwell on technicalities, overlooking the 'big picture'.

Source: Adapted from Belbin (1994).

Plant would equally be lost without a Completer and vice-versa. Belbin (1996) in *Management Teams* develops further the results of experiments carried out using different combinations of roles.

It is suggested when selecting a team an analysis of the types of role necessary to most successfully achieve the team's goal should be undertaken so the most appropriate team members can be selected if available. In doing this the team builder can then highlight the weaknesses and strengths of the team as a whole so the strengths can be exploited for the good and the weaknesses realised so the objective is not let down as a result.

The titles of the team-roles owe something both to historical factors and the need to avoid the preconceptions associated with the established alternatives. These could not be entirely overcome and can therefore be misread. For example the Chairman team-role refers to the characteristics of Chairmen found in winning companies. In fact some successful Chairmen of industrial commercial groups do not themselves adopt a typical Chairman stance, but make their marks as Shapers, where sharp or rigorous action is the order of the day; or as Plants, where the Chairman's role is basically strategic.

Company Worker or Team Worker team-roles have tended to be undervalued because of their titles. The former has been replaced in some firms by the title Implementers, but there is much to be said for Company Worker with its flavour of someone who acts as the backbone of the company. As an alternative to Team Worker, the term Team Builder has also been used. In practice some typical Company Workers and Team Workers in the business and industrial world become chairperson of their firms. In these cases some aspects of their Chairmen behaviour are learned, although other aspects of their style are likely to reflect their primary team-role.

4.6.1 Managing multicultural teams

As the world gets smaller and businesses work across national boundaries, and as society changes as a result of immigration, there needs to be an understanding of the effect these changes have on managing a team. The author remembers when working in the Far East for several years being advised of the differences between the British and the locals. True, there were differences, some significant, but in reality there were many more similarities. This is important to understand, otherwise there is the danger of overcompensating for each other's cultural needs. While not trying to give a comprehensive and authoritative list, the examples below are meant to give an indication of the kinds of issues that could impact on running a multicultural team.

- *Hours of work*. Some societies believe it is a sign of incompetence to work excessive hours, others think working long hours is necessary to show one's enthusiasm and hope it will impress their manager. Others have siesta periods during the day because of the climate.
- *Mode of dress*. In the United Kingdom it is expected that people in many white-collar jobs should be dressed in suits irrespective of the temperature. In Scandinavia it is more likely to find people in similar roles dressed more casually.
- *Method of address*. In some countries first names are commonly in use, whereas in others the surname is more likely to be used, especially as a sign of respect.
- *Respect and face*. In China age carries respect and often goes with position in the hierarchy of the organisation. The culture of 'face' is extremely complex. Full comprehension is very difficult to achieve. For example, if you make someone lose face, then you also lose face. This has much to do with the emphasis on compromise in negotiations and why the signing of a contract is not always easy to achieve as their word is expected to be accepted, because if they do not meet the agreement, they lose face. Even small details such as handing one's calling card are an issue. It should be handed with two hands rather than with one.

- *Authority.* Some cultures expect to be told what to do and will not question the instruction and feel uncomfortable being asked their opinion on how to carry out the task.
- *Religion.* This is a significant issue for many reasons. Different faiths have different modes of dress, attitudes to women or need specific prayer times set aside. There are the different holidays such as Christmas, Ramadan, Passover, Diwali, Chinese New Year and so on, non of which coincide, and there are issues associated with fasting, dietary requirements and days of worship.
- *Language.* With the increasing use of foreign nationals in the workplace, the lack of ability of some to communicate well in English needs to be considered, especially for health and safety issues.

4.7 Management of change

While there are good reasons for understanding the processes of managing change, in many instances, if it is necessary to manage change, then it is too late and the opportunity has probably been missed. It is argued an organisation should be set up and managed in such a way that change can naturally be accommodated by the culture of the people within the organisation.

However, there are circumstances where it is inevitable and in particular when personnel are promoted or have a change of job. In these cases, the individual is put through a transition stage that if not managed correctly can often lead to an unsatisfactory outcome or failure. This can often be seen and is referred to as the Peter Principle, which states that 'everyone rises to his or her level of incompetence', or put another way, 'everyone rises to at least one level above that at which they are competent'. The effect of this is either the business does not operate to its full potential, because of incompetence, or there is an unacceptable turnover of staff due to their frustration or stress.

4.7.1 The transition process

This process is divided into several stages many of which the reader may have already experienced, when changing schools, going to university, moving to another job, getting married and having children.

Stage 1 – Immobilisation or shock

When first going into a new work scenario there is often a sense of being overwhelmed. The previously held view does not match reality, hence the feeling of shock. Before starting the new job, the vision of the job is too positive and it is only when seeing the reality that one sees the limitations and difficulties.

Typical statements indicating this might be: What am I doing here? This isn't the job I thought it was. Did I really apply for this job? I want my mum!

Stage 2 – Denial of change

This is a temporary retreat into false incompetence where one tries to minimise the change or trivialise it. There tends to be a reversion to previously successful behaviour.

If this goes on too long then the problems start to arise, as the previously used behaviour becomes dominant. If they stay at this stage then the Peter Principle comes into play as they perform badly because their behaviour is inappropriate in the new situation.

Typical statements are: In my last job we did it this way. This situation is very similar to my last job.

It is not that previous experience is not valuable, and maybe the way it was done in the last job might not be wrong, but it is the clinging on to past experiences as a determining factor in solving new problems that is symptomatic of the reluctance to move on.

Stage 3 – Incompetence and depression

There is a lot of frustration at this stage. The individual feels as if they are floating, because of the realisation of change and finds it difficult to know how to cope with the new situation or relationships.

This is a very important stage in the process as without realising that change is necessary, they cannot move on and develop. Current values, attitudes and behaviour need to be challenged to permit people to cope with the new situation. There is a danger the individual bottles this up and often organisations don't offer to help or even appreciate there is a problem. After all, they have employed the person to do the job. They offer a 'sink or swim' approach that does not help the situation. Indeed it can stop the real learning process and hence the development of the individual and the acquisition of knowledge.

However, this 'suffering' is important as the individual learns from it. There is a school of thought that suggests that unless one has a trauma or a major setback, the chances of being successful in life are reduced. The parallel being that the jolt to the system in both cases motivates the individual to strive for recognition or success.

Typical statements are: I'm not sure what to do. I'm confused. Can someone tell me what is happening here?

Stage 4 – Accepting reality

Prior to this point everything has been about hanging on to past values, attitudes and behaviours. At this stage there is a realisation of the actual situation and the relief and pleasure generated as a result of letting go of the past. Also there is a desire and readiness to experiment with change and thus optimism for the future and one's ability to cope with it.

Typical statements are: I can see where I was going wrong. I will try again and get it right. Why didn't I think of that before?

Stage 5 – Testing

At this stage the individual tries new approaches and behaviour as a means of resolving problems, looking for new ways to overcome the transition. They become much more active than before, showing high levels of energy, but also anger and frustration as the testing process progresses.

Of course as a result of this new-found enthusiasm, but relatively little experience, many more mistakes are likely to be made. Many of the experiments will turn out to be blind alleys. This behaviour needs to be encouraged and the mistakes not penalised, otherwise the individual will once again retreat to earlier stages. It is only by experimenting that new and effective ways can evolve.

Typical statements are: We should be doing it this way. It didn't work, let's try another way.

Stage 6 – Search for meaning, internalisation

At this stage the individual is looking to find the meaning or reason for why things are different, trying to understand the activity and causes of anger and frustration. It is only when they have experienced the anger and frustration they can attempt to make sense of it and why it has occurred. This is usually quite a reflective period in the process. These reflections establish the real meanings of the events that have occurred and the values that have lasted or those that need to be discarded.

Typical statements are: I now understand why that worked and this didn't. I see where they are coming from now.

Stage 7 – Integration

This is the final stage. The transition is now over. Conditions around appear stable as the changes are integrated into the experience of the individual. The individual will have developed newer and better ways of dealing with the job and confidence and self-esteem will have arisen, resulting in increased performance.

Typical statements are: I feel like a new person. Well I'm glad it's over, but I learned a lot. I understand where I was going wrong, but have also learned where my true strengths lie.

The above are generalisations as to the way individuals go through the change process and of course there are overlaps between the stages, some progressing much faster than others. It is however important to realise that individuals have to go through these seven stages. Further, if management is aware of the implications of these stages and at the same time watches for them, anticipates and takes the appropriate action, the speed of these changes and hence higher levels of performance will be realised sooner and the likelihood of staff leaving is also reduced.

4.7.2 Team development

It is suggested that there are four distinct learning stages in the development of a group of individuals to form a more cohesive group. These are known as forming, storming, norming and performing.

Stage 1 – Forming or testing

The forming stage is in effect a testing stage when individuals in the group are being acquainted with each other. They tend to feel inhibited so are polite with each other, impersonal in conversation, guarded in disclosing information (some will be reluctant even to offer opinion) and they will be watchful of each other's behaviour and conversation. They are testing each other out.

Stage 2 – Storming or infighting

During the forming stage fundamental differences between the individuals are unlikely to emerge because of testing each other out and because the group will be tending to concentrate on completing the tasks at hand. However, once the members of the team have gained in confidence, then a certain amount of infighting is likely to occur.

The leadership of the group may come into question. The initial leader, appointed at the forming stage, may be less dynamic and skilful than expected. The group members may demand a change of leadership to maintain a high level of performance, which the leader may resist. The change of leadership may also be requested not because of their skills, but because their style, values etc., are considered unsuitable or a more powerful member of the group may be making a bid for leadership. This in turn may split the group into different factions.

This effect of this is infighting, even if the leadership issues are resolved one way or the other, is that there will still remain tensions and differences of opinion resulting in feelings of animosity and bitterness, making it difficult for the new leader to function. Also, as the group is storming, different members may prefer different approaches to completing the task. This may mean certain members may feel unwanted and redundant.

Personality differences also have an effect at this stage, increasing the likelihood of tension. If these tensions are not resolved, the conflict and confrontation could become the normal behaviour of the group. The problem with this is that even though the personality clash may only affect two or three members of the group, the tensions created can be disruptive for all members. If these tensions are not resolved then confrontation will become the norm.

Whatever the cause of the infighting, the behavioural pattern of the group is characterised by conflicts and divisions. Individuals who feel demotivated leave the group and some of those left feel trapped. Others may opt out of the group entirely, just concentrating on the tasks set rather than the relationships within the group.

Stage 3 – Norming or doing

The speed at which the storming stage takes place largely depends upon the style and personality of the group leader as well as pressures on the group to produce results. The sooner the group tries to fulfil certain task goals, the sooner it will break out of the infighting stage.

By doing tasks, norms of behaviour and professional practice begin to be established and barriers between members of the group begin to disappear as a result of exchanging views, ideas and experiences. However, tensions will still remain so the group, or part of it, may fall back into stage two. This can occur several times and in extreme and unsuccessful groups they never progress past this stage.

Stage 4 – Performing or identity

Breaking out of stages two and three is no easy matter. Much depends on the effectiveness of the leader who needs interpersonal, counselling and listening skills so they can act as a third party (consultant) to warring group members. It is important group members identify with the group mission or purpose. A skilful leader is able to shape a meaningful identity for the group, which is essential to move into stage four.

Once in stage four the group becomes cohesive. They are more supportive of each other, sharing information and ideas and tolerating each other's differences. They use each other's strengths and talents to a greater degree, are more resourceful and flexible in their approach to problem solving and task performance. They begin to perform as a genuine team. Problems can arise and a drift back to stage three if not properly led, especially as time goes on and new members replace other members of the group.

Bibliography

Bartol, K.M. and Martin, D.C. (1998) *Management*, 3rd edn. New York: McGraw-Hill.

Belbin, R.M. (1994) *Team Roles at Work*. New York: Butterworth Heinemann.

Belbin, R.M. (1996) *Management Teams*. New York: Butterworth Heinemann.

Campbell, A. (2015) *Winners*. London: Hutchinson.

Dyer, W.G. (1995) *Team Building: Current Issues and New Alternatives*, 3rd edn. Reading, MA: Addison-Wesley.

Freeman-Bell, G. and Balkwell, J. (1996) *Management in Engineering*, 2nd edn. Englewood Cliffs, NJ: Prentice Hall.

Gray, J.L. and Starke, F. (1988) *Organizational Behaviour: Concepts and Applications*, 4th edn. Columbus, OH: Merrill Publishing Company.

Mullins, L. (2010) *Management and Organisational Behaviour*, 9th edn. Harlow: Pitman.

Peter, L.J. and Hull, R. (1996) *The Peter Principle*. New York: William Morrow.

Robbins, S.P. and Coulter, M. (2012) *Management*, 11th edn. Englewood Cliffs, NJ: Prentice Hall.

Thomas, W.T. (1997) 'Toward Multi-Dimensional Values in Teaching: The Example of Conflict Behaviours'. *Academy of Management Review*, 2, 484–490.

Thomson, R. (1993) *Managing People*. New York: Butterworth-Heinemann.

Part III

Financial control of construction projects from conception to handover

5 Sources of finance

This is not meant as a definitive list of sources but to give the reader an indication of the kinds of organisations that can be considered, and to indicate which type of development they are most likely to be appropriate for.

The finance required for development falls into two basic categories. First, 'development finance' or 'short-term money' that is needed to cover all the costs incurred in purchasing the land, the development, design and construction processes. This may be paid back shortly after completion if the development is sold on and a profit made. Second, 'funding' or 'long-term money' that is used to cover the costs incurred when holding on to the completed development as an investment. This money is repaid using the revenue generated from renting the property. The monies for each may not come from the same source.

Some developments are financed entirely from the developer's own capital, but this is the exception rather than the rule. Most developers will go to one or more of the financial institutions to seek funding as this spreads the risk in the event of the project failing. However, if they have surplus cash it may be in their interest to invest it in property, especially if the expected returns from rental growth are good.

5.1 Insurance companies and pension funds

The pension funds and insurance companies take a long-term view, as they need to achieve capital growth to pay out the agreements made with pensioners and policyholders. They tend to be more cautious and conservative than other lenders as a result of their responsibility and in recent years have reduced the percentage of their portfolio from investment in property to equity shares, but are still major investors in property. However, since the fall of the stock market in 2002/3 they have moved from equity to safer havens and property is often considered to be a safer long-term investment. They will have a clearly defined set of criteria that have to be satisfied before being prepared to invest. As a result they expect to have significant control on the ways monies are to be spent, the quality of the building, its location and financial status of the tenants. The outcome of this is that developers are inclined to produce development schemes that satisfy the financial institutions rather than the future occupants.

They tend to only invest in large-scale projects, especially those in prestigious sites and occasionally become the developer taking on the risk. Generally, their aim is to minimise risk and maximise future yields; yield being the annual income received from the asset expressed as a percentage of its capital cost or value (see 6.2.11).

5.2 Banks

The banks are approached if the development is not acceptable by the insurance and pension companies or the developer cannot or will not provide the necessary financial guarantees. The banks will use the property asset as security for the loan. It is attractive, as it is a large identifiable asset with a resale value.

5.2.1 Clearing banks

The clearing banks normally offer short- and medium-term loans, although this is not always the case. Much depends on such factors as how buoyant the market appears, current government policy and the track record of the developer. They are more flexible in their approach to loans than the insurance companies, especially for refurbishment and development of older buildings. They also take an interest in the business of those of the prospective occupier and of the property development company as well as the property itself. If the loan is high, the bank may secure an equity stake in the project as they are exposed to a greater risk. Residential developers tend to go to the clearing banks as their requirements are only for a short-term loan.

5.2.2 Merchant banks

Merchant banks are more entrepreneurial than the clearing banks and for that reason are likely to take a greater risk. At the same time they will tailor-make a financial package for the individual developer. They may also divide the funding into smaller packages and obtain funding for each package from other banks. The cost of borrowing from them is usually higher than from the other main sources of funding. Other products, provided by specialist banks, include Mezzanine and Monoline loans that carry higher interest rates.

5.3 Private persons

Any individual who purchases a property is an investor and many in the United Kingdom have entered the housing market as a means of increasing their wealth and as a hedge against inflation, although this is not always a successful short-term strategy if the market value falls and the owner enters negative equity. Without these types of investors the speculative housing market would not exist in its current form and housing development projects could well be different. The majority of these investors purchase the property they live in, but others purchase with a view to refurbish and sell on at a profit, while others accumulate property and earn rental income. An interesting new development inspired by the use of social media is 'crowd funding'; an example being the Peckham Coal Line urban park developed in London based on individuals within the community working together to improve the local built environment.

5.4 Building societies

Building societies lend primarily to the domestic property market because they rely on the income from the mortgagers to supply finance on a regular repayment that enables money to be lent to others.

5.5 Government and EC funding

Generally speaking, government has too many pulls on its purse strings to give out money for development and this is in part why they support the various Public Finance Initiatives. However there is funding available for certain initiatives especially if they wish to encourage development in an area that might otherwise be unattractive to developers, usually in the form of subsidies or reduced rates normally for a fixed period. They may contribute to projects that could have national significance such as buildings for an Olympic bid. They also indirectly fund projects with money raised on the National Lottery; projects such as the Millennium Dome, Wembley Stadium and the Lowry Arts Centre.

The Department of Trade and Industry may offer capital grants towards the costs of factories, and local authorities may also be willing to contribute to the development of local amenities and tourism.

The UK Government has also created Enterprise Zones as a means of job creation. Originally established in 2012 for a period of five years, this scheme has now been extended until 2021. Companies working in these Zones are given tax benefits and reduced business taxes as an incentive to encourage development.

The European Regional Development Fund assists in the funding of projects in the more deprived areas of the Community with the aim to improve the economy of the area. This is a very complex area of funding that is continually changing. Most of the funding is directed for a particular purpose and not normally directly to the developer. The developer can, however, work in partnership with the eventual end user to attract funds to enable the project to come to fruition. At the time of writing to determine what happens after Brexit would be supposition.

6 Valuation and development appraisal

6.1 Introduction

A simple definition of development valuation 'involves the calculation of what can be achieved for a development once completed and let, less what it costs to create' (Ratcliffe and Stubbs, 2009).

The developer needs to establish whether or not the project is likely to be viable. At this stage the information available is limited, and this chapter aims to indicate the various factors and the methods that can be adopted to inform the decision as to whether or not to proceed, and concentrates on valuation required for new projects and not the valuation of existing property.

The costs involved in a development include the land, design, construction and all the fees to the others involved in the process as outlined in 6.2. There are various questions the developer needs to answer depending upon the information available. These are:

- If the development costs and profit margins are known, what is the maximum cost of land that can be absorbed to make the project viable?
- How much profit can be made if the costs of the land and construction are known?
- How much rental income is required to justify the development?
- If the land value and profit are known, what is the most that can be spent on the construction costs?

6.2 Cost elements

There are many cost elements the valuation surveyor needs to put a value against in order to calculate answers to the questions raised above. Following are typical examples to be considered, although not all will be used in every case.

6.2.1 Cost of the land

This may be the price, after negotiation, the landowner is asking for the land, or the price the developer is able to offer after taking into account the cost of the development and the returns expected after selling or leasing and the profit required. The value of the land to the developer is not necessarily the asking price but the difference between the cost of the building to be built on the land and the market price of the finished development including the land. If this value is less than the market price for the land, then the development is not feasible.

Sometimes the client owns the land. In the case of government and local authorities the land cost may not be included in the overall budget calculations. In others, such as speculative housing, the developer will wish to pass on the current value of the land to the purchaser to release funds to purchase more land for further work. The value of the land will be related to supply and demand and in certain cases may only be available on lease – often 99 years – and may incur ground rent, which can have a significant effect.

The value of the land will be affected by several factors such as:

- its geographic location, for example land in the centre of London will be some of the most expensive in the world whereas that in the Highlands of Scotland may be very cheap;
- its proximity to the transport links, especially to road and rail, but also air, the latter being good for business but not so for residential;
- its proximity to other development and local facilities;
- the topography of the land – whether it is level or hilly and whether it has a high or low water table;
- the level of contamination – this can be a very expensive to remedy costing many millions as shown in 6.2.4;
- public rights of way through the land that may have to be moved or could result in possible protests;
- restrictions on its use either required by the seller or as a result of planning restrictions. The former can be very restrictive in the case of leasehold land. Some residential leasehold agreements restrict the parking of caravans, erecting of sheds and greenhouses and in some cases satellite television dishes. However, the owner of the property has the right to buy the freehold after three years;
- the state of the national and regional economy.

6.2.2 Legal and professional fees in acquiring the land

At the feasibility stage the developer requires to know what the developed value of the land is as described above. To obtain this a valuation surveyor will be engaged. Once the sums have been completed and it is agreed the land is to be acquired, land agents and lawyers fees will ensue. The valuation surveyor may also be asked to value different plots of lands for the development. For example, if it has been decided to relocate a government department to an area, the valuation surveyor will look for plots of land that can accommodate the development and produce a report outlining both the costs of the development (building and land) on each of the plots along with the advantages and disadvantages of each location. An important consideration at this stage is tax especially Stamp Duty Land Tax.

6.2.3 Total building floor area and lettable area

This is a function of the area of the site that can be practically built upon and local planning policies that may place restrictions on the density of the development, the maximum heights of buildings and so on. There also may be conditions laid down over means of access, numbers of car parking spaces permitted, landscaping, protected trees and views. Taking all these into account will determine the size of the building and the maximum amount of usable space. The method used for measurement is normally the RICS code of measurement practice or the new international property measurement standards.

6.2.4 Construction costs

The construction cost is the greatest outlay of all at a time when no revenue is being generated. It is important that this figure is as accurate as possible. This is a function of how far the design process has progressed. There are certain elements that need to be looked at in particular as the costs of these can be much higher than predicted if not based on some detailed investigation and analysis. Examples of these are:

- *Demolition of existing structures*: the cost of demolition of existing properties will vary dependent upon:
  - the value of the materials in the existing building in terms of the likelihood of recycling;
  - the volume of material to be disposed of, because of landfill charges;
  - the distance from the tip;
  - the ease or otherwise of dismantling;
  - whether or not it is in a busy urban area, because dust and noise become a greater issue, as do difficulties associated with transport and access.
- *Decontamination of land*: this can be very expensive depending upon the area and depth and type of pollution. Contamination is usually found in brown field sites where there has been industrial use. It can be as a result of petrol spillage or in certain parts of the country from mining activities, notably coal. The latter can cause polluted watercourses and water tables. Another major source of contamination occurs when building upon landfill sites. It is said that the decontamination costs of the site upon which the Millennium Dome was constructed, was of the order of £200 million.
- *Archaeological finds*: if an archaeological find is discovered on the land, then work must stop to permit an archaeological dig to take place. The time this takes depends on the significance of the find. It may be that the finds are just recorded and the construction can then continue or in extreme cases the finds may be so significant, as for example in the Shakespearean Rose Theatre in London, that the development may have to stop totally. While finds cannot always be predicted, the likelihood of finds in certain cities such as Chester are more probable than in others. Searching archives for information could be worth the effort in these cases.
- *Previous building work below ground*: it is not unusual when constructing on ground previously built on to find an array of potential problems that have not been recorded. Examples of these include cellars and redundant underground services the statutory authorities have no record of or that are in a different place than thought. The more detailed the search, the more likely these problems can be anticipated, thereby reducing delays and associated costs.
- *Ground conditions*: the type of subsoil and level of the water table can have significant impact on the costs of the foundations. The ground below the building has to support the weight and the activities that occur within the building. Generally speaking the weaker the soil the more expensive the foundations will be. Undetected variances in the ground type may delay the construction work

while redesigning takes place. Hence the more detailed knowledge there is about the ground conditions, the clearer and more accurate the costs will be.

- *Building services*: the costs of the provision of the building services can exceed 60 per cent of the cost of the completed building. This means it is often the largest cost element of the building. Thus careful scrutiny is required to ensure an effective integrated solution is produced that takes account of both the capital costs of installation and the running costs when the building is in use. It cannot be stressed enough the importance of considering this aspect of the design as early as possible, especially in terms of the strategic design decisions effecting energy consumption.

The total cost of the building is not just the cost of the materials, labour and plant used to construct it, but also the contractor's site and head office overheads, profit and all the temporary works, such as scaffolding, formwork and soil support. There are also the planning costs and fees such as the community infrastructure levy, and Town and Country Planning Act section 106 agreements, for example social housing, where the developer may be required to contribute to other facilities in the area such as roads and amenities as a condition of planning permission.

6.2.5 Professional fees

These fees are usually expressed as a percentage of the total costs of construction and vary from 10 per cent if the building is simple with lots of repetition, to as high as 17 per cent on very complex refurbishment projects. In practice the majority are around 12.5 to 14.5 per cent. These fees can be found in the recommended fee structures by the appropriate professional institutions and the RIBA Handbook of Practice Management. However, while it is useful to use these figures for development valuation, when it comes to engaging the profession, it is also possible to negotiate a fixed fee. It is argued by some that, other than the professionalism of the person engaged, there is no incentive to produce a cheaper solution if one is paid as a percentage of the final cost of the construction work.

6.2.6 Development and construction period

Throughout both the development and construction processes, costs accrue that have to be financed. It is therefore essential to know how long these will take so that the necessary funding can be put in place. While on the surface it would be reasonable to expect the time to construct the building to be longer than that of the development, this is not always the case. Much depends upon the complexity of the project, the time spent on the planning processes – especially if it goes to an appeal – whether there are several landlords involved, and obtaining the finance depending upon the current economic climate, the purpose and location of the development.

The process of obtaining planning permission can be relatively simple and cheap whereas if it goes to public enquiry it can be drawn out and expensive. In extreme cases, it may even have to go before parliament. There can be considerable costs incurred in these processes, both in terms of the fees charged by the consultants and the delays inherent in the system.

6.2.7 Costs associated with facilities management

If the project is to include a facilities management contract, the developer will also need to look at the capital costs of providing materials and equipment otherwise provided by the prospective client, including the costs of maintaining and replacement. Examples are:

- *Furnishings and fittings*: these include furniture, carpets, blinds and curtains.
- *Equipment used for the business as distinct from running of the building*: these include kitchen equipment and computer facilities.
- *Costs of managing, running and maintaining the building including energy and waste disposal*: if the project is to be sold on this does not apply. However if the building is leased part or all of these will have to be costed and paid for by the developer. The building will need maintaining and repairing as parts and components wear out. Decoration is required from time to time and the building will have to be kept clean both internally and externally. For multi-use lettings this service charge will have to be divided between the tenants.
- *Business rates and insurances*: rates bills may be passed down to the tenant, but if there is not full occupancy then this proportion of the rates will have to be borne by the developer. The building structure and fabric will have to be insured against damage and fire.
- *Security*: not only does the capital cost of security systems have to be considered as part of the capital cost of the building but also the manpower required to supervise and control the security of the building. In some cases, depending upon the use of the building and the implications of multi-occupancy, this can be an unexpectedly high sum since it can be required for 24 hours of the day, every day of the year.

6.2.8 Disposal costs

The three types of agency agreement most likely to be used when selling or letting a development are:

- *sole agency* – an agreement with a single firm. The expected fee would be between 1.5 and 3 per cent of the agreed selling price or approximately 10 per cent of the annual rental value;
- *joint agency* – where two agents are instructed by the developer to jointly sell or let the property. This type of agreement is used when, for example, a mixed development may require specialist knowledge by the agent or if the property is being advertised nationally. The developer will have to pay up to 1.5 per cent times the normal fee;
- *sub-agency* – where the agent may employ another agent to assist in the marketing of part of the development because of their specialist expertise. In this case fees are likely to be similar to a sole agency agreement.

In certain cases it may be decided to dispose of the development by auction.

6.2.9 Cost of finance

As has been indicated in Chapter 5 the developer will obtain finance from a variety of sources so as to distribute the risk. In return the developer will be charged interest and this has to be costed into the development calculations. This is a complex issue outside the remit of this text, however it is important the reader understands the issues involved. Finance has to be raised for the cost of the land, the design and construction processes and the period from handover of the building to the receipt of the first income from rents or from selling the completed development. All of these include professional fees as discussed before. The finance required is needed at stages over the development, as these different expenditures have to be covered as and when they occur.

The simplest way to estimate the finance for construction is to either take half the interest rate over the total construction period, or the full interest rate over half the building period as demonstrated in 6.5.2, the residual method. Both methods give different figures. To be more accurate the costs of construction can be broken down into monthly expenditure and the interest calculation conducted on the amounts of finance raised as and when it occurs as shown in 6.5.3, the discounted cash flow analysis. Compound interest is usually applied to the finance acquired for the purchase of the land and is normally over the entire development period.

The rate of interest paid depends upon the source of finance available to the developer, but for these calculations it is normal to apply a figure a few points above the current base rate.

6.2.10 Rental income

The calculation of rental income is complicated by the fact that it is obtained over several years in which time interest rates can change, demand may vary, not to mention inflationary influences. It is probably the most important cost item in the calculation and yet will rely heavily on experience and prediction rather than hard evidence. It could be argued that the most successful developers are those who have a natural ability to get this right.

The evidence most likely to be used is comparisons with rents obtained on similar properties in the area, adjusted to take account of the nature and location of the property, car park provision, available floor area and the sizes and shapes of accommodation in the development. The issue is further complicated by having to take into account the running costs of the development. For example an older property may have higher heating and lighting bills than a new purpose-built one. They can also have different natural daylight and ventilation characteristics, and standards of insulation.

It is also necessary to assess the likely speed of obtaining near to or full occupancy, as this will also affect the revenue stream. Developments with pre-letting agreements with tenants increase the certainty of the calculation. Often in major retail developments, development does not go ahead until some of the major high street retailers are signed up. The prospective tenant may be offered a fixed rent for a given number of years to act as an incentive.

6.2.11 Yield

A term commonly used is 'yield' which is an important concept. In essence it means the level of earnings the investors obtain from their investment, the higher the yield the more attractive the investment is. So a 15 per cent investment on £100 will pay £15 per annum whereas a 5 per cent will only produce £5 per annum. However, a word of caution, usually the higher the yield the greater the risk is. The developer has to take account of this risk when making the decision as to whether or not to proceed with the project. The quality and reliability of the tenant may also impact on the projected yield.

Some examples of the yield expected are shown below:

- shops (prime; excellent locations) 5–6 per cent;
- shops (fair; good locations) 6.5–8.5 per cent;
- offices (prime; excellent locations) 6.5–9.5 per cent;
- offices (secondary: let in suites) 8.5–12 per cent;
- industrial (prime; excellent locations) 8.5–10.5 per cent;
- industrial (secondary; poorer locations) 13–15 per cent;
- residential property 12–14 per cent (this will vary as market rents change).

6.2.12 Administrative costs of leasing

If the building is leased to one or more tenants, then rents have to be collected or chased, dilapidation schedules have to be produced and, when tenants leave, the building has to be surveyed and comparisons of its state made against the agreed dilapidation schedules. It then has to be re-advertised and let.

6.2.13 Contingencies

This is a percentage allowance made to cover for any unforeseen circumstances. Typical examples of uncertainty at various stages of the development are unforeseen ground conditions affecting the cost of the foundations and the costs of resulting delays caused by redesign, costs of planning gain and special needs of major tenants. How much is allowed for depends to a large extent on the quality of the information available at the time of the calculation of the development costs. The more accurate the information then the more certainty and hence less contingency provision is required. Typical figures are 5 per cent on the construction costs or 3 per cent on the gross development costs used. On refurbishment work this figure will normally be higher because of increased uncertainty. Rather than have a contingency item, some developers increase their expected profit by about 2 per cent.

6.2.14 Required profit

Any development carries with it a potential risk and the amount of return developers expect will take this into account. However a figure of between 15 and 17 per cent is generally the minimum return a developer would expect on the cost of the development.

6.3 Risk analysis and sensitivity analysis

The problem with any method of valuation is the number of variables that are involved, most of which are based on estimates. Add to that the fact that applying interest rates can be done in different ways, it can be seen that there is a significant opportunity for error to occur in the calculations. The purpose of this section is to comment on the potential for error, but not to enter into statistical probability calculations. If the reader wishes to develop this further, several of the texts referred to can assist, notably Ratcliffe and Stubbs (2009).

- *Land costs*. Sometimes the land costs are known because the developer owns the land, but if not then an offer has to be made. If the land is on the market a value will have been assessed by the vendor and the purchaser can then negotiate accordingly. However, if the land is not on the market, then its price and the time taken to purchase the land will be less predictable, as the vendor is usually in a stronger position during the negotiation.
- *Purchasing fees*. Since these are normally a percentage of the cost of the land the risk is directly proportional.
- *Usable area of the building*. The topography, shape of the site and planning restrictions can all affect the footprint and height of the building that can be built on the site. This in turn may determine the amount of circulation space required, which has to be deducted from the overall floor area to establish the useable area for a specific use. It should be noted that the developer might not have decided what this use might be so this can be significant. Knowledge of the future use and a certain amount of design work will increase the accuracy of this assessment.
- *Cost of the building*. The accuracy of this depends on how far the design has progressed, the quality of the site survey and the type of contract being negotiated with the building contractor. It is possible to agree a fixed price with the contractor taking the risk. For example the Lowry on Salford Quays had to be designed and built to a fixed budget, as this was all that was available. However this is unusual and normally the developer would be expected to take part of the risk and extra costs might be incurred as a result of ground difficulties, archaeological finds and other problems.
- *Professional fees*. These can either be a fixed fee or a percentage of the cost of the building, in which case the risk is proportional.
- *Time for design, building and letting*. Any delays that occur during any or all of these processes have a direct effect on the costs of the interest on the finance obtained for the project, but in times of high inflation or shortages of the particular type of property, it has been known to be lucrative for the developer. This would be a very risky strategy to plan for.
- *Rental value*. Establishing the likely rental value is one of the most sensitive factors. One is predicting the rent that might be collected a considerable time into the future. Since this is a function of supply and demand at the time, there are many unpredictable factors that can come into play, notably world and national economics that can be affected dramatically by unforeseen events such as September the 11th, the credit crunch, or war. This is why developers aim to obtain major tenants in advance of completion and even before commencement of the development. This means a hedge against a changing market place and also

reduces the risk of not letting smaller parts of the development, especially in the retail sector, as potential small tenants know that the larger ones will attract customers.

- *Finances for the project.* Before the development starts, all the finance for the design, construction and letting periods will be in place. This can be on either a variable interest rate leaving the developer exposed to fluctuations or on a fixed rate. However, in both cases there may be penalties if the project is delayed and the loan period has to be extended.
- *Investment yield.* This is determined by the market place and carries the same uncertainties as rental value unless the project is pre-sold.
- *Marketing and selling costs.* The costs of the agent's fees are expressed as a percentage of the selling price of the building or the annual rental fee and are therefore linked to any extra costs incurred in either of these. The developer may also spend extra money on promotion especially if the project is not being let as quickly as anticipated.

It is important that the developer looks at all these factors and analyses the risks involved and does not rely entirely on instincts and experience, although these two issues should not be ignored. Since each of the factors outlined carries different potential risks, the range of probability needs to be considered and statistical analysis applied. This is referred to as a Sensitivity Analysis.

6.4 Mathematics of valuation

6.4.1 Formulae

There are several formulae needed to calculate how money invested will change over the years as a result of interest made or lost. This is important for life cycle costing calculations and discounted flow calculations used for property valuations.

6.4.2 Compound interest or amount of £1

If money is invested over a number of years it will earn interest at the end of the first year. This interest will be added on and at the end of the second year this new sum will be used to calculate the interest for the second year and so on.

The formula for this is:

$$\text{Compound interest} = (\text{£s invested} + \text{interest})^{\text{number of years}}$$

or for every pound invested

$$CI = (1 + i)^n$$

So if £1000 were invested at 4 per cent over five years, the calculation would be as follows:

$$1000\,(1{+}0.04)^5 = 1000(1.04)^5 = \text{£}1220$$

Therefore £1000 invested would yield £1220 after five years. Since this formula is the basis of the rest the derivation is shown here:

Let the interest per annum on £1 be i.

The amount at the end of one year will be $(1 + i)$

The amount at the end of two years will be $(1 + i) + i(1 + i) = 1 + 2i + i^2 = (1 + i)^2$

The amount after three years will be $(1 + i)^2 + i(1 + i)^2 = 1 + 3i + 3i^2 + i^3 = (1 + i)^3$

Hence after n years will be $(1 + i)^n$

And the total interest paid on £1 in n years will be $(1 + i)^n - 1$.

6.4.3 *Future value of £1 invested at regular intervals*

If, rather than investing a lump sum one invested the same sum annually, then the calculation would be different. At the end of each year the total in the fund would comprise all the previous annual investments, plus the compound interest earned on them and the next annual contribution. The formula for this is:

Future Value = [(£annual investment + interest)$^{\text{number of years}}$ – £annual investment)] / interest

If £200 was invested annually over a period of five years at 4 per cent interest, then

Future value for one pound invested = [(£1+0.04)^5 – £1] / 0.04 = (0.22)/0.04 = £5.50

Therefore for £200 invested per annum the yield after five years would be £1100.

6.4.4 *Present value of £1 (PV)*

The question being answered here is what would £1 in a few years' time be worth today. In other words: what sum needs to be invested at the present time at a given rate of interest to accumulate £1 by the end of a given period of time. This needs to be addressed as for example a developer may wish to make a return of so much in n years' time and therefore needs to know what has to be invested now to make this return. In effect the opposite of the compound interest calculation.

The formula for this is:

Present value of £1 = £1 / (£1+ interest)$^{\text{number of years}}$

If a return of £1000 pounds was required in five years' time discounted at 4 per cent per annum, what would have to be invested today?

Present value of £1 = £1 / (£1+0.04)^5 = £1 / 1.22 = £ 0.819

Therefore the amount required to be invested to day to yield £1000 in 5 years = £819.

6.4.5 Present value of £1 per annum/year's purchase

This is asking what is a series of annual payments of £1 discounted, at a given rate of interest, worth today? Put another way, the present value of the right to receive an annual income of £1 at the end of each year for a given number of years, each year's income being discounted at a given rate of compound interest. This is needed, as the developer wants to know if the total received rental annual income, discounted at a fixed interest rate over the life of the building (in perpetuity), is sufficient to make a profit after deducting all the costs of the proposed development. The formula for this is:

Years purchase in perpetuity (YP) of 1 = 1 / interest

If the annual rental income is £1000, discounted at 4 per cent per annum, what would the gross development value be?

Years purchase in perpetuity = £1 / 0.04 = 25

The gross development value would be 25 × £1000 = £25,000.

6.4.6 Present value of £1 payable at regular intervals

This would be useful for calculating the capital equivalent of regular outgoings such as maintenance, wages, rates or rents. The formula for this is:

Present value of £1 payable at regular intervals = $[(£1 + \text{interest})^{\text{no yrs}} - £1] / [\text{interest}(£1 + \text{interest})^{\text{no yrs}}]$

What is the present-day value of £200 paid annually for five years assuming an interest rate of 4 per cent?

Present-day value of £1 = $[(£1 +0.04)^5 - £1] / [0.04(£1 + 0.04)^5] = [1.22 - 1] / [0.04(1 + 0.04)^5]$
= [0.22] / [1+ 0.0488]
= £4.51

Therefore the present-day value of investing £200 per annum for five years = £902.

6.5 Development valuation techniques

There are various methods of evaluating the viability of a development, including the comparative method, the residual method and the discounted cash-flow analysis. There are other methods in use such as the investment method and an interested reader may refer to such texts as Scarrett and Osborn (2014) and Ratcliffe and Stubbs (2009).

6.5.1 The comparative method appraisal

This is a simple method by which the valuer looks at similar properties in the area where the values are known, for example, they have just been sold. Then, taking account of any differences, the valuer makes a value assessment. Estate agents selling residential properties use this method. The differences will include its architectural design and appearance, location, state of repair, quality of internal finishing, such as bathrooms and kitchens, and any other additions made. This method is effective when the market is stable, but more difficult to use when the market is volatile and house prices are rising almost daily, which is why it is essential when selling a property to obtain several valuations. For commercial and industrial properties the valuer would extend the list of comparables to take account of the specific use and needs of the client.

6.5.2 The residual method

This method has its limitations, but it does give a quick indication as to whether or not the development might be viable and is a fall-back position if comparisons cannot be used. It should only be used for a quick approximate answer. There have been serious criticisms of the method by the Lands Tribunal and its use as a technique for valuation in the 1970s' property boom also brought it into disrepute. The method is not precise enough as in practice expenditure is assessed during the construction phase at monthly intervals and in the case of the design, fees are often paid at the end of each phase of their work in line with the RIBA Plan of Work (Chapter 7.1). Its use can produce inaccurate forecasts affecting the developer's prospective profit. It can be a useful tool for the developer when considering several alternative projects in eliminating the clearly non-profitable ventures.

In essence the residual method is based on a simple equation:

Residual value = Gross development value – (Costs + Profit)

As indicated in the introduction the developer will most likely wish to find out how much he can afford to pay for the land and still make the profit required; or if the cost of the land and building works are known, establish the likely profit from the development. The figures used in the examples below are symbolic rather than true to life, but are used to demonstrate, in Tables 6.1 and 6.2, the calculation in two different examples.

Establishing the value of the land

The total floor area of the proposed office building is 3000m^2 of which 2400m^2 is lettable, the rest being circulation areas etc. The initial development period is estimated to be 12 months followed by a 10-month construction phase and an allowance of four months to let the property. Construction costs are estimated to be £1000 per m^2. It is expected to obtain rents of £200 per m^2 on the lettable space. Finance has been arranged at 12 per cent per annum and a 6 per cent yield is predicted. The developer wants to make 15 per cent profit on the capital value (see Table 6.1).

Establishing the likely level of profit

Using the same information but making an assumption that the land value is £2,500,000, the question posed is what profit will the developer make? See Table 6.2.

Table 6.1 Establishing the value of the land

Capital value after development		£	£
Estimated rental value per annum	2400m² @ £200	480,000	
YP in perpetuity at 6%		16.67	
Estimated gross development value or capital value			8,001,600
Development costs			
Building costs	3000m² @ £800	2,400,000	
Cost of finance on building costs (interest charges)			
– (12% pa for construction period on half cost of building)	2.4m × 0.5 × 0.12 × 10/12	120,000	
– (12% pa for letting period on full cost of the building)	2.4m × 1 × 0.12 × 4/12	96,000	
Professional fees	12.5% of building costs	300,000	
Cost of finance on professional fees (interest charges)			
– (12% pa for construction period on half cost of fees)	300k × 0.5 × 0.12 × 10/12	15,000	
– (12% pa for letting period on full cost of the fees)	300k × 1 × 0.12 × 4/12	12,000	
Subtotal		2,943,000	
Contingency 5% on costs including interest	2,943k × 0.05	147,150	
Promotion and marketing	estimate	60,000	
Agents' fees – 10% of annual rentable value	480k × 0.1	48,000	
Sale costs if property sold when fully let			
– (3% of capital value)	8,001,600 × 0.03	240,048	
Subtotal		495,198	
Net development costs		3,438,198	
Developer's profit 15% on net development costs		535,730	
Total development costs		3,973,928	
Residue to buy land	8,001,600 – 3,973,928	4,027,672	
Residual land value			
Let land value = v	v		
Cost of acquisition @ 2.5%	$1.025v$		
Cost of finance for land (interest charges)	$1.025v \times 0.12 \times 26/12 = 1.64v$		
(12% per annum for 26 months)			
Profit on total land cost @ 15%	$1.64v \times 1.15 = 1.88$		
The price developer can afford to pay for land	4,027,672/1.88	**2,142,000**	

Note interest added – not compounded for simplicity

Table 6.2 Establishing the likely level of profit

Capital value after development		£	£
Estimated rental value per annum	2400m² @ £200	480,000	
YP in perpetuity at 6%		16.67	
Estimated gross development value or capital value			8,001,600
Development costs			
Building costs	3000m² @ £800	2,400,000	
Cost of finance on building costs (interest charges)			
– (12% pa for construction period on half cost of building)	2.4m × 0.5 × 0.12 × 10/12	120,000	
– (12% pa for letting period on full cost of the building)	2.4m × 1 × 0.12 × 4/12	96,000	
Professional fees	12.5% of building costs	300,000	
Cost of finance on professional fees (interest charges)			
– (12% pa for construction period on	300k × 0.5 × 0.12 × half cost of fees)	15,000 10/12	
– (12% pa for letting period on full cost of the fees)	300k × 1 × 0.12 × 4/12	12,000	
Subtotal		*2,943,000*	
Contingency 5% on costs including interest	2,943k × 0.05	147,150	
Promotion and marketing	estimate	60,000	
Agents' fees – 10% of annual rentable value	480k × 0.1	48,000	
Sale costs if property sold when fully let – (3% of capital value)	8,001,600 × 0.03	240,048	
Subtotal		495,198	
Net development costs		3,438,198	
Land cost		2,500,000	
Acquisition cost @ 2.5%	2,500,000 × 0.025	62,500	
Cost of finance on land (interest charges) – (12% pa for development, construction and letting period)	2,562,500 × 1 × 0.12 × 26/12	664,250	
Total development costs		**6,664,948**	
Profit 8,001,600 – 6,664,948	1,336,752		
Developer's percentage profit	(1,336,752 × 100)/ 8,001,600	**16.60%**	

6.5.3 Discounted cash flow analysis

This is described as a method of evaluating an investment by looking at projected cash flows and taking into account the consideration of the time value of money (i.e. interest and depreciation). While still not completely accurate, this method more truly reflects the actual expenditure incurred by the developer. Its weakness is that it is still vulnerable to any delays that might occur and if expenditure varies from that predicted, the applied interest rates can distort the final picture.

Taking a simple example:

- The cost of the land is £400,000.
- The fees for acquiring the land are £20,000.
- The building costs are £2,000,000.
- The design team fees are £300,000.
- The agency fees are £60,000.
- A contingency of 5 per cent is added to the building and design costs.
- The development is sold for £4,000,000.
- The present value of £1 is 6 per cent.
- The design period is six months.
- The construction period is 18 months.
- The development will be sold during the following six months.

While this is fine, it is more usual to set out this type of appraisal with columns for income, expenditure, net cash flow, discount factor (PV) and discounted cash flow; the NPV being the final total. Time periods are typically reflected in the rows.

Table 6.3 demonstrates six-month periods for simplicity, but could in reality be broken down further into either three- or one-month periods. The figures in brackets are in debit and the others in credit.

As a learning exercise, this information can be entered into a standard spreadsheet, then, by altering the land costs, the building costs or the PV of £1 by 1 per cent

Table 6.3 Cash flow table

Item/Months	*0*	*6*	*12*	*18*	*24*
Land costs	(400,000)				
Purchase costs	(20,000)				
Building costs		(600,000)	(700,000)	(600,000)	
Design team fees	(100,000)	(50,000)	(50,000)	(50,000)	(50,000)
Agency fees					(60,000)
Contingency 5%	(26,000)	(32,500)	(37,500)	(32,000)	(2,500)
Sale proceeds					*4,000,000*
Cash flow per period	(546,000)	(682,500)	(787,500)	(682,000)	*3,987,500*
Cash flow cumulative	(546,000)	(1,228,500)	(2,016,000)	(2,698,000)	(1,389,500)
PV of £1 @ 6%	1	0.971	0.943	0.916	0.890
Net present value (NPV)	(546,000)	(662,707)	(742,612)	(624,712)	*3,548,875*
Cumulative NPV	(546,000)	(1,208707)	(1,951,319)	(2,576031)	*972,844*

Table 6.4 Disadvantages and advantages of the two methods

Residual method	
Advantages	*Disadvantages*
It is simple to use	Land Tribunal has been critical of the method
Can be used for all sizes of developments	The timescale of the project and the finance is very approximate
The residual can be the profit if the site price is already fixed	A small change in one of the factors considered can change residual out of all proportion
In simple terms it mimics the market approach. i.e.	
Gross Development Value – Costs = existing value	Many assumptions have to be made on the input data
Discounted cash flow	
Advantages	*Disadvantages*
More scientific and there is greater analysis	It is considered too academic by some
Detailed analysis of expenditure and income	Can produce over-analysis
A sensitivity analysis can be carried out	Increases the opportunity for input errors
It demonstrates the cash flow requirements	At early stages of development there is insufficient firm data leading to inaccuracy
Can include possible inflation influences on larger schemes	
Can more readily assess the contribution of the various elements to the scheme	

increment, and then different combinations of all three, it will be seen the affect uncertainty can have on the final cash flow sum.

The net present value (NPV), as used in this example, is what the sum of money will be worth at that point in time compared with the present time because it has been discounted, in this case at 6 per cent.

6.5.4 Comparison of residual method and discounted cash flow

There is no doubt that there is a role for both methods. The residual method is the more appropriate for an initial appraisal as it provides a more immediate view on the project's viability and remains the most common in practice in spite of the reservations of the Land Tribunal. The discounted cash flow method allows for more fine-tuning, but both have their limitations because of the problem of ensuring the data input is accurate. See Table 6.4.

Bibliography

Ashworth, A. (2010) *Cost Studies of Building*, 5th edn. Long man.
Ashworth, A. and Perera, S. (2015) *Cost Studies of Building*, 6th edn. London: Routledge.
Baum, A. (2011) *The Income Approach to Property Valuation*, 6th edn. EG Books.

Baum, A., Mackmin, D. and Nunnington, N. (2011) *The Income Approach to Property Valuation*, 6th edn. London: Routledge.
Ferry, D.J. and Brandon, P.S. (2007) *Cost Planning of Buildings*. Oxford: Blackwell Science.
Jaggar, D., Ross, A., Smith, J. and Love, P. (2002) *Building Design and Cost Management*. Oxford: Blackwell Science.
Millington, A. (2001) *An Introduction to Property Valuation*, 5th edn. London: Routledge.
Ostine, N. (2013) *RIBA Handbook of Practice Management*, 9th edn. London: RIBA.
Ratcliffe, J. and Stubbs, M. (2009) *Urban Planning and Real Estate Development*, 3rd edn. London: Routledge.
Scarrett, D. and Osborn, S. (2014) *Property Valuation: The Five Methods*, 3rd edn. London: Routledge.

7 Introduction to design economics

7.1 RIBA Plan of Work 2013

Before investigating this topic it is important to see the context in which it occurs. In the UK work is traditionally carried out to the RIBA Plan of Work. It is often used to determine when and how much the architect and others in the design team are paid.

This document is a very succinct way of understanding the process and permits the user to focus on the issues relevant at the time. It comprises eight stages of work: strategic definition, preparation and brief, concept design, developed design, technical design, construction, handover and close out, and in use. Each of these stages is divided into eight subsections: core objectives, procurement, programme, town planning, key support tasks, sustainability checkpoints, information exchanges and UK Government information exchanges. Further direction is then given for each of these subsections for each of the eight stages of work.

Take for example environment design decisions. It is important the design team consider strategic environmental decisions such as energy strategies and the use of natural light and ventilation at the concept design stage as this will determine the orientation and footprint of the building.

As indicated in 6.2.5 professional fees paid to architects, building service engineers, consultant structural engineers, quantity surveyors and others, vary from 10 per cent to 17 per cent of the final cost of the building, depending upon its complexity. This figure is divided between the design team according to their contribution. Architects can be paid as they complete various stages of the RIBA Plan of Work. The others involved would be paid in a similar manner depending on when and how much effort would be needed to comply with a particular stage of the work.

While many are still paid this way much has changed in the last few years, because of the different ways that work is procured and the need to fast track projects. It is clearly in the interests of the members of the design team not to quote a lump sum for the work unless the brief and service to be provided have been very clearly defined, which explains why there is a preference for the fee to be linked with the final cost of the project.

7.2. Factors affecting the cost of the building

7.2.1 Introduction

The following sections are written with the assumption that the decision has been made to progress the development and a thorough site survey has been carried out to reduce uncertainty. It should be noted there are often overlaps between the various sections.

Clients, especially public bodies, used to have two different cost budgets for buildings, namely the capital cost of the building and the maintenance and running costs. These were almost invariably kept separate with the result there was no incentive to design buildings that were efficient during their life span. This has over the years gradually changed so that now emphasis is placed upon making design decisions that take account of these issues. More detailed analysis is given in Chapter 9 Whole life costing and life cycle analysis.

7.2.2 Energy and carbon foot print

In essence this is establishing the amount of carbon generated both during the construction and use of the building, and developing ways of reducing it as much as possible. It commences by first establishing whether or not the building is located in the best place, bearing in mind such issues as the energy consumed in the travel of personnel, visitors, and the import and export of materials from the premises. The orientation and footprint of the building can affect the ability to harness natural resources such as sunlight, heat and natural ventilation and at the same time alleviate problems caused by wind chilling and there are design solutions to improve this further such as automatic shutters and solar shading. The selection of materials for the construction of the building is crucial as many materials have a high embodied energy and may also have to be transported long distances from their place of manufacture and their life expectancy needs to be considered. Insulation issues need to be addressed and consideration given to providing higher standards than those laid down in the Building Regulations. The building can be used to generate energy using solar panels. The selection of energy efficient plant is important. The use of heat pumps, tri-generation and district heating schemes especially on large projects is often feasible.

7.2.3 The site

No two sites are the same and some of the issues have already been raised in 6.2.4, such as construction costs.

- The shape and area of the site is the significant determining factor as this sets the parameters as to the possible footprint in terms of shape and orientation, and the number of storeys that will be required. Restricted sites also affect the amount and way materials can be stored during the construction process and limits the type and positioning of accommodation for site personnel. It may be necessary to find alternative accommodation close to the site.
- The topography of the site is relevant, as sloping sites generally will mean more expensive building.

- Poor ground conditions will often mean more expensive foundations and temporary works during construction required either to support the ground during excavation or in keeping water at bay if the water table is high.
- Other influences on substructure costs include the cost of dealing with contaminated ground, hidden services, underground structures and archaeological finds.
- The location of the site can have a significant affect on the construction costs. In a congested urban area there may be restrictions on the hours of working, the times vehicles can off-load, difficulties in access, levels of noise permitted and trespass problems associated with tower cranes passing over neighbours' properties (see 19.2). Rights of way, such as pavements, may have to be temporarily diverted or covered to protect the general public from falling objects. Sites situated in rural surroundings may incur extra transport costs and access routes may be too weak to accept the heavy transport without some widening or strengthening.
- Surrounding buildings may have to be protected from damage from the construction work and existing buildings on site demolished.

7.2.4 Size and scale of the project.

While 7.2.5 to 7.2.8 can distort the cost implications of this section, generally speaking the larger the project for a particular use, the cheaper the unit cost becomes. The reasoning behind this is as follows:

- Larger projects can be more efficiently managed and completed faster as it becomes more economical to use sophisticated plant and equipment, subcontractors put in more competitive bids because of the size of their contract, and materials can be obtained cheaper due to quantity.
- As the size of a project diminishes, the design cost as a proportion of the overall cost increases as the time taken for the design is disproportionately higher.
- Providing the overall unsupported floor span does not become excessive, the larger the floor plan area, the cheaper the building becomes. This is because of the costs of providing external walls and, if load bearing, their foundations, is relatively expensive. This is referred to as the wall-to-floor ratio.

7.2.5 Usable and non-usable space

The term non-usable space is used to describe areas of the building that are primarily non-productive. This usually means circulation spaces that include corridors, lifts, stairwells, escalators and floor areas within rooms/spaces needed for the passage of people and goods. The developer will obtain the greatest financial return from the building if this non-usable space can be kept to a minimum. Other issues are as follows:

- Often open-plan design offers a lower ratio of non-usable space.
- With the exception of very small buildings, such as a private house, some protected circulation space will be mandatory to permit safe evacuation from the building in the event of fire.
- Circulation spaces used in the event of fire will have to be of a minimum size to comply with fire regulations.

- A case can be made in some buildings to have a larger than required circulation space, especially in entrances and lobbies, so as to portray a grand image to visitors and users.
- Complicated and irregular footprints usually require a greater proportion of non-usable space. This is why, for example, hotels tend to use simple long rectangular accommodation blocks with a corridor running down the centre, with the bedrooms opening from either side thereby reducing the proportion of non-usable space.
- In complex situations space planners (see 1.8.2) are brought in to assist in optimising the remaining usable space.

7.2.6 Plan of building or footprint

This is a very complex problem. Much of the footprint may already be determined either by the site itself, such as a confined city site, or by the very nature of the business such as a supermarket or car manufacturing plant. Environmental issues may also come into play. If the developer requires an energy-efficient building using predominantly natural light and ventilation, then the distance from the external wall to the back of the room is limited to approximately 6 to 7 metres. This means, if using a central corridor for circulation, the maximum distance from the two external walls is approximately 15 metres. Those working in the space furthest from the windows may need some artificial lighting depending upon the nature of their work. If the footprint of the building were in excess of this then atriums could be considered as an alternative to providing artificial lighting and ventilation.

Generally, because of the wall-to-floor ratio, a square or rectangular building is the most economical to build as any change of direction complicates the roof details and sometimes the foundations. A complex footprint can slow down the production process of cladding, especially if constructed of brick and block, because of the number of changes of direction. Square buildings are not an economical solution if the site has a significant slope and in this case, it is generally accepted rectangular buildings following the direction of the slope are more economical.

If the development comprises more than one building, there are economies if the buildings can be linked in some way as this reduces the amount of external wall and foundations that have to be constructed. For example it has been estimated that for the same comparable area a semi-detached house is 6 per cent more expensive than a terraced house.

7.2.7 Height

Generally the taller the building, the more expensive it is compared with low-rise construction. However, it may be necessary to build high because of limited site area, scarcity of land and/or the land costs are so high they dominate the cost equation making it more economical to reach for the skies. This is often the case in major cities throughout the world. The key reasons for this extra cost are as follows:

- The structure and the foundations have to be more substantial than low-rise. This is because the structure has to support the extra height and also resist the horizontal wind forces.

- In order to overcome the problem of disturbing the structural equilibrium of adjacent existing buildings it may be preferable to create basements. This creates extra floor space that may be of benefit to the developer.
- Circulation spaces have to be increased in area to evacuate the greater number of building users and to satisfy fire regulations.
- Lifts will have to be provided.
- The building will be more highly serviced using more sophisticated and larger equipment, which in turn has to be installed by specialist contractors.
- The construction costs are higher because of the need to provide cranes and hoists to lift and transport materials and labour.
- Long-term maintenance is likely to be more expensive because of the extra risk of deterioration at high levels, difficulty of access to carry out the works and more complex equipment installed for the building services.

In spite of these reasons there are occasions where buildings of a few storeys high offering the same usable area may be more economical than single storey as some costs go down or remain constant as the footprint diminishes and the numbers of storeys increases. Examples of these are as follows:

- The roof covering diminishes significantly, though not completely in proportion, as circulation spaces increase. For example a-two storey building only requires a roof of about 50 per cent of the single storey building for the same usable area.
- The increase of the foundation design may not be substantial initially, until the weight of the multi-storey solution requires a different design solution.
- Other than for disabled access and the movement of heavy or bulky goods, lifts will not be required.
- Internal finishes remain similar irrespective of height.

7.2.8 Storey height

The floor-to-floor dimension is determined by several factors, such as minimum heights laid down in the building regulations, the ability to accommodate tall pieces of equipment and plant required for the user's business, space to cater for building services such as ventilation ducts and raised floors for Information Technology services. Other buildings such as sports centres, swimming pools, tiered lecture theatres and conference facilities will require extra height because of their specific design requirements. Any increase in storey height adds to the overall cost of the building as it both increases the quantities of materials used and adds to the weight of the building which in turn has to be supported by the structure and foundation. There are also energy implications in running the building, as there is a greater volume of air to be heated or cooled, not to mention the thermal capacity implications depending upon the materials selected for the construction of the building.

7.2.9 Buildability and construction

This is in part linked to value engineering (see Chapter 10), but in essence is to do with how the design has taken into account the problems of construction and produced solutions to assist the construction process, thereby making it more productive and hence more economical. Some examples are:

- to design so standard-sized components can be used without adaptation on site or alternatively agree with the manufacturer to produce the required size;
- to have as much repetition as possible, rather than many different sizes, without compromising aesthetic considerations;
- to consider the sequence of operations in terms of productivity and continuity of work;
- to ask the question 'How is what I want produced in practice?' and if the answer proves to be 'With difficulty' then 'Can I modify my design accordingly?';
- to ask whether it is possible to modify the design and methods of fixing so that all exterior work can be carried out from within the building so no scaffolding is required on the outside of the building;
- if possible to bring suppliers', subcontractors' and the contractor's production expertise as early as is practical into the design process;
- to consider the use of off-site production. Sometimes the components produced can be more expensive, but not always, but savings can be made in the increase in the speed of construction that results;
- to ascertain whether there is a freely available source of both material and skilled labour of the kind required to complete the building;
- to determine whether provision has been made for safe working conditions for the maintenance of the building after completion.

7.2.10 Deconstruction

Increasingly, consideration is being given to ease of dismantling the building so as much of the material used can be recycled when the building comes to the end of its useful life. There are some serious implications for the designer in this concept. Not only has the designer to consider how components and structure are taken apart, but that many of the newer construction techniques designed to combat skill deficiencies and improve productivity make segregation of materials more difficult, such as gluing architraves and skirting. Equally many of the completed multi-material components are not readily dismantled. At first glance, this could mean increased construction costs, unless alternative design solutions are provided, but this has to be balanced against the longer-term issues. It may be that as designers consider facilities management issues in more detail, the need to replace, repair, maintain or alter the building economically may also have a positive knock-on effect on the economics of deconstruction and recycling.

Bibliography

Ashworth, A. and Perera, S. (2015) *Cost Studies of Buildings*, 6th edn. London: Routledge.
Ostime, N. (2013) *RIBA Handbook of Practice Management*, 9th edn. London: RIBA.

8 Approximate estimating

8.1 Introduction

These techniques can be applied for several reasons in the development of the project. First, as a feasibility study to determine whether or not the project is financially viable. Second, for budgeting the project as a means of establishing the financial implications throughout the project from design to completion, so that the client is aware of the financial requirements, known as the cost plan, and also to assist in the obtaining of funds to finance the project; and third, for estimating in providing information to assist management in deciding whether or not their company wishes to tender (see 13.4).

8.2 Methods

Four methods are available, the accuracy of each very much dependent upon the accuracy and quantity of information and data available. These are:

- functional unit method or unit of accommodation method;
- superficial (floor area) method;
- elemental cost plan;
- approximate quantities.

8.3 Functional unit method

This method is largely used in the public sector where significant sums of money are involved, but very little, if any, design work has been executed.

Examples of a functional unit are the number of hospital beds, student residential accommodation places, and high- or low-risk prisoners, hence also the term 'accommodation method'. Having established the number of places required, this number is then multiplied by a figure based on the cost of providing such a place as shown in Table 8.1. This cost is calculated from previous provision and adjusted, making allowance for inflation, changes in specification and design, the market forces at the time, land costs, etc. It can rarely be accurate as a method because of the many intangibles, not least the condition of the land it is being built on and availability of services and infrastructure. Indeed the accuracy could be as much as 25 per cent. However, it does give a ballpark number for the setting of outline budgets. It can also

Table 8.1 Functional unit costs

Functional cost	*Area* m^2	*Cost £1000,s*
District general hospital	65–85 / bed	65–96 / bed
Theatres		7.5–11 / seat
Secondary school	6–10 / child	4.5–8.5 / child
Large student residences (200 plus rooms)	18–20 / bedroom	11–20 / bedroom

be used to establish approximately how much provision can be made if there is only a certain amount of money available for the purpose.

8.4 Superficial area method

This is an improvement on the previous method, but relies on some elementary design work being carried out. This is a method readily understood by developers, designers and builders alike.

It is necessary to know the superficial area of the building (the footprint) the number of storeys and any significant differences on any floor level. It also requires the type of usage for the areas involved if they are different. No deductions are made for stairwells, lift and circulation spaces. This superficial area in square metres is multiplied by an approximate figure based upon the usage of the building. This will vary dependent for example, on the amount of services, quality of finish etc. that the building type may require. As can be seen in Table 8.2, the range for each building type is very large, so further adjustments to this multiplying factor will be made to take account of such variants as location, quality of specification, complexity of shape, number of storeys, and suspected ground conditions including substrate and topography.

Table 8.2 Superficial area costs

Building type	*Costs per square metre*	*Building type*	*Costs per square metre*
Hospitals	£700–1400	Schools	£400–1000
Swimming pools	£900–2000	Student residences	£600–1400
Public houses	£500–1400	Hotels	£300–1000

8.5 Elemental cost plan

This can be produced from the designer's preliminary drawings. To do this, a list of the elements of the building is drawn up and an elemental cost added. This later information comes from measuring the size/amount of the element and multiplying by a unit rate. This rate will be based, in the case of the contractor, on the experience of the company and estimating staff employed and for the Quantity Surveying practice on their previous expertise and standard works, such as Spons, Laxtons and the BICS Standard Form of Cost Analysis or the BICS online (www.bics.co.uk).

Table 8.3 Elemental cost plan

Element	*Elemental costs £/m²*	*BUILDING A*	*Elemental costs £/m²*	*BUILDING B*
Substructure*	68.1	197490	95	365,750
Steel Frame	68.81	199549	67	257,950
Upper floors	36.21	105009	36	138,600
Roof	50.43	146247	50	192,500
Stairs	14.22	41238	14	53,900
External walls	62.5	181250	63	242,550
Windows and external doors	46.55	134995	47	180,950
Internal walls and partitions	31.03	89987	31	11950
Internal doors	6.59	19111	7	26,950
Wall finishes	13.36	38744	13	50,050
Floor finishes	61.21	177509	61	234,850
Ceiling finishes	17.67	51243	18	69,300
Steelwork (fireproofing)	3.88	11252	4	15,400
Fittings and furniture	9.48	27492	9	34,650
Sanitary appliances	5.6	16240	6	23,100
Disposal installations	4.4	12760	4	15,400
Water installations	6.03	17487	6	23,100
Heat source and space heating	52.16	151264	52	200,200
Ventilation and cooling	20.69	60001	21	80,850
Electrical installation	103.45	300005	103	396,500
Lift installation	0	0	13	50,050
Security alarms	27.28	79112	27	103,950
Fire alarms	10.73	31117	11	42,350
Builder's work in connection	7.03	20387	7	26,950
Minor works	3.1	8990	3	11,550
Total elemental cost	*728.53*	*2112737*	*768*	*2,956,800*
Preliminaries for building	58.31	169099	58	223,300
Net Building Cost	*786.84*	*2281836*	*768*	*2,956,800*
Site works	105.39	305631	105	404,250
Drainage	24.18	70122	24	92,400
External services	7.16	20764	7	26,950
Preliminaries for external works	8.21	23809	8	30,800
Net scheme cost	*931.78*	*2702162*	*912*	*3,511,200*
Design fees	44.83	130007	45	173,250
Statutory fees	11.03	31987	11	42,350
Overheads and profit	55.17	159993	55	211,750
Contingencies	34.47	99963	35	134,750
Budget total	*1077.28*	*3124112*	*912*	*3,511,200*

Alternatively, rates used on similar buildings are proportioned accordingly as shown in Table 8.3, which is adapted by permission of the Chartered Institute of Building, from *Code of Estimating Practice* (1997), endorsed by the CIOB. In this case Building A has already been built and its overall area was 2,900m^2. The proposed building, B, is 3850m^2. Prices can be adjusted to take account of obvious variances between the two buildings. For example in this case in the Substructure*, building A used reinforced concrete pad foundations and the case of building B, piled foundations.

8.6 Approximate quantities techniques

Clearly accurate quantities are to be preferred, but this may not always be possible, as the design may not have reached a sufficient development to provide this detailed information. In such cases, approximate quantities can be used. Examples of where they may be used are where speed is the key issue and all the detailed production drawings cannot be produced in time and where the ground surveys have not been completed or there are other unknown factors.

There are three different ways of achieving a detailed measure of the actual quantities used in the construction:

- The work is measured at the end of the contract.
- The work is measured as soon as the production drawings are completed and then substitution bills are produced.
- The work is measured as it is constructed.

Approximate quantities give a more accurate estimate of the likely cost of the building than the previously described methods, but can only be used if there is sufficient detailed design completed to carry out the calculations. Similar in approach to the production of the bills of quantities, the sequence of abstracting the information is usually the same, as far as it can be, to the methods of measurement (see Chapter 12), but in this case the unit of measure is a composite of several smaller units and only measures the main or significant elements of work to be constructed. For example: brick manholes, described as: 'Excavation, 150mm concrete bed, 215 class B engineering brick walls, 100mm clayware main channel, concrete benching, step irons and cast iron manhole cover'; and strip foundation, described as: 'Excavating, 225mm concrete, brickwork, dpc, common bricks.'

In the case of approximate quantities estimating, because the design is further advanced than in the previous three, if the design is being developed using building information modelling it may be possible to take advantage of this in this estimating method.

Bibliography

Brook, M. (2016) *Estimating and Tendering for Construction Work*, 5th edn. London: Routledge.
Cartlidge, D. (2013) *Estimator's Pocket Book*, London: Routledge.
Chartered Institute of Building (1997) *Code of Estimating Practice*, 6th edn. London: Longman.
Greenhalgh, B. (2013) *Introduction to Estimating for Construction*. London: Routledge.

Laxtons Building Price Book. London: Butterworth & Heinemann.
Smith, A.J. (1995) *Estimating, Tendering and Bidding for Construction Work*. London: Macmillan Press.
Smith, R.C. (1999) *Estimating and Tendering for Building Work*. London: Longman.
Spon's Architects' and Builders' Price Book. London: E & FN Spon.
The Aqua Group (1999) *Tenders and Contracts for Building*, 3rd edn. Oxford: Blackwell Science.

9 Whole life costing (WLC) and life cycle assessment (LCA)

9.1 Definitions

There can be confusion in the way the terms life cycle costing/analysis and whole life costing are used. The author uses the following definitions for the purpose of this text.

> *Whole life cost* (WLC) is a tool to assist in assessing the cost performance of construction work, aimed at facilitating choices where there are alternative means of achieving the client's objectives and where those alternatives differ, not only in their initial costs but in their subsequent operational costs.
>
> (from the developing ISO Standard 15686–5, Buildings and constructed assets – Service life planning)

> *Life cycle assessment* (LCA), sometimes referred to as life cycle analysis is, 'a method to measure and evaluate the environmental burdens associated with a product system or activity, by describing and assessing the energy and materials used and released to the environment over the life cycle.
>
> (Edwards et al. 2000)

9.2 Introduction

Traditionally buildings were looked at financially in two distinct ways: capital costs and maintenance costs. The effect of this was that many buildings, especially in the public sector, were built as cheaply as possible within the budget allocated for this purpose, often using inferior materials and components that needed replacing at frequent intervals, thereby raising the maintenance budget.

Now the industry looks at the costs incurred throughout the life of the building, as it is argued that the client, although initially needing to find more capital money, would, in the long term, pay out less for the building over its lifetime as a result of lower maintenance and running costs. For example by increasing the insulation and using the lowest energy lighting available, the capital costs of proving this combined with the energy consumption during the life of the building would be less than building to minimum insulation standards and using less efficient bulbs. There would also be a further saving, as the boiler used to drive the central heating would be of smaller capacity.

The problem with WLC calculations is that the number of variables that have to be considered inevitably make the outcome figures suspect, especially when predicting costs over a long timescale. For example who can predict what interest rates are going to be 15 years from now? However, it is important at the design stage to be making environmental decisions that both impact on the outline and detailed design of the building, considering such issues as the orientation and footprint of the building, the energy strategy and selection of materials with an eye on both the impact of the building itself and the impact resulting from running and maintaining the building. The developer will want to know whether or not the decisions made for sustainability reasons stack up financially; they probably will.

This subject area has become more relevant with the increasing trend towards developments such as Public Finance Initiatives (PFI) that require not just the building to be constructed, but also to be maintained by the contracting organisation for 25 years or more – known as facilities management.

It is also important to relate these decisions with value engineering (see Chapter 10), which can be used as a tool for WLC and LCA. This is because any value engineering applied to a building at the design stage will have possible implications for the long-term maintenance and use of the building.

The first part of this chapter considers the cost effectiveness of a building in terms of its capital cost, maintenance and running costs, i.e. WLC, but it should also be remembered that cost effectiveness to the user/owner includes other issues besides the cost of the building. For example, an airport is also about the number of long-term parking spaces available, ease of setting down and picking up passengers, the speed at which passengers can be processed, speed of baggage handling, security checks and processing and so on. In other words the cheapest building is not always the most effective in use.

9.3 Cost centres

In a whole life costing exercise it is necessary to consider all the expenditure that may be required throughout the life of the building. All of these are affected by the design decisions. There are various approaches to categorising these. The RICS proposes:

- capital costs
- financing costs
- operation costs
- annual maintenance costs
- intermittent maintenance, replacement and alterations costs
- occupancy costs
- residual values and disposal costs.

Kelly and Male (2003) suggest:

- investment costs
- energy costs
- non-energy operation and maintenance costs
- replacement of components
- residual or terminal costs.

9.4 Period of analysis

The life expectancy of a building both in terms of its physical state and its usefulness needs to be considered by the designers. After World War II, prefabricated houses were produced with only a 5- to 10-year life expectancy. Some of these are still in use and indeed a few have become listed buildings. In the early 1960s many clients believed buildings should only be designed with a 20-year life expectancy after which they would be demolished. On the other hand the cathedrals built over history, including the more recent examples, such those at Coventry, Guildford and Liverpool, have been designed with a very long life expectancy.

Another approach is to consider building with a more flexible design so that changes in use could occur. Historically this has happened already. For example, large Georgian terraced houses have subsequently been converted for uses other than housing, such as offices, hotels, nursing homes, small schools and hospitals. While there remains an attitude of designing and building for the initial purpose intended, some developers are thinking ahead and looking into the future about revised usage.

The period of analysis is also a function of how long the investor is interested in the building. For example, a facilities management contract on a PFI contract may be 25 or 35 years. The developer/investor may not be concerned about what happens after that, although the final owner such as the government may have a different view on this. The developer will also be influenced by whether or not they are developing with a view to sell or rent or use themselves.

9.5 Factors affecting life expectancy

9.5.1 Physical deterioration

The physical state of any building will decline over the passage of time. The fewer preventative measures taken, the quicker this will take place as can be seen in the rapid state of deterioration once a building is left unoccupied and nature takes its course. However, if a building is well maintained, it will normally last until it is demolished. The rate at which buildings deteriorate depends upon many factors, including:

- the developer setting inappropriate standards of performance specification;
- a lack of appreciation of the causes of degradation;
- poor design detailing;
- inappropriate selection of materials and components;
- poor construction practice;
- inadequate maintenance regimes;
- damage and vandalism;
- a different use by the owners and occupiers from the original design.

9.5.2 Economic obsolescence

This is in part related to some of the following causes of obsolescence (9.5.3–9.5.7) as economic analysis will be a key determining factor in accepting the need to vacate or demolish the building.

In Hong Kong in the 1970s in some cases when buildings were worth less than the land they were built on they were demolished and a taller one built. This is an example of economic obsolescence. Here the potential value of the land and new development is worth more than the existing rental income from the existing building. In other words more profit can be obtained from the site by demolishing and building something else that offers a greater return.

9.5.3 Functional obsolescence

Buildings are usually designed with a specific purpose in mind. If that use changes or ends then the building may not be appropriate for another use. The Millennium Dome is a classic example of how difficult it can be, remaining empty for many years. It is not always the case, however. Manchester, along with other cities, has and is continuing to convert many Victorian industrial buildings and retail stores into residential accommodation, bringing life back into the city by attracting tens of thousands of tenants or owner-occupiers. If no alternative use can be found, then the buildings will eventually be demolished.

9.5.4 Technical obsolescence

As with most items produced, technology moves on and the product becomes technically dated. Buildings are no different. Some components wear out faster than others and have to be replaced. New technologies for components become available that make the building more efficient and/or easier to maintain. For example, single glazed untreated softwood timber window frames are replaced by double-glazed uPVC frames.

The building usually will have been designed for a specific purpose, but as the user's technological requirements alter as a result of the latest plant and equipment, different handling requirements and the development of information technology systems, the building may be ill-equipped to cope with these changes.

9.5.5 Environmental obsolescence

If the building is old the insulation standards may be such that the cost of heating the building has become prohibitive, especially if the processes now being used do not emit the same levels of heat as previously. This coupled with other factors could make the building obsolete for the owner.

9.5.6 Social obsolescence

Some buildings such as many of the multi-storeys built in the 1960s to resolve the post-war housing problems became socially unacceptable and many have now been demolished and replaced with low-rise housing. Buildings cannot readily be moved from one location to another and many churches and chapels have become redundant either because of falling congregations or changing populations. For example, many immigrant populations of the Muslim faith have gradually filled inner town and city areas previously populated predominately by Christians who have moved out into the suburbs. Some of these buildings are now being used as mosques, but many have been sold on and converted for uses other than for worship.

9.5.7 Legal obsolescence

This occurs when the building is unable to meet the current regulations without substantial costs being incurred that outweigh the benefits to the building owner. This could be a structural defect making the building unsafe, the discovery of substantial amounts of damaged asbestos that has to be removed or sealed in, not satisfying the current fire regulations and, more recently, legislation concerned with disabled access.

9.5.8 Aesthetic and visual obsolescence

Public taste changes over a period of time. What was once acceptable visually now becomes unacceptable and sometimes vice versa. Some buildings seem to transcend time and are always liked. The 1960s was a period in UK history where there was an explosion of ideas and experiments right across the arts, and architecture was no exception to this. Unfortunately, whereas in the other art forms mistakes could be hidden or destroyed and then forgotten about, disliked buildings remain around for a much longer time because of economic considerations. However, often in the end public pressure can have an influence on the decision to demolish the building.

9.6 Data for whole life costing

9.6.1 Historical data

Those concerned with running and maintaining a building collect data over a number of years. This can include alterations, replacement, general maintenance, energy consumptions and so on. If the data is collected conscientiously, then this can be a very useful source of information. The problem is the data may have been collected for different reasons. For example, the service engineers may be collating an energy audit, cleaning may come under a different section and alterations and building maintenance under small works. The way costs are presented may also be unhelpful. It is more useful when calculating the WLC, to know the cost of painting a square metre of wall rather than the total cost of decorating the executive suite (for instance), so that this data can be used for other applications. It is important that from the point in time it has been decided that data should be collected for future WLC purposes, the method of collecting is standardised so that it is user friendly.

9.6.2 Experience

Never underestimate the value of experience. It is not always quantifiable, but it is not unusual for the experience of a key person in a section who has been there for a long time to often reliably predict the life of a component, piece of equipment etc., and in particular how long operations may take to complete. This experience was tapped into in 1992, when a number of building surveyors were asked to fill in a questionnaire estimating, using their experience, the life expectancy of a variety of components. The results were published as RICS (1992) *Life Expectancies of Building Components*.

9.6.3 Manufacturers' literature

Manufacturers will often offer guarantees as to the life of their product. If they are a reliable and well-established company, they will not offer guarantees of life expectancy below that which they, from their experience, can offer as this could result in court action against them with excessive damages being awarded. Their guarantee can therefore be taken as a minimum life before any remedial or replacement work will be needed. How much is added to this relies to a certain extent on knowledge gained from experience and reading trade literature and magazines. Using guarantees from small organisations is dangerous as the owner may be out of business before the guarantee expires.

9.6.4 Research databases

Increasingly, informed databases are being produced by various organisations such as the Building Research Establishment, but these are only accessed if paid for. However, the costs are relatively small compared to the savings to be made on a significant project.

9.6.5 Calculations

An example of the use of calculations is the annual amount of energy needed to heat/cool and light a building once the basic footprint and construction of the exterior elements has been decided. This calculation would also have to take into account the usage of the building in terms of days per week/year, the costs of reheating after cooling down over a weekend/holiday, likely solar gain, the type of work being carried out within the building and the numbers of people using it. This would not take account of plant efficiency, selection of fuel, changing fuel prices and so on, but serves as a useful guide as to the likely costs.

9.7 Issues to be considered at the design stage for whole life costing

9.7.1 Maintenance

This is to do with maintaining the physical integrity of the materials, components, plant and equipment and should not to be confused with operating costs such as cleaning (9.7.3). It can be described in two categories:

- *Planned maintenance*. This is where the maintenance is carried out on a predetermined timescale similar to the maintenance logbook provided with a new car. It may be for aesthetic and comfort reasons such as decoration, to ensure the highest possible efficiency from plant or to protect the basic fabric from deterioration and eventual failure. A good example of the latter being the replacement of flat roof coverings before the roof leaks.

- *'Crisis' maintenance*. This is dealing with the unexpected such as a burst pipe or the expected but unpredictable, such as light bulb failure. The term is sometimes applied to maintenance protocol that has no planned maintenance, but just reacts as needed.

Planned maintenance can be costed relatively easily using today's prices. It is then necessary to apply the appropriate formulas to estimate the future costs in real terms and then it can be established what finance is required to pay for it. The latter can be estimated, but relies much more on experience and historical data for buildings of a similar nature. Whether or not crisis maintenance is cheaper in the long run than planned is a different debate, but the advantage of planning is that the disruption likely to be caused to the users of the building can be included in the thinking processes. It is important that the designer takes maintenance into account, especially ensuring the replacement of components can be easily and rapidly executed so as to reduce costs and disturbance.

The costs of maintenance vary depending upon the use of the buildings. Buildings that are highly serviced such as hospitals will need more maintenance than others. It has been suggested that this can vary from between 5 and 30 per cent of the occupancy costs.

9.7.2 Energy

There are several factors that impact on the energy costs of the building:

- the initial capital costs of the plant and equipment that consume energy such as for heating, cooling, ventilation, de-humidification, lighting, lifts, escalators and kitchen equipment;
- the efficiency of the plant – some plant is more energy efficient than others;
- the staff required to operate plant and equipment such as boiler men – these will have to be employed throughout the life of the building or until replaced with automated plant;
- the staff required to maintain the equipment – this service could be outsourced, especially with the developing expertise in facilities management companies. Much will depend upon the frequency of maintenance and the skill required to carry out the work as against the convenience of having in-house staff;
- the cost of parts and replacement equipment – the design decision is whether or not it is worthwhile spending more on a piece of gear now and replacing it less frequently than selecting a cheaper one requiring more frequent changes. It should also be noted that equipment technology continues to make advances, so changing certain plant at more frequent intervals might offer significant energy savings, which would compensate for the replacement costs;
- the cost of providing access to maintain/replace the plant, which can be a significant issue. If the plant requires a large amount of space for this to occur, but the event only occurs once every 20 years, it might be cheaper not to provide this costly space, which will be redundant throughout this period. Users/owners will then have to accept the disruption and extra cost of perhaps having to remove elements of the building to carry out the work;

- the selection of fuel – a very difficult issue, for example in the late 1960s, the tendency was to use oil for heating, as it was readily available and very cheap. Then in the early 1970s when OPEC was formed oil prices rocketed and alternatives were sought, notably gas. With the deregulation of the energy suppliers and the ability to source energy from different suppliers and hence more competition, it has become more difficult to estimate what costs will be in the future;
- the footprint and orientation of the building – extremely important as the former will determine the amount of natural ventilation and lighting the occupiers can use as against requiring artificial lighting and mechanical ventilation, and the latter determines the effects, or otherwise, of solar gain and wind chilling;
- the location and siting of the building and exposure of the building to different climatic conditions – windy sites will have a cooling effect on the building whereas well-protected sites may require mechanical ventilation and air-conditioning especially in the summer. Temperature variations will also occur depending upon the height above sea level and geographical location;
- the energy consumption – a factor of many of the above plus the design of the outer fabric is primarily about the amount of insulation provided, elimination of cold bridging and control and utilisation of solar gain. It is also a function of the use of the building. For example, the greater the number of people and amount equipment in the building, the more heat generated with potential overheating problems.

The energy costs of a building are in the order of 20 to 35 per cent of the occupancy costs, but there is no reason why this figure could not be reduced below this with sensible design decisions and innovatory solutions.

9.7.3 Cleaning

Cleaning is an important issue for both the general welfare of the staff and the company's image. Few like working in untidy dirty conditions and an unkempt working environment gives off the wrong signals to potential and existing clients. The design process very largely determines the cost and method of cleaning.

A significant issue, often neglected, is that the selection of a particular material at the design stage determines what method of cleaning is utilised which in turn may determine the material used for cleaning, some of which have potential hazards associated with them (see 21.9.3)

The selection of materials or components also affects the frequency and ease of cleaning. Can the windows be cleaned from inside or are ladders, cradles etc., required to carry out the work from the outside? How easily does dirt become engrained in a surface and can the surface be washed, dried and polished in one operation?

A final consideration is whether in the future, the use of the space will change either by function or numbers of occupants (see 9.7.4 also). The costs of cleaning are between 10 and 20 per cent of the occupancy costs.

9.7.4 Communications

Telecommunications provision has revolutionised the way building users function, especially in commerce, industry, education and health and is set to continue. It is suggested that not only will the equipment become more sophisticated, but it will also change the way we work with increasingly more people working from home for at least part of the week. All this means that systems will be replaced at regular intervals with more up-to-date equipment and the way office space is utilised will also change.

There is the possibility that as a result of home working, more 'drop-in' office space will be needed, where employers do not have their own personal space, but just find an available desk. In turn this means that storage of personal items and files etc. will have to be considered.

Currently the use of raised floors in open-plan offices permits minor alterations to office layouts to occur relatively easily provided fire exit routes are incorporated or maintained. However, if substantial changes are required as a result of the previous comments, the costs of modification would increase.

9.7.5 Alterations and flexibility

Following on from communications, it can be expected in the life of most buildings, there will be a lot of alterations. The more potential for flexibility that is built in, the less cost will be incurred later and the greater the opportunity to change the use of the building for new tenants/users.

9.7.6 Security

This is an uncertain area as the need for security systems has only become apparent in recent years. A key issue here is to allow for systems to be updated as and when required without disrupting the building users. However, it should be noted that especially since 11 September 2001 it could be that changes of layouts and means of access also have to be considered. This will inevitably be an extra financial burden.

9.7.7 Financial advice – rates and taxes

This is the province of specialist advisors. Tax law changes each year as a result of the Chancellor of the Exchequer's budget. It can affect areas as diverse as capital gains, maintenance depreciation and incentives for employment in the area the building might be located.

9.8 Is whole life costing effective?

If capital costs and maintenance are separated because the developer is selling on, then it is likely that as the developer is only concerned with the profit made on the capital cost of the building, WLC issues will not be at the top of the agenda. On the other hand if the contract includes both construction and facilities management the attitude will be different.

WLC has its limitations because of the high number of unpredictable variables used in the equations especially that of anticipating inflation and interest rates. Long-term investments are also at the vagaries of the stock market. Since the life of the building is expected to be anything from 30 to 100 years this makes cost predictions difficult. Therefore, if negotiating a facilities management contract, it is necessary to have the opportunity to renegotiate new rates every few years.

It has been argued in some quarters that volatile inflation is a problem, but this argument is suspect as interest rates tend to track those of inflation, the difference between the two remaining relatively constant. For example, if inflation is running at 15 per cent, interest rates might be say 18, but as inflation falls to 0 per cent, interest rates drop to 3 per cent. Hence in real terms there is little difference. However, irrespective of this, not all building costs increase at the same rate as inflation. This is because products may come from parts of the world with different rates of inflation. Some building work, such as maintenance, that is labour intensive, may be carrying a higher rate of inflation than where there is a high material content.

It is difficult to predict the future. For example poor design detailing and workmanship will demonstrate failure earlier than anticipated. The projected planned maintenance schedule is based upon the assumption that the building owners will keep to it, but they may not, initially saving money but potentially building up a backlog of work that may be more expensive later.

However, one can get tied up with the academic rigours of precise and accurate calculations at the expense of raw common sense. Unless, after calculation, differences between two solutions are marginal, the fact that one solution is clearly an improvement on the other is surely all that matters. While the figures produced may be incorrect in reality, if there is a clear indication which of the two (or more) solutions is the most cost effective, then the selection becomes obvious, assuming all the variables have been considered. Using different likely variables and seeing how the sums work out can be tested mathematically. It is a very inexact science and there is much debate as to the validity of the calculations and the reader must draw their own conclusions. This naturally leads into including the environmental issues. This is referred to as life cycle assessment. It is argued that following this route rather than the economic debate is more fruitful, especially when investigating long-term sustainability issues.

9.9 Life cycle assessment (LCA)

LCA is an analytical tool that attempts to assess the material content and the majority of environmental impacts of any manufactured item. In construction it can be used to assess materials, sub-components, components and the whole building. To understand the full impacts of a particular component or material to be used in a building requires a considerable amount of research and is normally outside the time constraints of routine building design and material selection.

There have been calls for manufacturers to publish LCA data on their products, but this is not required of a manufacturer at this time. There is also no international agreement over LCA techniques, which makes comparisons difficult especially since many products and components used in construction are imported and used directly or incorporated into other products, or exported and then re-imported after being processed elsewhere. There are various LCA systems in use and under development such as the BRE material analysis computer based tool, ENVEST 2 and IMPACT. The aim of IMPACT is to integrate LCA, Life Cycle Costing and BIM.

However, in spite of these developments there are still a number of questions that need to be asked.

- What level of detail should be included? Are nails and screws included for example?

- The impact various materials have over time in relationship to the whole building. For example, the building's life is 60 years, the roof tiles 30 years and the boiler 10 years.
- The relative weighting of the various impacts. How seriously is global impact such as climate change compared with a local impact such as river pollution or ground contamination?

To give a further indication of the problem, Table 9.1 considers the various impacts materials can have.

9.10 Whole building environmental assessment

Life cycle assessment is aimed at materials and components whereas whole building environmental assessment is concerned, as the name implies, with the whole building. LCA is more comprehensive if conducted thoroughly. Methods of assessing the overall environmental performance of a single building have been developed, notably BREEAM. These are attempts to formalise environmental impact analysis of a building usually to ensure conformity of approach in order to grant a certificate of label. This is very necessary where claims of 'greenness' need to be substantiated and compared. These types of tools usually attempt to assess environmental performance in terms of the environmental issues such as demonstrated in Table 9.2 which along with Table 9.1 is extracted from Curwell *et al.* (2002).

This is often achieved via indirect factors, e.g. the operational energy consumption of the building as an indicator of CO_2 emissions. In terms of building materials and components, the early systems usually consisted of a series of checklists of materials

Table 9.1 Key factors affecting environmental assessment of building materials

Process	*Issues*
Upstream – extraction and manufacturing	Energy involved in both extraction and manufacturing processes Transportation from source to manufacturing plant Depletion of resources – how much reserves remain The amount of despoliation caused by the manufacturing process Quantity of waste generated by both extraction and manufacturing Quantity of pollutants generated during these processes Proportion of the product made from recycled materials
Construction	Energy involved in the construction process The distance the material has to be transported to site How much waste is generated and how much is, or can be recycled? What pollutants are generated during the process?
Buildings in use/maintenance	Durability of the material in a specific application Life expectancy
Downstream – demolition, disposal and recycling	Pollutants caused during demolition Pollution as a result of disposal The volume of waste disposed of The distance material has to be transported either to tip or recycling point What proportion is recyclable or reusable? Ease of disassembly

Source: Curwell *et al.* 2002.

Table 9.2 Current environmental issues

Global issues	Global warming Acid rain Ozone depletion Deforestation Loss of biodiversity Resource depletion	*Local issues*	Contaminated land Solid/liquid waste and landfill External air quality (nitrous oxide, sulphur dioxide, particulate matter etc.) Water quality (drinking water and watercourses Ecology and biodiversity, flora and fauna
Health issues	Sick building syndrome Asbestos/fibrous materials Indoor air quality Volatile organic compounds Drinking water quality Radon Electro-magnetic radiation		Desertification Deforestation Access to 'green' space Noise (from building process and from adjoining owners or users) Radon Electro-magnetic radiation

Source: Curwell *et al.* 2002.

to be avoided such as such as volatile organic compounds (VOCs) in paint. In some cases limited performance dimensions such as embodied energy were included.

These methods struggled to adequately address complex issues such as the contribution they might make in enhancing the performance of the whole building. An example of this is thermal insulation, which may cause pollution or is energy intensive, but reduces energy consumption throughout the life of the building.

The latest versions of BREEAM overcome some of these problems with a more intensive assessment of materials used in the elements of buildings and the BRE ENVEST2 programme demonstrates what can be done when LCA data underpins a whole building performance tool.

9.11 The future and sustainable building development

To place the discussion above in context, it is important to view it in terms of sustainability issues. The development, use and maintenance of buildings result in a consumption of 6 tonnes of building material per person per year in the UK (BRE 1996) and it is estimated it is 10 tonnes in the USA.

Predictions of the resource efficiency gains that need to be achieved for society as a whole vary from Factor 4 (75 per cent reductions) to Factor 20 (90 per cent reductions). This may seem impossible, but projects have shown some elements such as energy or water efficiency can achieve Factor 4.

The BRE Environment Office has achieved 30 per cent energy reductions and has a 90 per cent recycled content. The Vales have built buildings with an 85 per cent reduction in energy use from that used in buildings built to the Building Regulations prior to the recent revision and the house they built for themselves in Nottinghamshire is autonomous from building services.

There is no doubt that we have the technology and expertise when applied together to construct buildings to Factor 4. Whether the political will exists is another story. The other problem is what to do about existing building stock?

Bibliography

Ashworth, A. and Perera, S. (2016) *Cost Studies of Building*, 6th edn. London: Routledge.

Building Research Establishment (1996) *Buildings and Sustainable Development*, Information Sheet A1, Garston.

Building Research Establishment (BRE) *Environmental Profiles of Construction Materials and Components*. Available at: www.bre.co.uk/Environmental_Profiles.

Curwell, S., Fox, R., Greenberg, M. and March, C. (2002) *Hazardous Building Materials: A Guide to the Selection of Environmentally Responsible Alternatives*. London: E & FN Spon.

Dickie, I. and Howard, N. (2000) Digest 446: *Assessing Environmental Impacts of Construction*. BRE.

Edwards, S. Bartlett, E. and Dickie, I. (2000) Digest 452: *Whole Life Costing and Life-Cycle Assessment for Sustainable Building Design*. London: BRE Press.

Flanagan, R., Norman, G., Meadows, J. and Robinson, G. (1989) *Life Cycle Costing: Theory and Practice*. Oxford: BSP Professional Books.

Kirk, S.J. and Dell'Isola A.J. (1995) *Life Cycle Costing for Design Professionals*. New York: Kingsport Press.

Kelly, J. and Male, S. (2003) *Value Management in Design and Construction*. London: Taylor & Francis.

RICS (1986) *A Guide to Life Cycle Costing for Construction*. London: Surveyors Publication.

RICS (1992) Life Expectancies of Building Components. RICS Research Paper No. 11. London: RICS.

Stern Review on the Economics of Climate Change (2006) London: HM Treasury.

Vale, B. and Vale, R. (2000) *The New Autonomous House*. Thames and Hudson Ltd.

10 Value engineering and management

10.1 Introduction

Value engineering is about taking a second look at key design decisions usually concerned with making savings in the initial capital cost of the building and it does this by looking at the function of an element or component and looking for alternative methods to achieve the same functional performance. With the trend to new types of procurement routes such as construction management and management contracting coupled with a desire, post Latham (1994) and Egan (1998), to cooperate and work as a team, value engineering has become a valuable tool in improving the final design and construction solution for the building. Related to value is the ability to draw comparisons and look at other means of obtaining value. The last sections of this chapter include key performance indicators and benchmarking, building information modelling and lean thinking.

Although it often is, value is not necessarily about money or based upon quantified data as is illustrated in 10.3.4, but more about perception of value. What is valuable to one person is not necessarily valuable to another. For instance the first present a husband bought for his wife may have little monitory value but sentimentally is irreplaceable. A passionate environmentalist will be more concerned about the energy saved using low energy bulbs, than the type of light it emits. Further, there can be several parties involved with conflicting objectives such as hospital trust managers, the doctors, the nurses and the patients. It is the job of the value engineering manager to resolve these difficulties.

10.2 Functional performance

McGeorge and Palmer (2002) identified four levels of function:

- *Defining the project as a whole.* This is asking very fundamental questions about the purpose of the project. For example the proposition is to widen a road in the centre of a city to relieve congestion. Superficially the problem is the congestion and widening the road is but one solution. Diverting the traffic onto another route, introducing a one-way system or introducing congestion charges could also resolve the problem. However the fundamental question is why is there congestion?
- *Defining the spaces within the project.* The function of the foyer area of a theatre might be to allow the audience to meet and congregate before the performance, stretch their legs during the interval, have a drink and use the toilet. However,

it might also be used for art exhibitions, a meeting place, providing light refreshments during the day, and workshops.

- *Defining the function of the elements.* What is the function of a window? Initial thoughts are that it is to act as a ventilator or to admit natural light. However, both of these functions can be achieved by alternative solutions. Light might be provided using a roof light and ventilation provided artificially. However, it is also there because building occupiers wish to see the world outside and indeed there is evidence to demonstrate that lack of this external link can be mentally unhealthy if continuing for any length of time. The window is also part of the exterior wall and its contribution to the overall thermal performance of the outer skin has to be taken into account and so on. Construction methodology should also be considered. Can the window be fixed in position from the inside of the building thereby eliminating the need for external scaffolding and does this have a knock-on effect in determining how the rest of the outer skin is constructed? What should its dimensions be and how high above floor level should it be positioned. For example in a hospital ward there are issues of privacy relative to people outside the building.
- *Defining the function of components.* This entails looking at the element's component parts and seeing how they function. Can the windows be cleaned from inside safely? Can they be opened readily? Are they child proof? What standard of security is required? What performance is required from the glazing? How is the window fixed to the rest of the element?

10.3 Methodology or sequence of events

Miles' (1972) way, or methodology, that value management can be used, has become widely accepted. It comprises seven stages adapted for the purposes of the construction and design processes named orientation, information, speculation, analysis or evaluation, development, selection and conclusion.

10.3.1 Orientation

After the design team has been selected, a meeting is called for all the interested parties. These include the client's representatives, project management and the design team. The prime aim is for all involved in the project to understand all the issues and constraints and set the scene. It also enables those involved to get to know each other on a personal level and exchange information and ideas. The objective of the meeting is to address four questions:

- What is to be accomplished?
- What does the client actually need? This is not necessarily what clients think they want.
- What are the key and desirable characteristics of the project?
- What is a priority list of the key characteristics?

10.3.2 Information

All the information about the project is gathered together. The accuracy and depth of this process is directly related to the quality of the decisions that can be made and is therefore extremely important. However, this process is also determined by the amount of time available to collect it. Facts are more important than assumptions. Initially the kind of information required is to provide answers to the following:

- *Client's needs*. These are the fundamental requirements that the project must satisfy. This may not just be the physical needs of the building in satisfying specific functions, but may also include any statement the building must make for the client. For example, it is the head office of a major company or organisation, a national symbol or the greatest shopping experience in the area. The National Westminster Tower, Lloyds of London, Coventry Cathedral, the Welsh Assembly building and the Trafford Centre are examples of these. In projects not value managed, it is not unusual for decisions to be made without detailed consultation with those persons who are actually going to work or use the space. At some stage in the process, this level of consultation needs to take place. In other words the client is not just the person or persons who commissioned the building.
- *Client's wants*. These are things the client might want, if there is enough money left, but are not essential to the function of, or the statement being made by the building.
- *Project constraints*. These are issues outside the client or designer's control such as planning restrictions, shape of the site and rights of way.
- *Budgetary limits*. The amount of money the client has to spend on the land acquisition and the building. This will also involve decisions about life cycle assessment issues.
- *Project duration*. This includes both the design and construction phases. A retailer may stipulate they must have the building by September, so that they can be in business for the Christmas rush.

What also should be happening at this stage is the commencement of team building as this is the first time the main participants are all brought together. If this is thought out properly, the advantages to the rest of the process and project will be greatly enhanced. As the process continues more detailed analysis can take place at the differing levels of function (see 10.2).

10.3.3 Speculation or creativity

Brainstorming is a popular way of creating new ideas and resolving problems, although not the only way. The group is asked to consider the problem and 'brainstorm' ideas. All ideas are recorded and considered even if, on the surface, they appear ill-conceived and inappropriate. What often happens is that an ill-conceived idea develops a different train of thought that leads to resolving the problem. The group has to be encouraged to generate a large number of ideas and not to ridicule any submitted.

An example of brainstorming in practice was in the manufacture of the Jespersen industrialised building precast floor components similar in design to today's Bison hollow core floor units. The horizontal holes were made using withdrawable steel tubes. This worked well except for the forms used in the mould ends when it was necessary to withhold one or more of the tubes used to form the horizontal holes through the concrete unit to, for example, stiffen the component and act as an edge beam. These formers kept vibrating out.

Staff were assembled to discuss the problem; one of the foremen from another department suggested that cardboard forms should be placed inside the mould to stop the concrete from coming out. The idea of cardboard was impractical, as it would have been difficult to remove from the cured concrete on de-moulding, as it would have adhered to it. However, the idea of blocking from the inside resolved the problem. The solution was to insert cylindrical timber bungs into the mould ends, with a small lug to stop the bung from sliding through the hole. These remained during the casting and curing process, and the problem was resolved.

Interestingly, evidence seems to indicate, as in the example, that it is the inexperienced member of the group, rather than the expert who is more likely to come up with the most original suggestion, probably as that person is not mentally restricted by tradition and convention.

10.3.4 Analysis and evaluation

The team looks at the various suggestions, eliminating the ones that clearly have no place because they may not fully satisfy all the functional requirements, are unsafe, impractical, or far too expensive. This is a relatively simple stage and just acts as a filter for the next. There are a variety of ways of doing this but a simple and often used technique is to rate each of design criteria for each of the appropriate solutions. The rating scale chosen for this example is: Poor = 1, Fair = 2, Good = 3, Very good = 4, Excellent = 5.

Applying these to various types of floor finishings might produce answers as shown in Table 10.1. The row entitled 'weighting' is the designer's/client's perception of importance of the various criteria. For example, if the view is that environmental impact is the most important then 5 might be used. Initial costs may be considered to be more important than aesthetics, whereas another client may think the opposite. Using the same rating scale, the figures in the rows showing the different finishing are the designer's assessment of the performance against each criterion. The figures in the table reflect subjective opinion and should therefore not be taken as definitive.

By then multiplying the weighting of importance by the performance rating given and then adding these up horizontally, a guide is given as to which is the most appropriate solution for the client's needs as shown in Table 10.2.

It can be seen that in this case that terrazzo tiles are the most favourable with the highest score followed by wood blocks, but it must be remembered that in this example the weighting is based on the writer's subjective judgement only. Further, both the

Table 10.1 Sample design evaluation of floor finishes

Type of finishing	*Initial cost (5=cheapest)*	*Aesthetics*	*Cleaning*	*Replacement frequency*	*Environmental impact (1= greatest)*
Weighting	*3*	*5*	*3*	*2*	*3*
Flexible PVC tiles	3	3	4	4	3
Terrazzo	2	4	5	5	4
Wood blocks	1	5	3	5	4
Nylon carpet tiles	4	3	3	2	2

Table 10.2 Sample selection evaluations

Type of finishing	*Initial costs*	*Aesthetics*	*Cleaning*	*Replacement frequency*	*Environmental Impact*	*Total*
Weighting	*3*	*5*	*3*	*2*	*3*	
Flexible PVC tiles	3×3	3×5	4×3	4×2	3×3	53
Terrazzo	2×3	4×5	5×3	5×2	4×3	63
Wood blocks	1×3	5×5	3×3	5×2	4×3	59
Nylon carpet tiles	4×3	3×5	3×3	2×2	2×3	46

weighting and rating scale of one to five is crude, but this can be improved by having a broader scale.

10.3.5 Development

This stage takes the solutions still remaining after analysis and evaluation and develops them in some depth, which also includes buildability issues. Outline designs and costs are produced for each design. Life cycle assessments will also be produced at this stage if it is appropriate to the developer's long-term plans. Those that do not meet the prime design criteria are eliminated. Some further thoughts on methodology are given in 10.5 and 10.6.

10.3.6 Selection

The various solutions that are left from the previous stage are now presented to the management team and client for a final decision to be made. Drawings, methodology statements, programme implications and costings will be used to support these solutions. A commentary on the benefits or otherwise can also be included.

10.3.7 Conclusion and feedback

The client makes the final selection from this information. What is also important is that on completion of the building works, a feedback loop is completed to enable lessons to be learned for the future.

10.3.8 When value engineering should be applied

The further down the design and construction processes value engineering is applied, the less likely it is that money will be saved as the potential for making change is reduced. The costs of implementing change also increase as demonstrated in Figure 10.1. A simple and obvious example to clarify the point would be that it is easier to move a column on the drawing than it is to move it once constructed. More significantly it could be that by applying value engineering to the footprint of the building or allocation of space within the footprint, significant economies can be made in terms of the capital cost of the building, staff resources needed in maintaining the building and life cycle costing issues.

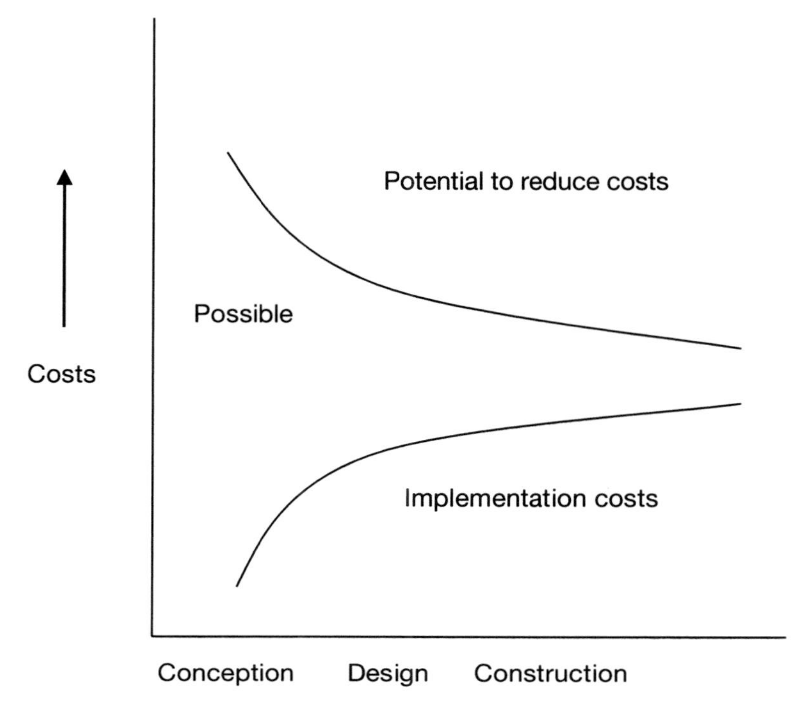

Figure 10.1 The timing of value engineering

10.4 Functional analysis method

In functional analysis the function of each component or process is examined by asking the question 'What does it do?' followed by a second question, 'How else can this be achieved?'

For example in terms of a process, 'How is paint applied to the wall?' Answer, 'With a brush.' 'How else can this be achieved?' Answer, 'Using a roller or spray.' These assume that the wall is built in situ. If the wall is a factory made component, then alternatively it could be dipped.

However, this does not go far enough because the function of the wall surface has first of all to be questioned. What does it do? Depending on the application, it could include a need to be:

- resistant to impact
- hygienic
- easy to clean
- decorative and colourful
- water resistant
- resistant to UV light
- possessing specific acoustic properties
- flame and fire retardant.

Deciding what the function of the surface is determines the types of paint available appropriate for the function which then might determine the method of application. The choice of surface finish might then determine the surface and material required to receive this paint and vice versa. This might be an expensive combination so it would be necessary to go round the loop again. In the end though the finish must satisfy the function desired from it.

10.5 Function cluster groups

There have been found to be benefits in dealing with clusters of functions. The Building Cost Information Service (BICS) standard elements have been used in Figure 10.2 (reproduced from Kelly *et al.* (2002) *Best Value in Construction* with permission of Blackwell Publishing).

The advantage of this approach is that it focuses the mind onto specific areas, readily understood especially by the client. The client may be particularly concerned with

Project mission and function of building

Cluster	Cluster Function	BCIS Element
STRUCTURE	Support load Transfer load	Substructure Frame and upper floors
EXTERNAL ENVELOPE AND WEATHER SHIELD	Protect space Express aesthetics	Roof External wall Windows and external doors
FUNCTIONAL SPACE, CLIENT ORGANISATION AND STYLE	Serve client function	Stairs Internal walls and partitions Internal doors Walls and finishes Floor finishes Ceiling finishes Sanitary appliances and plumbing Communications and data
INTERNAL ENVIRONMENT	Maintain comfort	Space heating Ventilation and AC Electrical
INTERNAL TRANSPORTATION	Minimise walking Raise/lower people and loads	Lift and Escalator
EXTERNAL INFRASTRUCTURE	Protect property Circulation/parking Remove waste	Site works and drainage external services

Figure 10.2 Function cluster groups

Source: Reproduced with permission of Wiley-Blackwell from Kelly *et al.* (2002), Ch. 5.

overall issues, such as the appearance of the building, the quality of the internal finishings or environmental impact of the building. In each of these cases the function cluster assists in this process. It can be from this approach that the value engineering can be directed into specific issues more readily than taking individual components.

10.6 Building information modelling

Building information modelling (BIM) is being advocated very strongly by the UK Government and in June 2011, it announced its BIM Working Party Strategy that requires all its construction projects to be using Level 2 (sometimes referred to as the third dimension) BIM by the year 2016. One of its main advantages is that all those involved in the design process have ownership of the total design and this is then handed over to the main contractor and subcontractor all of whom add their expertise into the model. This information is then passed to the owner/occupier of the building at the handover stage. This means there is no loss of information: not unusual on large projects. Government sees it as a means to obtain value for money and to reduce carbon emissions and improve sustainability.

Construction drawings have been traditionally two-dimensional showing plans, elevations and sections. While to the experienced eye, they are easy to interpret, there are limitations, especially in not identifying the actual size of certain components or where two or more are using the same place. For example, reinforcement steel is shown on a drawing as a line, but other than in the description, does not demonstrate the bar diameter size. The author remembers a structural engineering drawing of a large steel reinforced concrete beam that had so many stirrups that when it was prefabricated it became clear that there was inadequate space between each stirrup to allow the concrete to pass between them. Similarly, mechanical and engineering pipe work is also drawn using a single line, and as there are so many service pipes in a complex building, conflicts for space are almost inevitable. The situation is often exacerbated because valves used for controlling flow can take a lot of space.

Three-dimensional computer-aided drawings have been used now for a while and these have gone a long way to resolving the issues raised above. They have also, through integrated modelling, allowed the observer to have a 'building walk through' before the building plans have been approved by the client, or are to be used as part of the marketing process. BIM has taken this much further and has added the fourth dimension of time (the adding of scheduling) and the fifth of cost and it is further being developed to embrace maintenance and will eventually be extended to include demolition.

The modelling gives the designers, project managers and clients the opportunity to test solutions before committing to the construction process; this becomes a valuable tool in value engineering the project. The cost dimension allows the testing of alternative components and materials of different properties to obtain value for money. It also permits the testing of different solutions to inspect the carbon footprint of the building for embodied energy and for life cycle analysis.

The fourth dimension, that of being able to add scheduling, means that the project management team can experiment with access, site layout, phasing of works and selection of the appropriate plant and its location, for example, by positioning and selecting cranes and other fixed plant to ensure that those positions selected permit the building to be constructed most effectively. It can also look at safety issues both during the construction process and maintenance after the handover of the building, useful to comply with the CDM regulations (see 21.9.4).

The fifth dimension is the integration of the design, scheduling and cost estimating. This means that stakeholders can see what happens to the construction schedule and to the budget if changes are made to the design during the design or production stages. There is still much work to do in developing this so that it can be made available for all.

When fully developed, the sixth dimension is aiming to look at the effect of design changes on life cycle and facilities management issues and permit users to look at environmental impact and sustainability considerations.

The government's BIM strategy (HM Government, 2011) also included the requirement of the Construction-Operations Building information exchange (COBie) to be produced for each of their projects. A COBie is a database containing as much information about a building in as useful a form as possible. The idea is that key information is brought together in one format and shared between the client designers and construction team at defined stages in a project. At the end of the project it can be used for facilities management and management processes.

The benefits can be summarised as follows:

- Clashes of services can be avoided.
- Critical input from all involved in design and construction can be given earlier possibly permitting off site production to take place.
- Waste can be minimised.
- More environmentally friendly solutions can be sought.
- There is potential to reduce the design and construction duration.
- The building owner/occupier can use it to identify the cause of defects.
- There is increased value for money and cost savings.

10.7 Lean thinking

Womack and Jones (2003) developed the concept of lean thinking and the five lean principles to be used in any organisation. They had concluded that only a small percentage of time used in producing a product adds value to the customer. These five principles are: value, value streams, flow, pull and perfection.

Value is to thoroughly identify and understand the customer's needs and what they really want and are prepared to pay for. The customer includes both the owner of the building and the user(s), who must be satisfied the building matches their needs, including the handover date, and enhances their working. This should supersede everything except safety and environmental issues.

Value streams identify all the steps in the process and eliminate those that do not contribute value. These non-contributors include waste, unsatisfactory work and bureaucracy.

Flow refers to the aim to smooth out the flow of work so that any peaks and troughs are removed. This is not just in production but includes services and information. Examples of the obstacles to flow include waiting between processes, transportation, defects, overproduction, unneeded processes.

Pull refers to the rate of flow being determined by the customer so the speed of delivery is so that it arrives when required; sometimes referred to as just-in-time (see 23.6).

Perfection refers to continually striving to improve reducing waste at every level in the organisation and in the processes. It will not happen all at once. Attainable targets should be set and once achieved, the bar raised accordingly.

To be successful the key elements are:

- the sharing of information generated in the design, supply and construction processes;
- buildability: by eliminating complex components, reducing the number of different components by moving towards standardisation as much as possible;
- manufacturing off site where feasible;
- using the communications tools now available such as smart phones, tablet computers etc. to assist in making rapid decisions. For example, the design team can be shown a problem visually without having to leave the office;
- the use of building information modelling;
- using value engineering techniques;
- an equitable sharing of savings made;
- all participants being committed to the process;
- empowering people to make small changes every day;
- accepting that lean thinking is a long-term strategy and not a tactic for saving cost.

A very simple example of some of the principles in practice is the delivery of heavy prefabricated components such as precast concrete. The decision to use them is to speed up the construction process, but to maximise this components must be available so that the site crane is in use all the time. This requires the components to be loaded onto trailers by the supplier so that they can be taken off in the sequence required and at the same time in such a way that the remaining load on the trailer is even at all times to stop it overturning. Since traffic conditions vary there needs to be sufficient loaded trailers on site at any one time to permit continuity of construction. Delays can occur on site so there can be a build-up of trailers as a result. Also components can be broken on site or during the transportation. To overcome these points and to ensure a steady flow between manufacturer and site there needs to be very close cooperation and communication between the two parties. The manufacturer must also have very good control at their end to ensure the correct units are dispatched in the right order and of the correct quality.

Sir David Brailsford, who managed the successful cycle squad at the London Olympic Games, when asked how he was successful said, 'the whole principle came from the idea that if you broke down everything you could think of that goes into riding a bike and then improved it by 1 per cent, you will get a significant increase when you put them all together'; he went on to say that 'There's fitness and conditioning, of course, but there are other things that might seem on the periphery, like sleeping in the right position, having the same pillow when you are away and training in different places', and when asked, 'Do you really know how to clean your hands without leaving bits between your fingers?' his response was 'If you do things like that properly, you will get ill a little bit less. They are tiny things but if you clump them altogether it makes a big difference.' He called this approach marginal gain. The team won eight gold medals.

Bibliography

Egan, J. (1998) *Rethinking Construction*. London: Construction Taskforce: HMSO.
Harty, J., Kouider, T. and Paterson, G. (2016) *Getting to Grips with BIM*. London: Routledge.
HM Government (2012) *Industrial Strategy: As Government and Industry in Partnership*. Available at: www.bimtaskgroup.org/ (accessed 3 January 2017).
Kelly, J. and Male, S. (1993) *Value Management in Design and Construction*. London: E & FN Spon.
Kelly, J. and Male, S. (2002) *Best Value in Construction*, ch. 5. Oxford: Blackwell Publishing.
Latham, M. (1994) *Constructing the Team*. London: HMSO.
McGeorge, D. and Palmer, A. (2002) *Construction Management New Directions*. Ch.2, 2nd edn. Oxford: Blackwell.
Miles, L.D. (1972) Techniques for Value Analysis and Engineering. New York: McGraw-Hill.
Sanchez, A.X., Hampson, K.D. and Vaux, S. (2016) *Delivering Value with BIM*. London: Routledge.
Sherratt, F. (2015) *Introduction to Construction Management*. London: Routledge.
Syed, M. (2015) *Black Box Thinking: The Surprising Truth about Success*. London: John Murray.
Womack, J.T. and Jones, D.T. (2003) *Lean Thinking: Banish Waste and Create Wealth in Your Corporation*. New York: Free Press.

11 Procurement methods and types of contract

There was a time when this process was very much prescribed. The client would appoint an architect who would select the rest of the design team. This would include the quantity surveyor and structural engineer. If the client was experienced, they may have employed other members of the design team directly. Bills of quantities (B of Q) were produced as explained in Chapter 12. Contractors would tender for the work, and the lowest price, especially for public sector work, determining who would be awarded the contract. Contractors worked to the letter of the contract using every clause available to obtain extra payment. The result was often confrontation between the design team and the contractor, the client picking up the bill at the end.

This has changed over the last few decades with a multitude of different ways being developed for clients to procure work. Many of these changes have occurred because of clients' dissatisfaction at the way the industry was operating. A significant turning point was the publication of the report *Faster Building for Industry* (1983) that investigated different types of procurement then in use and analysed the effectiveness of each on a large number of contracts. The conclusions were interesting in that the traditional methods of procurement were not as successful as others in use, especially those that were more client orientated. This report anticipated many of the later findings of Latham (1994) and Egan (1998).

11.1 Traditional procurement

11.1.1 Competitive tendering

This method is outlined in the introduction and is sometimes referred to as single-stage selective tendering. A significant amount of design has to be completed before the quantity surveyor can produce the B of Q and the contract is put out to tender. Some work is offered on open tenders permitting any contractor to apply, but this can mean far more contractors apply than is manageable. Pricing a contract is costly and this approach causes a considerable amount of wasted effort by the numerous tenderers. Much work is offered as selective tendering, when only contractors who have the prerequisites to carry out this work are invited to tender. Many clients operate a selective tender list that contractors can apply for inclusion. This will involve interviews and presentations allowing the contractor to demonstrate their ability to carry out work for the client.

On being awarded the contract, the main contractor puts the subcontracts and materials requests out for tender, unless the architect had already nominated them in

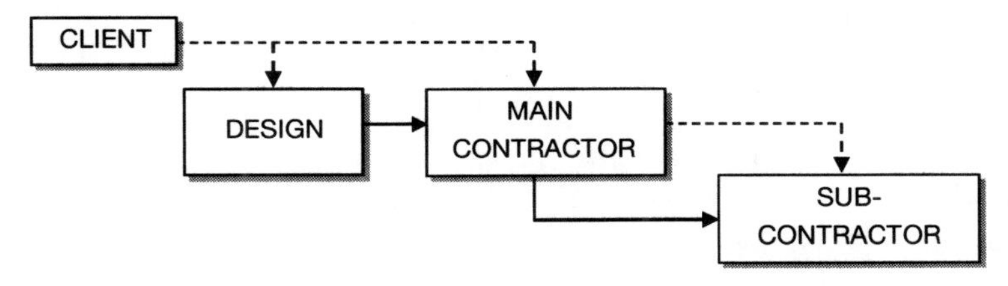

Figure 11.1 Traditional method

advance. This occurs occasionally when, for example, an order is placed in advance because the lead-in time for the manufacturer is such that delaying the placing of the order could cause delays during the construction process. As can be seen in Figure 11.1, the duration from appointment of the architect to commencement on site is lengthy.

The dotted arrows show the contractual links and the solid line the control, i.e. management links, between the various parties. The contract used in this case is almost invariably the JCT11 Standard Building Contract with Quantities. There can be other links between the client and specialist subcontractors if they carry out a design function. The B of Q is used as the basis for the financial administration of the project (see Chapters 16 and 18) and any errors in the pricing are borne by the contractor. Recompense for any delays or mistakes made by the design team is claimed by the contractor.

It is important to note that the main contractor has little or no effect on the costs of the site selection, design, life-cycle costs or impact on the environment as a whole so all their experience and expertise is lost. The competitive tendering process requires the production of a combination of the following information for the contractor to price the work:

- specifications and drawings
- performance specifications
- B of Q
- bills of approximate quantities
- schedules of rates.

11.1.2 Negotiated tender

In this method, the contractor is selected early in the process, which enables their expertise to be utilised in the design process. Often several contractors are approached and asked to submit costs for the management of the project, key items of work and profit margins. One of the advantages of this method is that the work on site can commence earlier than the competitive tender process permits. This is sometimes referred to as a two-stage process. Bringing the contractor in early enables it to make a technical contribution to the design, improve buildability, modify the design to suit specialist plant and equipment, order materials early to reduce delays and, above all, bring in subcontractors to contribute in a similar way. Finally they can bring their management skills into the process. The contract between the client and various parties can be adapted to suit the particular needs of all.

It is ideal when the client has a programme of new or refurbishment work to be carried out over a period of time as it enables all the same parties involved – the designers, client and contractors – to build up an understanding and work more effectively as a team. This is sometimes referred to as serial contracting when the contractor tenders in the knowledge that if successful more work will be forthcoming. A further variation on this is term contracts when the same process occurs, but the continuity is for a fixed term of say two years after which the client can either renegotiate with the contractor or start the process again. In these cases, cost forecasting is much more accurate for the client as the cost data is available even before the next contract commences. Time and expenditure are reduced as for each new contract the tendering process does not have to take place. Opponents of this method argue that the client doesn't obtain the lowest price, but then does the lowest price always provide the best solution for the client? Also there is less likelihood of claims being submitted at the conclusion of the contract. The contractual relationships remain the same as outlined in Figure 11.1.

11.1.3. Competitive tender based on approximate bills of quantities

Using the same contractual relationships, an alternative approach is to use approximate quantities. This method is used when the project has not fully been designed and a full B of Q cannot be produced in time for the tendering process to take place. As the contract proceeds, work is measured on a monthly basis and the monies paid are agreed based upon the approximate B of Q as far as is possible. This permits the construction work to commence earlier, but financial control can be less accurate than with a full B of Q.

11.2 Design and build

This is usually a competitive process, but can be by negotiation. The client invites contractors to tender not just for the construction work but also for the design, giving an inclusive price. This can be a fixed price, which clients find attractive as one of the major variables in their budget is eliminated and there can be no claims for late design information. However, there can be some difficulty in selecting a contractor because comparisons between the different designs, costs and time required for the execution of the design and construction may vary. It is important the client selects a contractor who specialises in the type of building required. After selection they also only have to deal with one organisation compared with the traditional method.

The client should write a comprehensive brief to suit their requirements, otherwise there is the danger that the contractor will produce a building more inclined to their production needs rather than those of the client. This of course places the inexperienced client at a disadvantage as they may find difficulty in doing this.

In the early days of design and build many contracts were for factory sheds; however, this has now developed into such common practice that major contracts are procured this way. The advantage of this method is the contractor has specialist knowledge that may well outweigh that of the architect normally employed by the client, resulting in more cost-effective solutions and fewer variations. Compared with the traditional method it provides a more rapid completion of the project as the design

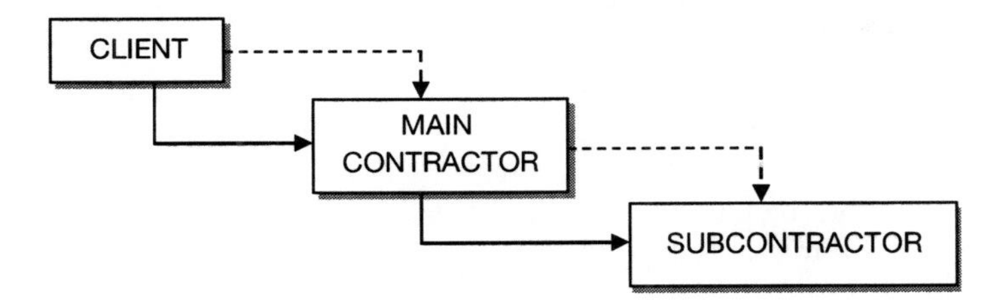

Figure 11.2 Design and build

and construction processes can overlap. The contractor can use its own design team or designers they regularly work with, and can tailor their production expertise and produce building solutions. In some cases the client will have produced outline drawings and performance specifications upon which the contractor will develop the initial design and then construct it. They may employ an agent to supervise the contractor, especially if inexperienced. The advantage of doing this is that the client maintains a design overview and can more significantly influence the final design outcome at a stage when it is easier to change.

Another form of design and build contract is turnkey where the builder not only designs and constructs the building, but also provides all that is necessary for the client to commence their activity. In other words not only is the building ready for occupation but is also ready for use. All the client has to do is to 'turn the key' in the front door. The contract normally used for this type of procurement is the JCT11 Design and Build Contract. In Figure 11.2 the dotted arrows show the contractual links and the solid line the control between the various parties.

The client loses some control over the project as the risk is being largely borne by the contractor. This means also that any alterations the client may require can be more expensive than using the traditional method, although contractors try to accommodate the client's needs whenever possible.

11.3 Prime costing

This is an extension to design and build contracts (11.3). Other than the work that has to be done initially prior to selection for which no fee is paid, this is a risk-free type of contracting and is therefore very attractive to the contractor. Usually about six companies are approached to produce a pre-qualification document. After consideration the client then reduces the number to about four companies. In competition the construction company puts together technical proposals, a design solution and an indication of the costs, which means a lot of assumptions have to be made since the final design has not been agreed. A preliminary cost plan is also produced along with an estimate of the preliminary costs.

After this stage, two contractors are selected for the final phase of the process. These two then make presentations to the client and after client evaluation, the preferred bidder is chosen. From this stage onwards the contractor is reimbursed for all work. What is important is that there is transparency on all costs and this includes the supply chain. So a cladding subcontractor would advise a price for fixing the cladding, but would itemise this into cladding as an item, the insulation added, the labour and any plant required. The contractor works closely with the client and the end users, if appropriate, to develop the design. Examples of end users in a hospital development would be the doctors, nurses, administrators and patients. It means fully understanding the work process in the proposed development, the number of people employed in each section of the process and any requirements they have. The design is then produced. This also involves working closely with the suppliers and subcontractors on their design if they have been commissioned to do this.

11.4 Construction management

There can be some confusion between the terms construction management and management contracting (11.5). So for clarity both sections commence with a definition for the terms used in this text. In this case the client employs the design team and then a construction manager is employed to programme and coordinate both the design and construction processes for which they are paid a fee. The role also involves improving the buildability of the design. Figure 11.3 demonstrates the contractual relationships: the dotted line contractual, and the solid control responsibility.

The construction management team does not carry out any direct work themselves, but divides the work into packages and sublets them to specialist subcontractors. The contract is between the client and the subcontractor even though the construction manager will have usually negotiated the work package on behalf of the client. This means the client has to be very knowledgeable; if not this is not an appropriate method of procurement. The design team will be outsourced totally, either selected in part or whole by the client, or totally by the construction management organisation.

The term design team is a loose description as within the process of development it may well include many other professionals such as lawyers and financiers (see Chapter 4). This is because the process of management can commence as early as the feasibility of the project. A good example of this is when a client is competing for a PFI contract and employs a management contractor to control the process.

The construction phase can commence prior to the design being completed and, because of the nature of the contractual relationships, the traditional adversarial conflict between the various parties is removed. It is also possible to change the design as the contract progresses, providing the particular work package has not been let and it does not have an effect on packages already awarded.

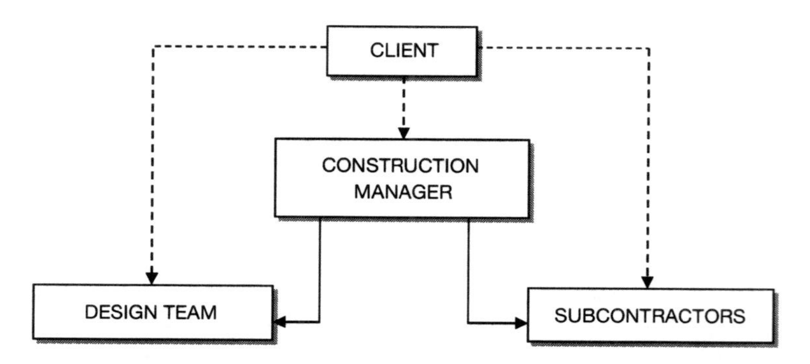

Figure 11.3 Construction management

Unlike design and build contracts, where the client knows reasonably accurately what the final cost of the building is likely to be, in this case until all the packages are let there remains uncertainty and projected costs rely on the elemental cost plan the private quantity surveyor (PQS) has produced. This means that the construction manager has to be very cost aware to be effective.

11.5 Management contracting

A contractor is employed by the client to manage the construction process and is paid a fee. The two main differences compared with construction management are first, the contract for the subcontractors is with the management contractor rather than with the client and second, while having a significant input into the design process, the management contractor does not have the responsibility to manage this part of the process.

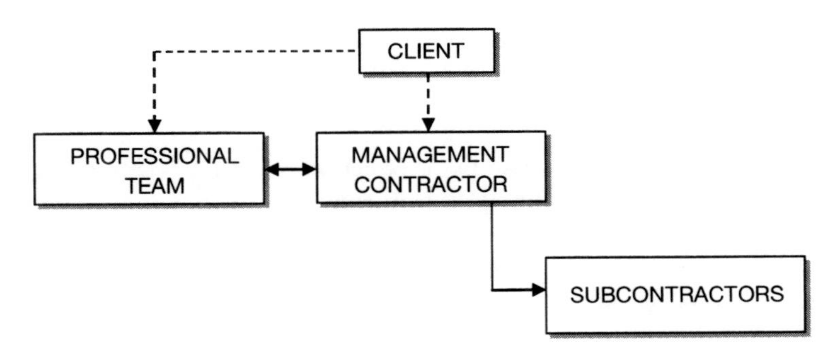

Figure 11.4 Management contracting

11.6 Partnering

There are two types of partnering agreements: that of strategic or multi-project partnering, and project partnering that is for one project. It may well be that a successful project partnership leads to a strategic agreement. The greater benefits accrue on multi-project partnering as this allows the partnering arrangements to be extended to the supply chain (Chapter 24). When this occurs it is a logical extension of continuity type contracts such as serial contracts (11.1.2). However it should be noted that for EC and UK public projects, partnering agreements can only take place after the contract has been awarded which makes strategic partnering difficult to use. This may change as a result of Brexit.

The essential aim of partnering arrangements is the achievement of trust and cooperation between the partners. It is a commitment of all the partners to make the project successful and it is this attitude rather than the contractual links that drives the relationship. It is argued that this commitment cannot take place unless the decision to enter a partnership is taken at the highest level in all the participating organisations. The partnering process is about team building, communication and team spirit so that all involved work together as one, towards the common goal as the prime objective, eliminating personal organisation interests from their thinking. Partnering

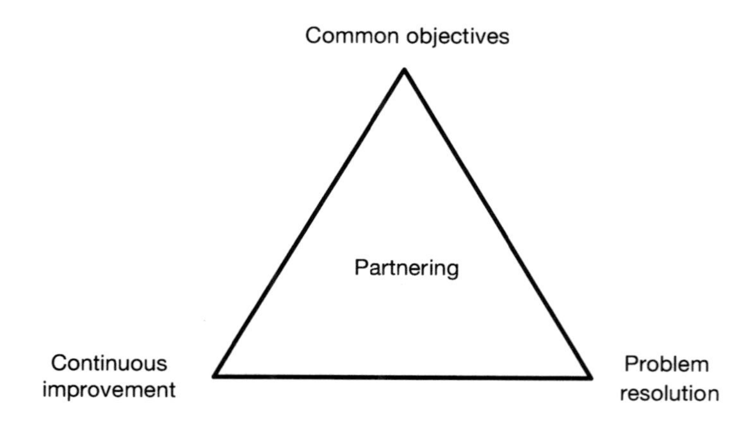

Figure 11.5 Partnering
Source: Adapted from The Aqua Group (1999).

should commence at the development and design stage and it is important that, if possible, the composition of the team should remain stable throughout the duration of the project.

In achieving the proper relationships, it is necessary to agree three key objectives as shown in Figure 11.5: mutually agreed objectives, such as improved performance and reduction in costs; actively looking for continuous improvement; and having a common approach to resolving problems.

It is important that an agreement is drawn up between the partners. This should cover the following:

- a statement of working in good faith;
- a method by which open-book costs of all the partners can be demonstrated and seen by all parties;
- a clear statement of the roles and responsibilities of those involved;
- how the lines of communication between parties will work;
- a procedure for resolving disputes. Hopefully this will not have to be enacted, but unfortunately the world is not perfect and therefore there needs to be a clearly defined way of dealing with differences.

The client in the partnership agreement is not necessarily from the traditional property development companies and one of the earlier examples of partnership in construction demonstrates this as described in the brief case study below.

The Barnsley Partnership

The Barnsley Metropolitan Borough Council (BMBC) made the decision to form a partnership. They were seeking an input from private enterprise in a rolling programme of development using sites owned by the Council. These were not to be selected because they were the most lucrative for the partner to develop. They then ran a competition to find a contracting partner and Costain Construction was the successful candidate. The partnership was formed in 1990.

All profits and losses were shared equally between the partners, which included a proportion of the contracting profits. The company offered a social dimension to the partnership by way of integrated training programmes for local people and also participated in local community activities. One example of this was that the company's human resources department worked up and sustained a youth centre in one of the deprived parts of the town; another was after having done a development for a supermarket to provide staff training. A demonstration of the success of this aspect of the partnership was the award of the Lord Mayor of London's trophy for participation in the community.

Costain was allowed, within the terms of the agreement, to negotiate on price for three out of five projects, allowing the BMBC to tender on the other two for the purposes of maintaining Costain's competitive edge. The negotiation was not a pushover for Costain in that the negotiator from the Council was not part of the partnership and was in close touch with the current construction costs, hence protecting the council's interest.

The expected output for the Council was to use the expertise of the private sector as well as the company's client base, improve the local economy and environment, dispose of some difficult sites and earn some extra money. On the other hand Costain expected to gain from the development profits, construction profits from non-competitive projects and get the opportunity to win similar types of partnerships in other parts of the United Kingdom. It was also felt that at that time doing business with a social dimension was a valued cachet, especially by some of the trade unions who had been influential both in the concept and in the award to Costain. It was also perceived the partnership would benefit by an understanding planning department in obtaining planning permission, albeit there would never be favouritism.

On the surface the idea that a strongly socialist council could have a partnership with a private company perceived as being more 'right wing' was an interesting concept. In practice clashes of culture were never allowed to interfere with the management of the partnership. The agreement was a tightly worded document and the management board was charged to ensure this aspect of working together would not be a problem.

A Costain employee was appointed as the chief executive of the partnership, and reported to the management board which included the council leader, the chief executive of the council and the director responsible for the valuation and planning department.

The partnership had its successes, but also disappointments. Benefits accruing from partnering in general include:

- by definition, elimination of the adversarial approach between the partners thus minimising the likelihood of incurring legal costs;
- the development of working relationships over the long period of association, although if not watched this can lead to complacency;
- in strategic partnering, reduction tendering costs since each new contract awarded is not done in competition with others;
- opportunities for innovation and value engineering;
- efficiency occurring as a result of familiarity with each other's systems which may also be modified to assist in supply chain management enablement;
- improvement in the anticipation and identification of risk.

The negative aspects if not properly controlled and monitored include:

- personal interests taking precedence over the common good;
- new personnel not having the same enthusiasm as those who were in at the beginning and hence not buying into the philosophy;
- in the event of dissolution of the partnership, parties resorting to type and looking out for their own interests at the expense of others;
- lack of competition potentially resulting in higher costs.

11.7 Private Public Partnerships (PPP) and Private Finance Initiatives (PFI)

The basic idea of the Private Finance Initiative (PFI) is that private companies build the project, such as a hospital, school or road using their own money normally provided by the banks (see 5.2). The government rents it back from them over a stated period such as 25 or 35 years after which the ownership of the building either returns to government or a further contract is negotiated. The PFI became very complex and in 1997 the Bates review made 29 recommendations to streamline the process. This became known as the Private Public Partnerships (PPP).

Some contractors have been put off the PFI because of the high costs of bidding and the inherent financial risk. This is because if, after the initial interest shown, the contractor is shortlisted, a significant team of expertise has to be put together all of whom generally work for no remuneration relying on their costs being recouped if they are successful with the bid. The bidding process can be very costly as it will usually include providing a design solution to meet the client's brief, carrying out value engineering on the solution and deciding on production methods, and then agreeing sources of finance. All of this takes time and money and on major contracts could run into millions of pounds.

There are several forms of contract for PFIs and PPPs, examples are shown below:

- Build–operate–transfer (BOT): used for infrastructure projects and PPP, the public body contracts with a private sector body to design and build the infrastructure, operate and maintain this facility for a fixed period. The private company raises the finance for the project and retains all revenues generated until the end of the period when it is transferred back to the public authority.
- Build–operate–own–transfer (BOOT): similar to BOT except that the private company owns the facility until transferred back to the public authority.
- Design–build–finance–operate (DBFO): similar to BOOT except there is no ownership transfer from the public authority. Used in particular for toll roads.
- Build–own–operate (BOO): in this case ownership remains normally with the private company.

11.8 Apportioning risk

The HM Treasury defines risk as 'the uncertainty of outcome, whether a positive opportunity or negative impact'.

In development, risk is the difference between the anticipated cost of the building, including all development, design and construction costs, and the final actual costs. Since a significant factor in the procurement of contracts is concerned with distribution

Table 11.1 Financial risk

Type of procurement	*Risk*
Traditional	The contractor takes a risk initially when tendering in competition. Once the contract is awarded, the client carries the risk
Construction management	Depends upon the form of contract, but generally the contractor takes the greater risk, but since packages are let closer to execution of the work, less risk is built into the package price by the subcontractor
Design and build	Contractor takes the risk and responsibility. If this is a fixed sum rather than having some form of inflation proof contract, the risk is greater
Management contracting	Same comments as construction management
Partnering	Partnering is not about allocation of risks but, as Latham asked for, 'a shared financial motivation'
Private Finance Initiatives	The unsuccessful bidders can take considerable financial risk, after which some of the risk is transferred to the client, the amount depending upon the wording of the contract
Prime costing	Other than during the pre-qualification period, the client takes the risk

of financial risk between the various parties involved, a summary of this has been produced in Table 11.1. It should be noted that the more the client apportions risk, generally speaking the less control the client has over the project. For simplicity the risk apportioning shown is between the client and the contractor. However, the client will have almost certainly laid off some of the financial risk with the financers of the project. It should also be noted that much depends upon the way the contract has been drafted so the table above is only a rough guide.

Bibliography

Egan, J. (1998) *Rethinking Construction*. London: Construction Taskforce: HMSO.

Faster Building for Industry (1983) London: HMSO.

JCT11 Design and Build Contract (2010). London: Sweet & Maxwell.

JCT11 Standard Form of Building Contract with Quantities (2011). London: Sweet & Maxwell.

Kelly, J., Morledge, R., and Wilkinson, S. (2002) *Best Value in Construction*. Oxford: Blackwell Science.

Latham, M. (1994) *Constructing the Team*. London: HMSO.

McGeorge, D. and Palmer, A. (2002) *Construction Management New Directions*, 2nd edn. Blackwell.

Smith, R.C. (1986) *Estimating and Tendering for Building Work*, 7th edn. London: Longman.

The Aqua Group (1999) *Tenders and Contracting for Building*. 3rd edn. Oxford: Blackwell Science.

12 Method of measurement and bills of quantities

12.1 Introduction

The purpose of this chapter is to introduce the B of Q and new rules of measurement so that their composition, application and usage are understood. It is not intended to teach the reader how to produce B of Q in practice.

In the past much of the contracting work was obtained via the traditional route as described in traditional methods of procurement (11.1) using B of Q. While in recent years the amount of work using this approach has declined, much of the other forms of contract between the client and the contractor is based upon the principles laid down in these two documents. The employer's quantity surveyor produces the B of Q – an itemised account of all the work that has to be carried out on the contract. The method and sequence of measurement in the bills is based upon the RICS New Rules of Measurement 2: Detailed measurement for building works (NRM2). The civil engineering standard method of measurement (CESMM4) is used in civil engineering projects.

12.2 The standard method of measurement and the new rules of measurement

The great advantage of the new rules of measurement – detailed measurement of building works (NRM2) and its predecessor the standard method of measurement, 7th Edition (SMM7) – is that it determines a uniform method accepted by all parties. Prior to these documents, estimators were often left in doubt as to what items in the then B of Q actually meant. This could lead to misunderstanding, inaccurate tendering and a basis for dispute when payments were made for work completed.

In 1912 a Joint Committee, made up of members of the Surveyors' Institution and the Quantity Surveyors' Association, was constituted, charged with the task of producing a standard set of rules for measuring building works. In 1918 four contractors nominated by the Institute of Builders and the National Federation of Building Trades of Employers were added to the committee. From time to time an input was sought from representatives of certain trades. The first edition published in 1922 was based upon the practice of leading London quantity surveyors of the time, modified to suit practices throughout the United Kingdom. Subsequent editions have been produced, refining and modernising the document to suit the practices of the day, culminating in the 7th Edition of the SMM7 published in 1988. In 2012

SMM7 was replaced with NRM2 to take account of modern building and procurement practice. SMM7 is still used in many parts of the world.

12.3 Composition of NRM2

The new rules of measurement are divided into different work categories as shown in Table 12.1. Each is a clearly defined area of work. All the categories with the exception of Preliminaries, have a physical presence in the final building. The Preliminaries are discussed in 12.4. It is important to note that the sequence shown, i.e. 1 to 41, is the sequence in which measured work is sometimes presented in the B of Q. This uniformity helps the user of the bills to access information.

Each of these sections is then subsequently broken down into further subsections. It is also useful to have it broken down in this way for the contractor, because if all like materials and processes are together it is easier to abstract information to send to suppliers and subcontractors for quotations. It can also be used to provide material quantities for the planner to produce the pre-tender programme and later on, for ordering materials.

There are then further subsections to produce further sub-classifications. For example in the case of masonry, the type of brick construction and whether or not the brickwork has face work on one or both sides and its overall thickness. At the same time it goes into further detail including whether measured as linear, metre super or metre cube, all of this with a view to clearly specifying the finished work.

Table 12.1 Detailed contents of NRM2

	Categories		*Categories*
1	Preliminaries	21	Cladding and covering
2	Off-site manufactured materials, components and buildings	22	General joinery
3	Demolitions	23	Windows, screens and lights
4	Alterations, repairs and conservation	24	Doors shutters and hatches
5	Excavating and filling	25	Stairs, walkways and balustrades
6	Ground remediation and soil stabilisation	26	Metalwork
7	Piling	27	Glazing
8	Underpinning	28	Floor, wall, ceiling and roof finishings
9	Diaphragm walls and embedded retaining walls	29	Decoration
10	Crib walls, gabions and reinforced earth	30	Suspended ceilings
11	In-situ concrete works	31	Insulation, fire stopping and fire protection
12	Precast/composite concrete	32	Furniture, fittings and equipment
13	Precast concrete	33	Drainage above ground
14	Masonry	34	Drainage below ground
15	Structural metalwork	35	Site works
16	Carpentry	36	Fencing
17	Sheet roof coverings	37	Soft landscaping
18	Tile and slate roof and wall coverings	38	Mechanical services
19	Waterproofing	39	Electrical services
20	Proprietary linings and partitions	40	Transportation installation
		41	Builder's work in connection with 38, 39 and 40

12.4 Preliminaries

This is the first section (1) of the B of Q. Preliminaries are concerned with the administration of the project and the provision of plant and site-based services such as accommodation and water, electricity etc. The content will vary considerably from contract to contract, but always needs careful attention. Generally, the value of the preliminaries can be from 7 to 15 per cent of the total cost of the project. Errors here can have major implications on the cost performance of the contract as a whole. For example, the salaries of staff employed on the contract are included here and if the contract takes longer to complete than programmed through no fault, there will be no more money left to cover these wages. Equally, if an extension to contract is agreed, for example to accommodate alterations made by the client, then these figures act as the basis for the calculation of recompense to the contractor and hence the need for accuracy. Note that NRM2 contains work sections covering Preliminaries for situations where the contract is based on a main contractor as well as individual work packages. The following is based on main contractor's preliminaries.

As the value of the preliminaries is so significant, if contractors were paid for them proportionately over the period of the contract, their cash flow would be severely distorted because of the initial outlay that has to be made to set up the contract for such items as site accommodation, installation of a tower crane and hoardings around the site. To overcome this, preliminaries are paid for in two ways:

- fixed charge for work that is considered to be independent of the duration of the contract;
- time-related charge for work that is considered to be dependent on duration.

NRM2 divides preliminaries into two main sections:

- preliminaries (main contract)
- preliminaries (work package contract).

In turn the two main sections are split into:

- Part A: Information and requirements
- Part B: Pricing schedule.

The complete list of items is shown in Table 12.2:

- Part A: Information requirements
- Part B: Pricing schedule.

12.5 Preliminaries classification expounded

The examples given for the contents of the various classifications following are only meant as an indication of what might be included. In reality, not all may be used and equally they may be expanded upon. Note the NRM2 always refers to the client of the project as the employer.

Table 12.2 Preliminaries classification list

Part A: Information requirements

	Classification		*Classification*
1.1	Project particulars	1.12	Employer's requirements: Operation/maintenance of the finished building
1.2	Drawings		
1.3	The site and existing buildings		
1.4	Description of the work		
1.5	The contract conditions		
1.6	Employer's requirements: Provision, content and use of documents		
1.7	Employer's requirements: Management of the works		
1.8	Employer's requirements: Quality standards/control		
1.9	Employer's requirements: Security/Safety/Protection		
1.10	Employer's requirements: Specific limitations on method/sequence / timing		
1.11	Employer's requirements: Facilities/Temporary works/Services		

Part B: Pricing schedule

	Classification
1.1–1.12	Employer's requirements
1.2.1–1.2.14	Main contractor's cost items

12.5.1 Preliminaries (main contract) Part A: Information requirements

1.1 Project particulars

- Name, nature and location of the project. This gives the contractor an immediate overview of the project and sets the work in context. The location is important to know in terms of how far it is from the office, which affects communications, the difficulty in recruiting sufficient local labour, which could mean transporting labour in, and whether or not it is familiar territory.
- The names and addresses of the employer (client) and consultants. This is important to know as the contractor may have worked with some or more of these parties before. The quality of this experience could affect the decision on whether or not to tender depending on whether the experiences were good or bad. If there had been an excellent working relationship and the previous contract(s) have run smoothly, then this could affect the tender price and produce a more competitive bid.

1.2 Drawings

- This is a list of the drawings from which the B of Q was produced. This gives the contractor an indication of the stage the design has reached and therefore the reliability of the data upon which the tender is being based. The less detail available then the greater the risk of the estimate being inaccurate. This increases the likelihood of claims for extra payments or extension to the duration of the contract being made, which can sour relationships with the employer as they see the cost of the project rising.

1.3 The site/existing buildings – read in conjunction with 1.4 Description of the work

- The details of the site boundaries are provided either by description and/or by reference to site plans or map references. The estimator will want to know, either by looking at these drawings or by site inspection, the road network around for ease or restrictions of access, which may affect the hours of work, ability to unload from the main road and/or times of delivery. From the drawings provided it can be seen what land, if any, is not being built on, allowing space for the provision of site accommodation and material storage. If the footprint of the proposed building takes up most of the site then the public will have to be protected against the possibility of falling objects, and perhaps footpaths diverted for safety reasons, all of which can be quite expensive. These drawings would normally indicate any existing trees both in terms of species and size, which if being left in place may have to be protected and could affect the building work.
- Details of existing buildings adjacent to or on the site. This could have an effect on the methodology and costs of work due to their proximity to the new construction. Examples include limitations in tower cranes over-flying other buildings (this is considered as trespass unless permission is sought and gained), or sensitive buildings, such as historical ones that may be readily damaged, or limited access. It is important to note that the contractor will be well advised to document the state of adjacent buildings once the contract is awarded, so as to avert claims for damage at a later stage. The usage of these buildings is also important. For example if the adjoining building is a dairy then the production of dust from the construction process would be problematic (1.9). If the site is approached from a main road, then the road and public footpath will be have to be kept free of mud during the course of the works.
- Existing main services (drainage, gas, electricity, water) both on and adjacent to the site should be identified so that the contractor is aware of them. This can affect the mode of work as well as offering the opportunity to connect to existing supplies in executing the construction work. It should be remembered that on old previously built-on sites, some of the services might not have been located and also may run elsewhere from that indicated. This should be remembered in the context of safety as, for example, an excavator cutting into a live electric main could have disastrous results.
- Any other information that is available about the site that may assist the estimator. This may include soil surveys and likely contamination implications. The detail of the soil surveys is important as the number and position of trial bore holes

drilled over the area of the site determines the accuracy of the overall site soil survey as it is easy to miss pockets of poor soil such as peat if inadequate or inappropriate sampling has taken place. As an example, a square in plan building was designed for a site on which an adequate number of trial bore holes were taken for its purpose, but subsequently it was decided to change the orientation of the building by rotating it by 45 per cent. Some of the trial bore holes were now outside the footprint of the building.

- Contamination can be identified from the soil samples taken, but also indications of problems can be identified by knowledge of the previous uses of the site. For example if a garage then there is a likelihood of petroleum spillage over a long period of time that would have polluted the sub soils. The presence of Japanese knot weed would be another important issue as it can be very expensive to deal with in its removal and preventing its spread by foot traffic.

1.4 Description of the work

- This is a general description of the work to be carried out, giving for each building to be constructed/modified/refurbished its main dimensions and shape. This would include the plan area and perimeter at each floor level, heights between floors and the total height of the building. It would advise of details of work being carried out by others that may impinge on the contract, and identify any other unusual features or conditions.

12.5.2 Contractual matters

1.5 The contract conditions

These are details of the form of contract that is to be used in the event of a successful bid. A typical example would be the JCT 11. The estimator will be especially interested in special conditions such as:

- how much retention is to be made at the end of the contract and for how long it will be held;
- the employer's insurance responsibility, any amendments made to the standard conditions, any performance bonds required and whether or not the contract is to be executed under hand or under seal. Under hand means that the contractor's liability for most defects is for a period of six years after completion, and under seal for 12 years. The type of contract utilised will depend on the method of procurement (see Chapter 11). Contract is a significant and complex subject in its own right and is not developed in this text. It should be noted that the JCT forms of contracts were designed to be fair to both signatories to the contract. Any onerous additions or removals of clauses, by definition, will favour one or the other;
- whether the contract includes for fluctuations, and if so at what dates these come into effect;
- any warranties that may be required for design work carried out by a subcontractor. There is a separate form for this named Collateral Warranty Form of Agreement;

- extra costs to the contactor for providing, for example, a guarantee bond that covers the employer in the event of the need to employ another contractor to finish off uncompleted work up to the value of the bond. This is not used on all contracts, but has been required by local authorities and other public bodies. The bond is usually 10 per cent of the value of the contract, taken out with an insurance company.

12.5.3 Employer's requirements

1.6 Employer's requirements: provision, content and use of documents

- This is concerned with who has the responsibility to provide drawings, i.e. the architect, consultants, contractor and subcontractors. It also defines what should be included so that nothing is overlooked, and how they should be used.

1.7 Employer's requirements: management of the works

This may include as examples:

- stating who has the responsibility for the supervision, coordination and administration of all the works. This may well differ depending upon the form of contract and method of procurement;
- the taking out of appropriate insurances to cover such risks as fire, vandalism, third party injuries and damage;
- the provision of a master programme for all the works and ensuring that progress is monitored. This may also require advising the employer accordingly at stated intervals. Monthly would not be unusual as the employer needs to be aware of any delays or if the contract is ahead of programme, as this can have a knock-on effect on their business. For example the need to bring forward or delay the installation of machinery or shop fit-outs. There will also be financial implications to their revenue stream. The type of programme might also be specified, identifying the amount of time allocated for inclement weather;
- collecting meteorological data on site. The site manager recording the basic information in the site diary often satisfies this requirement. It is important, as notwithstanding the comments in the item above, excessive inclement weather could be grounds for an extension to contract. The winter of 1962/63 in the United Kingdom was so severe, with sub-zero temperatures for approximately two months, much of the construction industry was brought to a standstill during this period;
- arranging site meetings at regular intervals;
- notification of when the work commences and finishes. This is not necessarily the whole contract but could be stages or phases of the work;
- notifying the Employer's agents (the quantity surveyor) before covering up work so that it can be measured correctly. This is most likely, but not always, to occur when working on the foundations and drainage, where work can be very rapidly built over or concealed;
- the keeping of site labour and plant records so that when calculating any claims or variations to contract, there is factual data;

- who owns unfixed materials stored on site. Until such a time is reached when a large proportion of construction materials arrive as and when they are required to be built into the building (just-in-time delivery), there will be significant quantities of building materials stored on the site waiting to be used in the building. Current contract almost invariably means that materials delivered each month and not built-in are measured and paid for in part or total by the employer. Who owns what is an interesting legal question as, if the supplier has not been paid for the material, they also have an interest in this.

1.8 Employer's requirements: quality standards/control

These can include as examples:

- The employer needs to be satisfied that the materials and components purchased by both the contractor and subcontractors meet the specification. This section gives the employer the opportunity to specify which samples of materials and components have to be obtained and what tests need to be carried out. The contractor can cost how much this will be.
- It is sometimes necessary for a sample of work such as a brick wall to be constructed so that a standard of workmanship can be agreed. This means that all similar work on site can then be compared against this example of workmanship. Alternatively, a sample of a component(s) can be erected on site showing the jointing between elements and fixing to the structure. This is done to show the employer that the solution works and what it looks like in practice.
- As a result of the debate on ownership of materials unfixed on site, the employer has an interest in the methods by which unfixed materials are protected and stored. Further, that it is the contractor's responsibility to replace damaged materials at their expense.
- If defective work or materials occur it is important that there is a procedure for sorting out the problem, since ineffective procedures could cause delays to the project. The employer needs to be satisfied that the procedures in place will not cause any problems of this nature.
- The contractor will be expected to clear up and remove all rubbish and surplus material as it accumulates and at the end of the contract or phase of work. This clause might be extended to specify not only that materials have to be disposed of correctly, but also have to be segregated and disposed of in an environmentally friendly way using recognised recycling organisations.
- Before the building is handed over, a detailed check, known as snagging, takes place to ensure all the minor problems are listed and remedial action taken. Examples of this would be touching up paintwork, ensuring all ironmongery works correctly, doors and windows are fitted properly, and nothing has been missed such as a screw out of a hinge or a shelf in an airing cupboard. This list can be very extensive depending upon the effectiveness of the quality assurance systems used by the contractor.

1.9 Employer's requirements: security/safety/protection

- Contracting by its very nature is noisy and dirty, but the emission levels can be controlled with careful planning and control. The employer can specify these levels.

The contractor can also be limited to the amount of noise made before or after certain times of the day. There may be a necessity to make provision to protect rivers, streams and aqua flows from effluent pollution. The Manchester Evening News Arena was constructed very close to the Boddingtons Brewery, which obtained its water from aqua flows below where the proposed building was to be built. Any pollution from the construction would have severely affected or indeed destroyed the brewer's business. One should remember that after the contractor has left the client still has to live with the neighbours.

- The public and private roads surrounding the site have to be maintained. This is in part a pollution problem resulting from mud carried off the site by vehicles, but also to do with damage from these and other heavy and tracked plant.
- Existing and surrounding buildings need to be safeguarded against damage by the main contractor or subcontractors, and any damage caused made good at their expense and not the employer's.
- The existing underground services will be marked on the site plan. The contractor must ensure these are not damaged in any way from excavation or overloading from either heavy traffic or excess demand. It may be specified that the contractor should take all reasonable steps to check that these drawings are correct and complete.
- Sites are vulnerable to theft, vandalism and intruders, which includes children who see the site as a gigantic playground, drug users and those sleeping rough. The provision of security on the site is not just to prevent the problems mentioned, but also to protect the same people against injury to themselves, as there is a duty of care. Employers wish to protect their reputation by not having such incidents on the site.
- The work needs to be protected as it progresses, especially completed work, from damage by impact, the weather and other trades.
- Protection of existing trees and habitats may be required by law as well as by the client. For example trees may have preservation orders placed upon them and other wildlife that includes bats, nesting birds and types of newts, not to mention certain species of flora that are also protected.

1.10 Employer's requirements: specific limitations on method/sequence/timing

These are issues that limit the contractor and subcontractors' method of operation, their sequence and timing. Examples include:

- restrictions on the working hours, which may include the start and finishing time, specific days of the week and public holidays. This may be due to the environmental impact to others adjacent to the site or because the client has particular views and beliefs such as not working on Sunday;
- restriction of access roads – the term used to describe both the temporary roads on the site and those public roads used to approach the site. The employer may wish to restrict access from certain roads on the grounds that the road is not substantial enough to take the loads or that it goes through say a residential area causing nuisance or a safety risk to children. Temporary roads on a restricted site may be relatively inexpensive but on a complex large site costly depending on ground conditions and distance travelled;

- any materials that are found on the site and have to be disposed of, excluding material that would be part of a bills rate such as excavate and cart-away or the normal wastage expected in other operations. There may also be environmental considerations stated here;
- design constraints, especially if the contract is such that the contractor is involved in or has some leeway in selecting materials. The employer may wish to ensure certain choices are made. Any planning restraints may also be identified here;
- the employer wishing to determine the method and sequence of work. This would be particularly important when working in an already occupied building or if the building is to be handed over in phases. It could well relate back to items such as noise and pollution as identified in 1.9;
- prohibition of the use of foul language and if occurring the offender should be instantly dismissed. It may seem strange on a construction site that such a condition be laid down but the contract awarded for the construction of the Mormon Tabernacle near Preston made that a requirement.

1.11 Employer's requirements: facilities/temporary works/services

- The employer who may have permanent or semi-permanent representatives on the site, such as members of the design team, the clerk of works and resident engineer, will need accommodation. The contractor will usually be expected to provide the facilities for them as requested and this may include furniture and equipment. This is additional to that required by the contractor (1.2.1).
- This provision could include such items as cleaning and the provision of sanitation, running water, heat, lighting, power and telephone.
- The employer may specify the location on site of both these offices and the accommodation of the contractor.
- Temporary hoardings around the site are often used by the employer to advertise themselves and they can lay down some strict conditions accordingly in terms of colour, quality and specification of material used and so on. Failing that, the contractors may paint them in their own livery colours and advertise themselves.
- Normally the employer, design team, contractor and subcontractors advertise themselves using name boards. There may be limitations in the overall size and position of the name boards laid down by the local authority planners or by the employer. It is not unusual for the dimensions of the individual boards to be controlled and specified and in some cases the employer may only want their name to be shown because of conflict of interests. For example if the contractor is also involved in property development, the employer may not wish this fact to be advertised.

1.12 Employer's requirements: operation/maintenance of the finished building

- This is the provision of facilities and services to be provided by the contractor and subcontractors on completion and/or after, to assist the employer to operate and maintain the finished building. This could include instruction and maintenance manuals, training for maintenance staff especially on the mechanical and electrical side. BIM may be one of the ways of doing this.

12.5.4 Preliminaries (main contract) Part B: Pricing schedule

1.1 Employer's requirements

1.1.1 SITE ACCOMMODATION

The type of temporary accommodation required by the employer and the employer's representatives, where separate from the main contractor's site accommodation, should include:

- offices for the management and administrative staff;
- stores for small and/or valuable components, materials and hand tools;
- canteen used for lunch breaks and a place to go during inclement weather;
- toilets, both male and female;
- drying room to hang wet clothes overnight;
- first aid facilities on a large site if a full-time first aider/nurse is employed;
- washing facilities;
- security cabins at the entrance and exit gates of the site.

The total costs of providing these includes erecting and fitting out, removal after the contract is completed, transport to and from the site, foundations and drainage, furniture, rates and insurance.

1.1.2 SITE RECORDS

Operation and maintenance manuals both hard copies and web-based information.

1.1.3 COMPLETION AND POST-COMPLETION REQUIREMENTS

Hand-over requirements including;

- training user's staff
- spare parts.

1.2 Main contractor's cost items

1.2.1 MANAGEMENT AND STAFF

The contractor's management team will vary considerably depending upon the size and complexity of the project. Most contractors will have a standard list of personnel likely to be used, which they can use as a checklist. It should be noted that the cost of staff is not just their salary, but includes:

- pension scheme (employer's contribution)
- annual bonus
- overtime
- car and expenses
- training levy.

The amount paid out will be different each month, as the staff required to manage the contract will change as the contract progresses. Typically the type of staff will include, on a medium to large contract, those shown in Table 12.3.

1.2.2 SITE ESTABLISHMENT

The type of temporary accommodation required by the contractor is dependent upon the needs in managing the contract itself and is in part determined by the numbers of staff decided on. What is provided will also depend upon the space available to site the accommodation. Note that this does not include any accommodation provided for the employer. On inner-city development the accommodation may also be off site, the contractor renting other premises.

Typical accommodation may include:

- offices for the management and administrative staff:
- stores for small and/or valuable components, materials and hand tools;
- canteen used for lunch breaks and a place to go during inclement weather;
- toilets, both male and female;
- drying room to hang wet clothes overnight;
- first aid facilities on a large site if a full time first aider/nurse is employed;
- washing facilities;
- security cabins at the entrance and exit gates of the site.

The total costs of providing these includes erecting and fitting out, removal after the contract is completed, transport to and from the site, foundations and drainage, furniture, rates and insurance.

1.2.3 TEMPORARY SERVICES

Temporary services include such items as:

- temporary electricity and gas supply: the availability of supply needs to be ascertained. Is it sufficient to run a tower crane and how far is the connection away from the site? The cost includes connection and disconnection charges as well as the running costs;
- temporary water supply: similar issues as with the electric supply, but if not readily available the use of bowsers can be considered;

Table 12.3 Site staff

Contracts Manager	Planner
Site Manager	Assistant Planner
General Foreman	Quantity Surveyor
Trades Foreman	Assistant Quantity Surveyor
Ganger	Safety Officer
Engineer	Secretarial Staff
Assistant Engineer	Security Staff
Chainman	Cleaners

- temporary telecommunication systems: the number of lines required for both calls and IT needs to be established, as this will indicate the costs of connection and the monthly costs;
- temporary drainage.

1.2.4 SECURITY

This has become a very important issue, not just because of potential theft and vandalism, but also the risk of terrorist activity; the latter depending much on the location and/or purpose of the building such as an airport terminal, government or prestigious building. There is security during the working day and outside working hours. In the case of the former, checking in and out procedures to ensure only bona fide people are on site, and in the latter patrolling the site. It will also mean providing material security items such as locks and alarms.

1.2.5 SAFETY AND ENVIRONMENTAL PROTECTION

Works required to satisfy requirements of CDM Regulations:

- The contractor will be expected to provide employees with certain protective clothing including safety helmets and goggles, safety equipment such as harnesses, barrier creams, signing around the site and training. It will also be necessary to provide appropriate protective clothing for visitors and often training before allowing them on site.
- Barriers and safety scaffolding.

1.2.6 CONTROL AND PROTECTION

This section gives the contractor the opportunity to price for providing:

- any necessary surveys;
- protecting the finished works prior to handover;
- the provision of samples;
- drying out completed buildings.

1.2.7 MECHANICAL PLANT

Small tools and plant could be a very long list, but Table 12.4 gives a flavour.

Large items of mechanical plant would include those shown in Table 12.5.

Some of these items may be included within the all-in unit bills rate (see 13.11) and would not be included here. This would be the case if a piece of plant is being brought in for a specific relatively short clearly defined operation as distinct from a dumper say, which may be on the site for the majority of its construction programme and used for a multitude of different bill items. Other equipment that is being hired may also include costs for the drivers' time if this is provided as part of the hire agreement such as with large excavation plant. The running and maintenance costs will be provided either by the contractor and/or the hirer.

Table 12.4 Small plant and tools

Barrows	Saws
Blocks and tackle	Water pumps
Steps	Concrete vibrators
Trestles	Concrete screeders
Buckets	Compressors
Pick axes	Generators
Spades and shovels	Fuel tanks
Drills	Etc., etc.

Table 12.5 Mechanical plant

Excavators and earth movers	Concrete mixers and Silos
Concrete mixers and pumps	Lorries
Dumpers	Hoists
Forklifts	Piling plant
Mobile cranes	Pavers
Tower cranes	Tractor and trailers
Compressors	

1.2.8 TEMPORARY WORKS

Some of these items may have been taken account of in 1.9 and 1.11 as employer's requirements, but will probably be priced by the estimator here.

- Besides satisfying the employer, the contractor is also interested in access roads both on and onto the site. Approach routes to the site are important to ensure that goods coming to the site have unobstructed access and to establish the best route in and out of the site.
- Strengthening of existing roads could have already been specified (1.9), but if the road is not substantial enough, then a cost will be incurred to carry out the necessary works, as it is the contractor's responsibility to protect the highway.
- It is sometimes necessary to divert public rights of way running through the site as well as re-routing public pavements adjacent to the site by providing temporary walkways.
- Access scaffolding is required on the outside of most building projects for access, platforms, towers and inside for stair- and lift-wells. It will also be used as guard rails to prevent people falling into holes, from the building or into a trench. This will also include scaffolding boards and toe boards used for walkways and for ladders for vertical access.
- Support scaffolding must be provided for formwork to stairs and soffits and propping once the formwork has been removed until the concrete has reached its design strength.
- If traffic from the site has to cross a highway or if the works partially obstruct the flow of traffic it may be necessary to include for the provision of some form of traffic control. This can be achieved by either automated lights or using labour.

Any other temporary structures that are required such as bailey bridges are priced here.

1.2.9 SITE RECORDS

The contractor should price for costs associated with any items not listed as deemed to be included site records.

1.2.10 COMPLETION AND POST-COMPLETION REQUIREMENTS

This section is concerned with costs associated with testing, commissioning and handover. The contractor should price for costs associated with any items not listed as deemed to be included.

1.2.11 CLEANING

Cleaning allows the site to be kept tidy and includes:

- cleaning of work in progress
- pest control
- final builder's clean
- waste management – providing skips and waste chutes to collect and to permit the segregation of waste to enable more of the waste materials to be recycled rather than sent to landfill tips.

1.2.12 FEES AND CHARGES

Fees and charges associated with the payment of fees for a range of items including:

- building control fees
- considerate constructors' scheme
- rates on temporary accommodation.

1.2.13 SITE SERVICES

Temporary works that are not specific to an element, for example:

- temporary screens
- support for unstable structures.

1.2.14 INSURANCES, BONDS, GUARANTEES AND WARRANTIES

The costs associated with the following should be included here:

- works insurance
- public liability insurance
- employer's liability insurance
- bonds

- guarantees
- warranties.

It is recommended by NRM2 that a priced cost centre schedule NRM2 Appendix C, together with all backup calculations, is submitted with the tender. Any items that are not priced are deemed to have no financial effect.

12.6 The bill of quantities

The main purposes of the B of Q are as follows:

- It enables all the contractors tendering for the contract to price on exactly the same information in a form that is standard.
- The risk to the contractor is minimised as each part of the work is clearly itemised, theoretically resulting in more realistic and competitive tenders. The proviso to this is that the completed bills total price may not be the same as the employer finally has to pay because of subsequent variations and claims.
- If a more accurate price is required then the employer and the design team will be encouraged to finalise the design and have produced a full set of drawings and specifications upon which the bills can be based.
- After the contract has been let, it acts as a satisfactory basis for the calculation of variations made as a result of design changes that occur during the contract period.
- It acts as a good vehicle to calculate the certified payments throughout the contract. (Certified payments are for the work done each month and any materials delivered during the same period, but not yet incorporated into the building.)
- As it is a quantified and described itemised list of all components and materials to be used in the building it gives the contractor the opportunity to negotiate and place provisional indicative orders of the bulk materials and components with suppliers.
- The same data can be used by the contractor's planner for programming the work and calculating the labour resource required.
- It can be used as a cost control system by the contractor to ensure the work is within budget.
- It is the basis for calculating and agreeing the final account after the completion of the contract.

Bibliography

Cartlidge, D. (2017) *Quantity Surveyor's Pocket Book*, 3rd edn. London: Routledge.

Civil Engineering Standard Method of Measurement (2012) 4th edn. London: ICE Publishing.

Lee, S., Trench, W. and Willis, A. (2014) *Willis's Elements of Quantity Surveying*. Chichester: Wiley-Blackwell.

RICS New Rules of Measurement 2 – Detailed Measurement for Building Works (2012).

Seeley, I.H. and Winfield, R. (1999) *Building Quantities Explained*, 5th edn. London: Macmillan.

Standard Method of Measurement of Building Works (1998) 7th edn. London: RICS.

13 Estimating and tendering procedures

13.1 Introduction

This chapter is primarily concerned with the invitation to tender using the traditional method of procurement, although much of what is contained here is also relevant to other methods. The term client is preferred to employer by the author as it is suggested it more properly reflects the relationship with the contractor, and reduces confusion when referring to employer–employee relationships within an organisation. It is interesting that in total quality management the term 'customer' is used (see 25.3).

It is important to note that the estimator's prime role is to estimate, as accurately as possible, the true costs of doing the work on site. This excludes any overheads and profit management may wish to add. This net cost is known as the estimate. The tender, being the price quoted to the client (the employer), is the estimate plus the overheads and profit. If the estimator has a proven track record of being accurate with the estimate, the more confidence management has in making strategic decisions as to the amount they should add to be truly competitive and surer of making their planned financial returns.

The more comprehensive and accurate the documentation provided by the client, the more likely the estimate will be close to actual. Equally important is that the likelihood of future claims, disputes, variations and delays in completing the contract will be reduced. This is very important as it is negotiations over these financial claims that can lead to confrontation between parties rather than the cooperation that is in everybody's interest for the smooth running of the contract.

Unfortunately, the documentation provided both at the tender and construction stage is not always accurate. Table 13.1 (adapted from the findings of the Co-ordinating Committee for Project Information's findings in 1987) identifies typical examples of deficiencies in the provision of information for both the tendering and construction processes and the reasons why. However, it should be noted that many of these are more likely to happen at the construction stage.

13.2 Selection and invitation to tender

Invitation to tender via the traditional and design and build methods of procurement comes about in primarily three ways:

1 The client or representative advertises in the press or trade papers and asks for contractors to tender. The contractor wishing to tender for the work, replies to

Table 13.1 Causes of problems with information

Cause	*Reason*
Missing information	It has not been produced or has, but has not been sent to the contractor
Late information	Has not arrived in time for the contractor to plan the work properly or is too late for materials to be ordered and delivered on time
Wrong information	The information is out of date, or includes errors such as dimensions and descriptions
Insufficient detail	Inadequate information for either tendering or the construction processes
Impractical designs	The design solution is either difficult to construct or will not work in practice
Inappropriate information	Either not suitable for the purpose intended or not relevant
Unclear information	Caused either by poor draftsmanship or the solution is ambiguous
Not firm	Provisional information which is indistinguishable from firm information
Poorly arranged information	Not produced in a consistent way, poor titling and generally difficult to read as a result
Uncoordinated information	Where it is difficult to read or relate one drawing with another
Conflicting Information	Where documents disagree with each other. This often occurs between say the architect's and the engineer's drawing

Source: Adapted from the findings of the Coordinating Committee for Project Information, 1987.

the advertisement and in return receives the contract documents. Sometimes a deposit is required, refundable on receipt of the completed bid so as to discourage frivolous enquiries. This process is referred to as 'open tendering'.

"Pre-Brexit" Public contracts in value have to be advertised across Europe as a result of a European Directive if in excess of approximately 5 million euros so that free trade can occur across the European Community. This figure is adjusted from time to time. The latest can be found in the *Official Journal of the European Union* (OJEU).

2 More generally in use, is 'selective tendering' when a list of perhaps three appropriate contractors is drawn up and they are asked to tender. This list is referred to as a preferred list.
3 In this case only one contractor is approached because of their track record in the particular type of work, and the design team wish to avail themselves of the contractor's knowledge during the design stage. It is highly probable that the client and/or design team has worked with this contactor in the past. This is known as a 'negotiated tender'.

In selective and negotiated tendering the experienced client will select contractors to tender based on the following:

- The size of the company may determine the range of value of work they can be expected to manage successfully. Large contractors often have difficulty in managing small contracts cost effectively as the overheads may be too high. Similarly, a small contractor may not have the ability to manage a very large contract. The quality of the management team is measured on how successful they are at maintaining relationships with subcontractors, and completing similar work well and on time.
- The experience the client has had with the contractor on previous work would reinforce the above knowledge.
- Their financial state is important. Indicators, although not always reliable, would be the length of time the company had been in business, financial checks carried out, bank references provided and insurance cover.
- Finally, knowledge of their current workload would be of interest. If overstretched, another tender may be too much or it may not be able to devote enough resources to the project.

Clients do not always select the lowest tender price. A client may prefer the security of someone it knows, or it is felt that the lowest bid is too low and could put the winner in danger of going bankrupt.

13.3 Tender documentation

When the tender documentation arrives, the estimator will check that all the information stated in the letter of invitation is enclosed, there is sufficient information to permit a realistic cost of the contract to be produced and that adequate time has been allowed to produce an informed and accurate tender.

Typical sources of information that come with the letter of invitation upon which this decision is made are:

- drawings that will be at least those used to produce the bills of quantities and may include others done since. There should be site layout drawings giving information about access roads, boundaries, topography, adjacent buildings, buildings and structures, services on the site, roadways, trees, restricted areas and the footprint of the proposed building. General arrangements such as plans, sections and elevations would be needed as well as any specialist drawings done such as specialist temporary works, and surveys of existing structures. Detailed drawings should include substructure, frame, floors, roof, cladding, internal structural and internal walls, internal finishes, and services;
- schedules of information including standard and non-standard joinery details, windows, doors, ironmongery, sanitary fittings, partitions, floor and ceiling finishes;
- specifications that includes performance specifications if appropriate;
- site investigation reports including soil surveys, water-table levels, and other technical reports such as the CDM health and safety plan, reports on existing asbestos and contaminated ground;
- bills of quantities;
- the form of contract and any supplementary conditions.
- programmes of work periods for the main nominated subcontractors if not in the contract form.

This information can and should be supplemented by carrying out a site inspection to understand the likely production problems due to space restrictions, traffic restrictions, state of existing or adjacent buildings, neighbours and effect of the impact on them of noise, dust and dirt, ease or otherwise of access, ground conditions, over-flying of tower cranes and other issues already highlighted in the employer's requirements and contractor's general cost items.

Other issues would include alternative solutions if there were insufficient space available on the site for staff accommodation required during the construction process, the location of the nearest landfill site and the charges, the local transport system, parking facilities and their costs, which can have implications in recruiting labour; also, what local facilities are available for food and accommodation, and whether the site is exposed to extreme weather conditions or the possibility of flooding.

The statutory authorities will be contacted to find out about supply and on controversial contracts it will be necessary to assess the likely disruption by protestors.

13.4 The decision to tender

On receipt of an invitation to tender, the contractor has to decide whether to accept or decline the invitation. If declined, the documents are returned to give the opportunity for another contractor to tender. The estimating procedure using full B of Q is very time-consuming and therefore costly. There is no point in expending this effort unnecessarily, so when an invitation arrives the contractor will consider the following, set against the background of the company's current workload:

- Is the size of the contract too small or too large? If too small, the culture of the organisation may be inappropriate to manage the works and overheads added will make the bid uncompetitive. If too large then the company's resources may not be able to cope. To go ahead in such circumstances would become a strategic decision because of the implications. The first M1 motorway project was divided into four separate contracts. John Laing bid for all four and although not the cheapest on each, the client asked them to consider constructing all four as one contract. The total value of the works was very high relative to their annual turnover. The board made a strategic decision to take on the work.
- Has the company experience of this type of work? Again this may become a strategic decision as an opportunity to break into this market. Equally it could be to far removed from the company's experience and the decision made to decline.
- Even though the company has experience in the type of work, have they the experience in the type of contract?
- How much competition is there? It is said that the success rate is about one in four to six bids and these odds can be built into the overheads calculations. If there is a lot of competition then the odds will lengthen, confusing these calculations.
- Is the proposed contract within the geographic boundaries normally operated within? If not, then tendering becomes a strategic decision.
- Does the company have the necessary resources, especially manpower, to cope? If not, the company will need to recruit the necessary staff. The implications of this can be considerable as it takes time to recruit and train suitable staff and, assuming they are available, they may expect continuous employment. This means

more work has to found, which means the company is looking at an expansion policy with all that entails.

- Has the estimating department the capacity and expertise to tender? The latter is important if the work is different to that normally tendered for.
- Are the documents provided comprehensive? If not then, the estimating department will have more difficulty in reaching a true cost. It also raises the alarm as how the contract might be serviced by the design team during the construction phase and increases the likelihood of claims and confrontation.
- Has enough time been allowed for to prepare an accurate tender? This can affect accuracy if insufficient time cannot be allocated to fully comprehend all the implications of constructing the project and to work out the best methods resulting in higher risks.
- What is the current workload of the contractor at the time? If the workload needed to meet the annual plan has been achieved, and management decides to tender, it can add lower than normal overheads to become more competitive. This is slightly more complicated than just how much work has been successfully bid for, as it must also take into account the chances of it being obtained, when it commences and over what duration.
- What is the financial status of the client and what is their track record for settling accounts?
- Supplementary contract conditions attached will either be insurances, warranties or performance bonds that are quite straightforward to price and will be common to all bidders. However, some non-standard conditions can be onerous and the contractor may wish to decline from tendering, renegotiate these conditions or add a premium.

Two issues occur regularly: strategic decisions and risk. In both cases the need for senior management involvement is identified. Strategic thinking is developed further in Chapter 27. However, management may also have statistical information about their previous performances against the current competitors (see 14.4), and this can be brought into the decision-making process. Risk can be managed with the knowledge that some of the burden of risk can be moved from the client to the contractor or that uncertainty increases the risk. In both cases the contractor needs to consider how much, if any, should be added to the tender price to take account of this increased risk.

13.5 Managing the estimating process

Once the decision has been made to tender for the project, the estimating department needs to control the work being processed through the section. To do this the company will usually keep a tender register that, on a monthly basis, gives a summary of the status of all contracts. This can be used to help manage tenders and as a record of past successes and failures in case any obvious trends have occurred. Typical headings are shown in Table 13.2.

The client column might be extended to name the architect, structural engineer and quantity surveyor used by the client. The percentages, low and high, in the results column are established, either from the client advising the contractor what the other

Table 13.2 Tender register

Tender register	*Period from:*		*Tender*		*Result*		*Comments*
Reference No.	*Client*	*Contract name*	*Submission date*	*Value £s*	*Yes% low*	*No% high*	

bids were, or down the estimator's grapevine. Irrespective of the source, this indicates the current market value as perceived by one's competitors and therefore acts as a benchmark for future bids. Although not shown in this table, the contractor should keep a record of all the competitors' bids for each contract to assist in developing a bidding strategy.

Many estimators will relate there is rarely enough time to complete the estimating process and as the final submission date approaches, it often reaches 'panic stations'. Therefore it is important that all estimates being processed through the department are programmed and monitored, as this demonstrates the overall workload. Each individual contract has its own programme. On the contract programme not only should each item of work be shown, but also which person in the department should be responsible for executing the task. Typical tasks that should be identified and entered on the programme, some as action dates and others as activities, are listed and briefly described below and summarised in Figure 13.1:

- the date the documentation was received from the client;
- selection of the estimator responsible to run the project. This might be after the decision to proceed, as the chief estimator may wish to check the documents to provide the information to decide whether or not to tender. This occurs shortly after the receipt of the drawings and must both give enough time for an informed decision to be made, and leave sufficient time, if the answer is yes, to permit the tendering process to be accomplished properly;
- organisation of a site visit, if not already done, as well as visits to consultants to inspect other available information;
- with all the data available, the methodology for construction is determined and a pre-tender programme produced (see 20.2.2). Much of the work contained in the bills will be subcontracted to others, so after analysis of the bills subcontractors are selected and invited to tender. Equally prices for materials not to be included in the subcontractor's quotation have to be obtained. Approximate dates and durations can be abstracted from the programme enabling realistic quotations to be submitted;
- a deadline by which all enquiries must be sent out;
- a deadline for receipt of all prices of both materials and subcontractors. The former is usually required earlier as this affects the pricing of items in the bills especially if the contractor is carrying out the work itself;

TENDER TIMETABLE

Description		Feb																				March						
	Keydate	1	2	3	4	5	8	9	10	11	12	15	16	17	18	19	22	23	24	25	26	1	2	3	4	5	8	9
Documents received	1 Feb																											
Select staff	1 Feb	■																										
Inspect documentation		■																										
Decision to tender			■																									
Site visit				■																								
Visit consultants					■	■	■																					
Determine methodology								■	■																			
Pretender programme									■	■	■	■	■	■	■	■	■											
Abstract, despatch subs.	5 Feb			■	■	■																						
Abstract, despatch mats.	10 Feb						■	■	■																			
Receive quotations mats.	24 Feb																		■									
Receive quotations subs.	1 Mar																					■						
Price labour and plant										■	■	■	■	■	■	■	■	■	■									
Price materials																				■	■	■	■					
Price subcontractors																						■	■	■	■			
Price preliminaries																		■	■	■	■	■	■	■	■	■		
Add overheads																									■	■		
Check																										■	■	■
Review meeting	10 Mar																											
Submit documents	11 Mar																											
Submission date	12 Mar																											

Figure 13.1 Tender programme

- commencement of the pricing of the bills while awaiting the return of this information, especially many of the preliminaries items;
- selection of the most competitive and realistic bid from material suppliers and subcontractors which is then entered in the bills;
- establishing a deadline date for completion of the pricing of the bills;
- checking calculations on completion of the priced bills;
- calculation of cash flow predictions;
- a date fixed for a meeting for the senior management to determine the overheads and profits;
- time allowed for preparing documents for submission and submitting on the date stated by the client.

13.6 Completing the bills

Many of the activities listed in 13.5 run concurrently, but for ease of understanding they are discussed separately. The estimator's tender programme (Figure 13.1) shows these overlaps; while the preliminaries items come first in the decisions on whether items such as plant should be included in the bill items or the preliminaries, which needs to be determined.

13.7 Building up the costs of an item/s

Section 3.3.3.13 of NRM2 states:

> In a bill of quantities, the following shall be deemed to be included with all items
>
> a) Labour and all costs in connection therewith.
> b) Materials and goods together with all costs in connection therewith.
> c) Assembling, installing, erecting, fixing and fitting materials or goods in position.
> d) Plant and all costs in connection therewith.
> e) Waste of goods or materials.
> e) All rough and fair cutting unless specially stated otherwise.
> f) Establishment charges; and
> g) Cost of compliance with all legislation in connection with the work measured including health and safety, disposal of waste and the like.

The establishment charges overheads, and profits will be excluded from the build-up of the unit rate as these can vary dependent upon the management decisions at the time of tender (section 13.22). The other items are labour, materials and plant costs.

Each item in the B of Q is based upon the NRM2. The process of calculating rates (13.8) does not occur every time a new tender arrives for pricing. A comprehensive library of rates based upon these calculations will have been developed over the years. However, changes will need to occur from time to time as costs increase, so they will be regularly updated. Occasionally new rates have to be produced in the case of some special requirement not covered in the database.

13.8 Calculating the labour costs

The costs of labour to the employer are not just the basic rate, but include many other items:

- the standard wage, which is the basic hourly rate multiplied by the weekly working hours agreed and published in the Construction Industry Joint Council (CIJC) Working Rule Agreement (WRA);
- overtime, which can be the hourly rate multiplied by 1.5 (time and a half) or 2 (double time);
- guaranteed minimum bonus, which is laid down in the WRA as a means of ensuring a bonus is paid if the operatives are unable to work as a result of, say, inclement weather;
- bonus, which is calculated based upon output (see 29.12.2);
- plus rate, either paid in lieu of bonus or to attract labour if in short supply. This is an hourly rate in addition to the basic wage rate;
- extra payments under the WRA for special skills such as driving and maintaining mechanical plant;
- travelling costs that can include both the time it takes to get to the site as well as fares. It must be remembered that the site may be a long way from where the employee was engaged;

- lodging allowances that may have to paid;
- Construction Industry Training Board (CITB) levy paid as a percentage of all employees' wages to fund training for the industry as a whole (see 29.9.3);
- public holiday pay;
- other holidays that operatives may be entitled to in their conditions of employment;
- employer's contribution to pensions, accidental injury and death cover schemes;
- sick pay, which has to be paid, up to a certain point, when operatives are ill or injured;
- National Insurance contributions by the employer;
- severance pay if the employee is dismissed prematurely;
- employer's liability and third party insurance;
- tradesman supervision. The trade foreman may either be employed as a full time supervisor or a part time one, the remainder of the time working on his trade.

From all this information an hourly rate can be calculated. This calculation needs to be done for both craft and general operatives because their basic wage rates are different. Adjustments can be made to the calculated rate to take into account:

- overtime working may well vary depending upon the time of year – for example it is more likely to be worked in the summer months when the daylight hours are longest;
- the amount of skilled labour available in the area, which can affect supply and demand or the need to bring operatives in from another area;
- what percentage of operatives are to be paid travelling expenses and how long it takes to travel in the particular area;
- the part of the country/world and the effect of climate;
- the part of the world and likely productivity and methods of work.

Organisations will do their calculation in slightly different ways, but the calculation shown is a typical method that is applied and has been adapted, with permission, from the Chartered Institute of Building *Code of Estimating Practice* (1997).

There are basically three components in the calculation: to establish the hours likely to be worked over the year; to calculate the annual wages; and to establish the average hourly rate to be used in the bills rate.

13.8.1 Average annual working hours available

Before reading on, the author wishes to issue a health warning about the figures used in this section. The hours of work, conditions of employment and the costs to the employer all change regularly as a result of legislation and nationally negotiated wage agreements. The figures, therefore, should only be used as a means of demonstrating the principles and not taken as actual figures.

An assumption is made that the basic week is 40 hours worked throughout the year (note at the time of writing the National Working Rule Agreement is 39). As the hours of daylight vary throughout the year, in the northern hemisphere the summer months provide more opportunity to work in natural light than in the winter. It would not be unreasonable to assume there are 30 weeks available for overtime in a given year. Not withstanding EC legislation on working hours, employers may negotiate special working conditions and hours of work on a specific contract because of the nature of the work, such as working shifts.

The employer pays the employees for public holidays and their annual leave. If 21 days of annual leave are paid for, it could be assumed that 7 days are taken at Christmas, 4 at Easter and 10 for a vacation. The latter two could occur during the summer period when extra overtime would be worked. In the UK there are three public holidays during the winter period – Christmas Day, Boxing Day and New Year's Day and five in the summer – Good Friday, Easter Monday, Spring Holiday, May Day and August Bank Holiday. The calculation for the total annual hours is shown in Table 13.3. For simplicity it is assumed in the calculations that the working day in the winter is eight hours and in the summer nine hours.

The total hours operatives are available to work in a year are 800 + 1179 = 1979. On average operatives are off sick usually in the winter for eight days, a total of 64 hours giving a net hours available for work of 1979 – 64 = 1915 hours.

A further deduction should be made for inclement weather, as it is almost certain that work will stop at some time during the year as a result of excessive rain, snow, winds or sub-zero temperatures. The amount deducted will depend upon the level of exposure of the site, its geographic position and time of year. Assumptions made on this may well change in the light of climate change. It is not unusual to assume a loss of 2 per cent, which in this case would mean losing a further 38 hours, making a figure of hours actually available for work as 1915 – 38 = 1877.

13.8.2 Calculation of the total wage costs

Having arrived at a figure for the estimated annual number of hours operatives are available for work, i.e. 1877, the actual annual cost can be calculated. This means taking the basic hourly rate and adding on all the other costs incurred by the employer in employing staff.

Basic wage

For the purposes of this calculation a figure for the wages has been assumed to be:

General operative: rate £332 for 40 hour week or £8.30 per hour

Craft: rate £442 for 40 hour week or £11.05 per hour

Table 13.3 Annual hours of work

Winter period		*Summer period*	
Working week	40	Working week	45
For 22 weeks	880	For 30 weeks	1350
Deduct		*Deduct*	
7 days leave*	56	14 days leave**	126
3 days public holiday*	24	5 days public holiday**	45
Total hours	800	Total hours	1179

* A 8.0 hour day, ** A 9-hour day.

Therefore the basic annual wages excluding overtime rates are:

General operative: 1915 hours × £8.30 = £15,894,50

Craft: 1915 hours × £11.05 = £21,160.75

Note this includes the 38 hours for inclement weather, as the operatives have to be paid even though they are not working. This is a simplification because there are other skill rates between operative and craft.

Overtime

Overtime is paid for any time worked in excess of the basic working week (40 hours). In the calculation shown in 13.8.1 no overtime is worked during the winter period, but a prediction of five hours per week is made for the summer period. This means that operatives working this amount of overtime are paid at time and a half. In other words they will receive 7.5 hours' pay for the five hours worked. While the operatives earn an extra 2.5 hours' pay in overtime, the contractor still only obtains five hours output so the overtime paid is, to them, non-productive time. The total cost of this non-productive time over the complete summer period is:

30 weeks less 4 weeks' leave and public holidays × 2.5 hours = 65 hours

The cost of this is:

General operative: 65 hours × £8.30 = £539.5

Craft: 65 hours × £11.05 = £747.50

Bonus payments

Operatives will also earn bonus at say one third of the basic week's wage, which is £2.77 for general operatives and £3.68 for tradesmen. This works out for the 40 hour week as:

General operatives: 40 hours × £2.77 = £110.8

Craft: 40 hours × £3.68 = £147.2

Sick pay

It was assumed that, on average, eight days' sick pay is taken on by each operative and this would occur in the winter period. No payment is made for the first three days (known as qualifying days), so the employer in this case will only have to pay for five days. The Department for Works and Pensions determines the amount of payment, which is on a sliding scale depending upon income. A basic sick pay of £23 per day is used for this calculation.

Trade supervision

Contractors and subcontractors will employ trade supervisors or foremen. The number of operatives they are responsible for will vary from company to company and the complexity of the works. Often these appointments are such that they spend half their time supervising and the remainder engaged in their trade. For the purposes of this exercise it is assumed that they are responsible for six tradesmen, are supervising for 50 per cent of their time and are paid an extra 75p per hour for the added responsibility.

The hourly cost for the gang is:

1 No. Craft foreman	£11.05 + £0.75 = £11.80
6 No. Craft	£11.05 = £66.30
	Total £78.10

However, the foreman is only working for 50 per cent of the time and supervising for the other 50 per cent, so the hourly rate of the gang in productive terms is:

£78.10 ÷ 6.5 = £12.01

The extra cost of supervision for the gang is:

£12.01 – £11.05 = £0.96

So the total cost per year of this supervision is:

£0.96 × 1915 (the available annual hours) = £1838.40

Working rule agreement allowances

For the purpose of this exercise there are none.

Tool allowances

Under the NWRA, joiners for example are entitled to approximately £2.00 per week and other trades £0.50 per week for providing and maintaining their tools. Although not part of the basic age, it is subject to NCI and tax deductions.

Construction Industry Training Board (CITB) levy

Construction industry employers are obliged to pay to the CITB a levy of 0.5 per cent onto the wages and salaries of operatives, clerical workers and management if taxed PAYE and 1.5 per cent on labour only subcontractors, so that training can be provided for construction skills, supervision and management. This figure is added to the labour unit costs.

National Insurance contributions

The amount of contributions paid by both the employer and employee is determined by the Chancellor of the Exchequer in the annual budget and therefore changes from time to time. At the time of writing the employer's contribution is linked to the employee's weekly or monthly wage and is currently 13.8 per cent of gross income.

Holiday credits and death benefit scheme

Holidays with pay used to be achieved by means of the employer sticking stamps each week to a card. The operative then could take this card to whoever was offering employment at the time, accumulating holiday pay that could be cashed at the time of going on holiday. The employer lodged the money with a management company who held the money until such time as required for holidays. This was accepted as being a good method of overcoming the mobility of workers moving from one contract to another. Resulting from European Union employment law this system was abolished and now the employer pays to the management company the equivalent of the daily rate of each operative's wages for each day of the annual holiday earned as a result of length of service during the working year.

A craftsman paid £11.05 per hour and working eight hours per day, earns a daily rate £88.40. So for the 21 days' holiday the employer needs to set aside £1856.40. The productive period of the year is 47.8 weeks (52 – 4.2), so the cost to the employer per week is £1856,40 ÷ 47.8 = £38.84

Severance pay and incidental costs

When employees either hand their notice in or are dismissed there are costs involved. In both cases it would not be unusual for there to be loss of interest in the job during the period of notice and hence loss of production. The employee could go absent, leaving the employer the cost of NIC, pensions and holiday pay. For the employee who has been dismissed there is the additional cost of severance pay. This cost will vary between companies depending upon their experience, opportunity and ability to maintain a constant labour force, but a percentage allowance for this of between 1 per cent and 2 per cent is not unusual.

Employer's liability and third party insurance

The costs vary from company to company depending upon their insurance record, the insurance company and the size and type of contract. A figure of an additional 2 per cent on the labour costs is not unusual. An alternative way is to express all insurances as a percentage of the tender price and include it as part of the overheads.

Summary

All the information above is summarised in Table 13.4 adapted from the sixth edition of the CIOB *Code of Estimating Practice* (1997). This spreadsheet this can be used by the estimating department to amend the inserted figures as necessary when changes occur such as an increase in wages and conditions agreed at national level or resulting from the Chancellor's Budget.

Table 13.4 All-in labour rate calculation

Description			*Hours*		
Summer period	No. of weeks	30			
	Weekly hours	45			
	Total hours		1350		
	Annual holidays (days)	14			
	Public holidays (days)	5			
	Total holidays (hours)		–171		
Winter period	No. of weeks	22			
	Weekly hours	40			
	Total hours		880		
	Annual holidays (days)	7			
	Public holidays (days)	3			
	Total holidays (hours)		–80		
Sickness	No. days (winter)	8	–64		
Total hours for payment			*1925*		
Inclement weather allowance		2%	–38		
Total hours available for work			*1883*		
		Craftsman	*Gen. ops*	*Craftsman*	*Gen. ops*
Earnings	Guaranteed min.wage /week	£442	£332		
	Bonus	£147.2	£110.8		
	Plus rate – attraction bonus	0	0		
	Total weekly rate	£589.2	£442.8		
	Hourly rate of pay (40)	£14.73	£11.07		
	Total Annual Earnings			£28,355.25	£21,309.75
Additional costs	Summer overtime hours @1.5	7.5	7.5		
	Winter overtime hours @ 1.5	0	0		
	Summer overtime total hours	65	65		
	Winter overtime total hours	0	0		
	Total annual cost overtime hours			£718.25	£539.50
	Sickness pay for 5 days	£23/day	£23	£115.00	£115.0
Trade supervision	No. tradesmen per foreman	6	0		
	Supervision plus rate/hour	£0.50	0		
	Time spent on supervision	50%	0	£1,838.40	0
* *note there could be an entry under Gen. ops for a ganger*					
	Sub total			*£31,026.90*	*£21,964.25*
Overheads	Training levy (based on LOSC)	1.5%	1.5%	£465.40	£329.46
	Employers NI Contribution	13.8%	13.8%	£4281.71	£3031.06
	Holidays with pay	£11.05	£8.30	£1856.40	£1394.40
	Public holidays 8 days	8 hrs/day	8 hrs/day	£707.20	£531.20
	Sub total			*£38,348.61*	*£27,250.37*
Severance pay	Sub total	*1%*		*£38,732.10*	*£27,522.87*
Employers liability		2%		£39,506.74	£28,073.33
Annual cost of operatives	Divide by productive hours – 1883				
Cost per hour				£21.55 ++	£14.90

Source: Adapted from CIOB Code of Estimating Practice, 6th edn. 1997.

13.8.3 Labour production rates

So far, the calculation of the labour rate is rather routine and mainly a case of inserting the current figures into a set of predetermined calculations. The real skill and expertise of the estimator is in assessing the impact of the work, site conditions and restrictions on production output and then deciding on outputs and the number of operatives needed to carry out the activity, i.e. the gang size.

The company will have acquired over the years a significant amount of productivity statistics from previous work and this is of immense value to this process. Where this is not available, a judgement has to be made from their own personal expertise and that of others both in and outside the company, supplemented by technical literature. It should be remembered that technical expertise from outside the organisation is often by its very nature biased in favour of their product or system.

There are very many factors affecting output so the estimator has to scrutinise the drawings and information provided to establish difficulties that might arise, at the same time ensuring the quality specified is attained and a safe method of work is provided. Typical production issues to be looked for include:

- the quantity of work, as large amounts might mean utilising larger more efficient plant and operatives become more productive as they learn all the nuances of the job;
- repetitive work, which is more productive than non-repetitive work;
- whether the operatives can get a good run at the work or whether the sequencing of other work will result in them continually being moved on and off the job;
- whether the work complex is either slowing down the process or requiring tradesmen with specific skills;
- how accessible the work is in terms of access, height, and depth – this can affect the ease of delivering the materials to the workplace, the type of equipment that might be used and the freedom of movement of the operatives in carrying out the task;
- heavy or irregular shaped cumbersome components that may cause loss of productivity;
- restrictions on working hours, shift patterns required, the need to use respirators or other restrictive protective clothing and extremes of temperature, all contributing to reductions in productivity;
- depending upon the activity, the time of year and associated weather conditions that may impact on productivity. For example, if it is very cold and windy, operatives working outside in exposed situations will inevitably spend a certain amount time trying to keep warm.

13.9 Building up material rates

30.9.1 Determining the amount of material required

Either the supplier or the contractor has to take into account the fact that the way materials are ordered is not always the same as the way they have been measured for the purposes of the B of Q, and a conversion has to take place. Examples of this are bricks that are mainly measured in m^2 but are ordered from the supplier in thousands, hardcore measured in m^2 or m^3 but ordered in tonnes and damp-proof membranes measured in m^2, but ordered by the roll and width.

The majority of items measured using the NRM2 are measured net. This ignores items such as overlapping tiles and roofing felts that require extra material than the net measure to complete the work. Materials such as concrete and hardcore lose some of their volume when vibrated or compacted. This extra material has to be calculated and taken into account when ordering. Waste also has to be accounted for and is covered in more detail in section 13.9.3.

When building up the unit rate for the tender, these processes have to be reversed and the quantities converted back to the nomenclature of the bill's descriptions.

13.9.2 Checking suppliers' quotations

While the estimator may well have experience about the costs of materials it is important to obtain quotations for the materials to be used on the specific contract, as the transport costs will vary and prices do fluctuate for the following reasons:

- the distance from, or time it takes to transport materials from source to the site;
- whether or not the quantity being delivered is sufficient to take up the full capacity of the vehicle;
- large orders tend to reduce the haulage costs as the number and frequency of loads increase. This means the haulage contractor has greater certainty on achieving full use of the vehicles and can therefore be more competitive.

Some of the bulk materials are relatively cheap, hence the percentage cost added for haulage is significant. The exceptions to this are supply and fix by subcontractors, and when the contractor's buyers have negotiated rates with a supplier of bulk materials such as aggregates.

When receiving a quotation, it is checked to see that the quotation is for what was requested. Most suppliers are honourable in intent, but for a variety of reasons, including making errors or misunderstanding the nature of the request, may not quote appropriately. Less scrupulous suppliers wanting to present a low and attractive price to obtain the order may 'confuse and confound' the inexperienced (see sections 13.15 and 13.16). Typical issues to be inspected are:

- Do the materials quoted for totally meet the specification stated in the enquiry? Sometimes the supplier proposes a lower or higher-grade material than is required, whether because that is all they have available or because the specifier is unaware of improvements in production and supply and has requested a lower specification than is normally available.
- If a section of the bills has been sent for a supply-only price, has every part been priced? An example of this is precast concrete. The bill sent might include kerbs and lintels as well as architectural cladding components. These are two different manufacturing processes and it would be unusual for the architectural cladding provider to price for kerbs and lintels and vice versa.
- Have transport costs been included? These can be high for the reasons outlined above if a supplier quotes the cost as ex-works.

- If a supplier has to deliver a part load only it usually costs more and their conditions may state that.
- It is expected that a delivery may arrive on site and not be unloaded immediately. This is all right to a point, but because of limitations to drivers' working hours, if this becomes excessive it has many implications to the haulage company. They may well lay down maximum waiting time after which a penalty will be invoked.
- Manufacturers require time from receipt of order before they can commence delivery, called lead-in time, and will normally state the duration. The estimator needs to check that, from when the order is placed, this lead-in time will comply with the site's programme requirements.
- The supplier will usually state, if not specified in the enquiry, the duration over which the components will be delivered. It is important that this complies with the site programme.
- Sometimes penalty clauses are introduced to cover for the situation when the site programme goes behind schedule. This is because the supplier may run out of space for storage or has to reduce the rate of manufacture to a less economic level.
- Some materials come in minimum sizes and are quoted at so much per unit, but the contract may require less than the amount quoted: for example, sheet material or lengths of pipe. This can either be taken into account in the price and written off as such, or taken into account in the waste calculation. If the amount not used is of significance then another use elsewhere could be considered. However, it should be noted that this could incur extra costs for transport and storage not to mention any environmental considerations.
- The costs may not include all other materials required: for example, fixings such as nails, screws and bolts.
- Who is responsible for unloading the vehicle? The driver may carry it out using unloading equipment attached to the lorry or the contractor may have to provide personnel to unload either by hand or aided by mechanical plant.
- If the material or component has to be tested, it is normally the contractor's responsibility and the costs incurred accordingly, but if it is the supplier, the estimator needs to be satisfied they have accounted for this.
- Is the price quoted a firm price or are fluctuation clauses involved?
- Discounts are often offered by suppliers normally dependent upon speedy settlement of the invoice by the contractor. The relevance of this is much to do with the current inflation rates and the effect on cash flow for both parties. Higher discounts can also be negotiated if the order is for full loads delivered directly from the manufacturer rather than builders' merchants.
- Where one party is concerned about the financial state of the other, pro-forma invoices may be used. This means that the invoice has to be paid prior to goods being delivered.

13.9.3 Assessing likely waste

NRM2 states what work is measured net and the contractor must take wastage into account. The amount of waste likely to be generated during the construction process is difficult to estimate (see Chapter 22). Most estimators will have their own rule-of-thumb method based upon their experience, but these tend to be general figures.

Different materials create different percentages of waste for reasons some of which are noted below along with other considerations.

- Some materials come in sizes not dimensionally compatible with the dimension required. Typical extreme examples of this are board materials, especially plaster board, pipes, timber reinforcement steel, bricks and blocks.
- Some work cannot be practically carried out to the dimensions specified because of buildability issues. The dimension of the excavator's bucket digging a strip foundation determines the width of the excavation rather than the stated dimension of the concrete.
- There is the likelihood of damage associated with unloading, storage, double handling, fixing and other trades working in the near vicinity of the product.
- There is risk of loss of materials through theft and damage.
- Poor workmanship may result in either remedial work or reconstruction.
- Materials may be used for the wrong purpose such as bricks in lieu of block work because it is easier for the bricklayer to do this than cut a block.
- The costs of disposal of the wasted material are rising as landfill taxes increase and segregation into different material types needs to occur before removal from the site.

With the pressures of both costs and environmental implications, there is a need to consider how waste material can be reclaimed, recycled or reprocessed, offering the opportunity to generate income to set aside against the cost of waste.

To improve the accuracy of these calculations, there is information available in research papers, price books and feedback from within the organisation. The problem with the latter is that in contracting most people move on from the contract to the next and are in a 'looking forward rather than back' mode, so it is often difficult to ensure that adequate feedback occurs in spite of the value of this source of information.

13.9.4 Storage, security and handling of materials

While covered in the preliminaries, the sites visit and site plans give the estimator an indication of problems and difficulties that might occur during the construction process. For example, a very restricted urban site may result in materials being delivered within a narrow time frame or there may be such a limitation on storage that part loads have to be considered.

13.10 The cost of plant

Plant development occurs all the time both at the larger end of the scale (cranes and excavating equipment) right down to hand tools, drills and saws. Some of these advances can have significant effect on productivity and costs. In small tools the development of battery-driven hand tools now having enough power and duration to be effective has eliminated the need for a supplied power source. On larger plant developments of hydraulic systems, laser guidance and control mechanisms have dramatically improved productivity. It is important therefore that the estimator is well versed in these so that the appropriate piece of plant can be selected for the work on the site.

There is an academic debate in many of the texts as to whether or not one should hire or own plant, but increasingly, unless one is in a specialist area such as piling, most medium and large contractors hire plant as and when required so as to offer the greatest flexibility and ability to use the most up-to-date purpose-made equipment. The other major exception to this is if the contract duration is of such a length as to make the economics of purchase and maintenance stack up.

On obtaining quotations for hiring plant, besides hire charges, there are several points to understand and questions that need to be answered:

- The cost of hire may commence when the plant leaves the plant depot and finish when it returns. These travelling costs can be significant as in the example of a large mobile crane that may take a day to get to the site, only work for a couple of days and then take a day to return to its depot. It highlights the need to consider continuity of work on site for expensive plant.
- It is usually on a weekly hire basis even though it is used for less. The exception to this is some of the large specialist equipment such as mobile cranes and day work lorries. Continuity of work is again an issue.
- There will be contractual obligations to ensure the plant is used, maintained and worked in a safe manner. These need to be scrutinised for any onerous clauses.
- Does the plant hire include the cost of the operator and if so what are the rates and conditions of pay? There might include limitations on driver hours.
- Even though the plant hirer supplies the operator, the contractor will have to take responsibility for any damage caused by the operator such as damage to services, unless it can be demonstrated that the operator is negligent, but even this may be excluded from the contract between the parties.
- Is the hirer or the contractor responsible for insurance? This may split responsibility – the hirer while in transit to and from the site, the contractor while on the site.
- The hirer will take care of breakdowns and maintenance. If no operator is provided, basic maintenance such as checking oil and water levels, keeping the plant clean, inspecting the tyres, provision of fuels and oils is the contractor's responsibility.
- The plant selected has to be available for when required on site.
- Is the quotation a fixed price for the duration of the hire or are there fluctuation clauses? On projects certain plant such as dumpers could be utilised for virtually the whole project, which could extend over two or three years.
- There may be a requirement to provide extra equipment to operate the plant. Examples are slings and lifting beams for handling components, concrete skips, specific types of excavator bucket and tools for compressors.

After this process the all-in rate for each item of mechanical plant can be calculated. In summary these are:

- the hourly cost of the machine, including the maintenance labour and parts, depreciation costs, which have to be paid for, either by the hirer or the contractor owning the plant;
- the amount of standing or idle time likely to occur, which must be estimated. While it is expected that especially heavy and expensive plant will be used to its full

Table 13.5 All-in plant rate calculation

Basic weekly hours	39		
Add 0.5 hours per day maintenance	2.5		
Total hours worked		*41.5*	
Standing time @ 1 hour per day	5.0		
Total productive hours per week	*41.5 – (2.5 + 5.0)*	*34.0*	
Weekly hire cost of plant say £200			
Productive hourly hire rate	£400 ÷ 34.0		£11.11
Fuel at 4 litres per hour @ 75p a litre			£3.00
Grease and oil @ 20p per hour			£0.20
Total weekly operator costs (41.5) hours)	£20.00 + 20p plant ops. plus rate	£838.20	
Hourly productive rate	£838.20 ÷ 34		£24.66
Total all-in cost per hour			£38.97
Total all- in cost for period on site	£38.97 × 4 × 41.5		£6269.20

capacity, in reality there will be down time resulting from mechanical failure, inclement weather and hold ups. This should be taken into account in the calculation;

- the all-in rate of the operator which is usually higher than calculated in 13.8.2 because the working week is longer due to extra tasks such as oiling, greasing, minor repairs and coming in earlier to start the machine and position it ready for the start of the normal working day. There may also a higher basic wage rate because of enhanced skill requirements;
- the fuels, oils and greases.

Decide whether or not to use this all-in rate for the plant in the preliminaries or in the unit rate in the bills.

A typical example of the all-in rate for a piece of plant is shown in Table 13.5. It is assumed that a 39 hour five-day basic week is being worked with overtime, the plant will be used on site for four weeks, delivery and retrieval charges are assumed to be inclusive in the hire rate. The contractor provides the driver. The driver spends 0.5 hours per day maintaining the plant that is over and above the working day. It is assumed the plant is idle on average for one hour per day due to stoppages caused by inclement weather, and other hold-ups.

The all-in cost per hour can be used for building up rates in the bills and that for the period on site if the plant is to be costed in the preliminaries. This calculation does not take into account productivity issues such as the output from a back actor excavating clay being different from that of shale or the output being a function of the bucket size.

13.11 Calculating and pricing unit rates

Previously it has been demonstrated how to calculate the hourly cost of labour, how to assess the quantity of materials required, taking account of waste and how to establish the hourly rate for a piece of mechanical plant. None of these calculations takes into account the different size factors that can determine whether or not output is high or low. In other words how long it will take to complete an operation. Other than the general comments made in section 13.8, it is not the purpose in this text to investigate in detail influences on productivity. The estimator will in practice take these factors into account when pricing the unit rate.

The estimator has at his disposal many sheets of data showing a wide range of outputs of various operations for differing circumstances. These cover every aspect of construction work from plant outputs in different materials to the coverage of different types of paint on various surfaces. Adapted with permission from Brook (2008), Table 13.6 shows typical data for fixing softwood skirtings, architraves, trims etc.

The next stage is to build up the all-in rate, which brings together the data for the labour, material and plant costs required to execute the described bills item. Table 13.7 demonstrates this for building a half-brick thick wall using cement mortar 1:3 mix, built in stretcher bond; as well as the rate to lay 1000 bricks and also the rate to lay one metre super of brickwork.

Table 13.6 Output rates for softwood

Size of member		*Nailed*	*Screwed*	*Plugged and screwed*	*Size of member*		*Nailed*	*Screwed*	*Plugged and screwed*
19	19	0.10	0.13	0.18	25	19	0.10	0.13	0.19
	25	0.10	0.13	0.18		25	0.10	0.13	0.19
	32	0.10	0.13	0.18		32	0.12	0.15	0.21
	38	0.12	0.15	0.20		38	0.12	0.15	0.21
	44	0.12	0.15	0.20		44	0.12	0.15	0.21
	50	0.12	0.15	0.20		50	0.12	0.15	0.21
	63	0.12	0.15	0.20		63	0.15	0.19	0.25
	75	0.15	0.19	0.24		75	0.17	0.21	0.27
	100	0.17	0.21	0.26		100	0.19	0.24	0.30
	125	0.19	0.24	0.29		125	0.23	0.29	0.35

The above outputs are for each joiner and the figures are hours/metre run.
Add 30% to the outputs for fixing hardwood.
Average waste allowance is 7.5%. Varies depending upon number of short lengths and mitres.
Add for cost of screws in lieu of nails.

Source: Adapted with permission from Brook 2008.

13.12 Costing the preliminaries

This section is written in general terms, although based on the NRM2 as, irrespective of which contract is being used, the preliminary items are generally the same. The references in parenthesis are for users of the NRM2.

General project details (1.1–1.4)

Since these are all to do with the description of the site and provide information for the tendering process, there will be no costs incurred.

Contractual matters (1.5)

Many of the clauses in this section of the contract will have no cost attached to them. The details are often complex and beyond the remit of this text, and are likely to

Table 13.7 All-in build up rate work sheet

Reference: *Date:* *Trade: Bricklayer*				*Bills item description: Common brick wall, 102.5mm thick, vertical stretcher bond, 215 × 102.5 × 65 mm*			
Item details				*Analysis*			*Net unit rate*
Description:	*Quant*	*Unit*	*Rate*	*Lab*	*Mat*	*Pl*	
Materials							
Deliver to site common bricks	1000	No.			150.00		
Wastage	5%				7.50		
Mortar delivered mixed	1	m³	£120				
Mortar 0.53m³ per 1000 bricks					63.60		221.10
Labour							
Bricklayers	2	hr	£21.56	43.12			
Labourer	1	hr	£14.90	14.90			
Gang cost				58.02			
Output 55 bricks/hour/brickie							
Cost = 1000 × 58.02/(2 × 55)							527.45
Total cost per 1000 bricks							*748.56*
59 bricks per m² for ½ brick wall							
Cost per m² £748.56 × (59/1000)							*£44.17*

change so the reader should refer to the JCT. Highlighted are typical issues most likely to result in costs to the contractor. This is not meant as a definitive list, as there is often ambiguity in that some can be costed either here or in other preliminaries items. These include practical completion and defects liability to cover for rectifying defects; the insurance of the work such as employer's liability, insurance, public liability insurance, joint liability insurance and covering the building works and materials against all risks; certificates and payments concerned with the percentage retention the client may hold from each monthly payment; and costs associated increases in national insurance, taxes and so on, that the contractor has no control over.

Employer's requirements (1.6–1.12)

Provision, content and use of documents (1.6) and Management of works (1.7). There is no cost to be entered here. The latter would be taken into account in the general overheads.

Quality standards/control (1.8). The provision of constructed samples and components can be costed in the usual way and entered here. If samples of concrete are to be carried out this can be readily calculated. For example, the numbers of core samples and tests, cube tests etc. can be ascertained from the instructions in the employer's requirements in conjunction with the amount of concrete ascertained from the bills, gives the data required. The local concrete testing laboratory will have a scale of fees for each test.

Security/safety and protection (1.9). Safety would not normally be priced here as it would be included elsewhere either as a overhead component, a staff component

Table 13.8 Security costs

Description	*Cost (£)*
Hoardings around the site	5,000
Repairs to ditto	750
Chain-link fencing around secure compound	600
Gate house and gates	2,000
Alarm system	700
Night security patrol for contract duration (15 months)	15,000
Total	24,050

in 1.2.1 or as part of the bills item. Security, such as the provision of hoarding, patrolling and alarms is normally found here. A simple example is shown in Table 13.8.

A sum may also be added to cover maintaining external roads and services, noise and pollution control, protection of trees on the site, protection of works in progress and repairs to surrounding buildings. How much, if any, is more a function of the risk the contractor assesses there to be.

Specific limitations on method/sequence/timing/use of site (1.10). These limitations are considered and allowances made for in the methodology of construction and the build up of the all-in rates. Where this is not possible a sum might be added.

Facilities/temporary works/services (1.11).

Operations/maintenance of the finished building (1.12). If this were solely the provision of plant manuals and operating instructions, there would be no need to cost it. However, if this involves the training of operatives in the use of plant, allowances are made here. Factors affecting the costs include the staff, the duration of the training, travel and subsistence.

Contractors' general cost items (1.2.1 –1.2.14)

This is where the serious money is spent in the preliminaries, so it is important that fixed and time-related charges are identified.

Management and Staff (1.2.1). Taking account of overall costs of employing staff as outlined in 12.5.4, the estimator needs to establish the cost of each member of staff working on the contract. This depends on the duration they are employed there. A standard format can be used for this, listing all possible members of staff, but for the purposes of this example, as shown in Table 13.9, only those employed are shown. Under column 'Number', the amount of time they will be spending on average each week over the duration of the contract is assessed. In some cases, such as the engineering staff, there may be weeks, especially at the beginning of the contract and when the structure is being set out and constructed, that they are employed full time on site, but then their presence is not required until the setting out of the site works.

While this gives the overall predicted costs of management and staff, which is fine for tendering purposes, it does not demonstrate when staff are required on the contract. Since staff costs are considered to be time related, this data needs to be calculated to demonstrate the monthly costs. For the purposes of this example it is

Table 13.9 Site staff costs

Staff description	*Number*	*Salary** (£/week)	*Total cost* (£/week)	*Expenses subsistence etc.*	*Cars £150/ week*	*Total* (£/week)
Contract manager	1	700	700	100	150	950
General foreman	2	500	1000	2 × 200 = 400		1,400
Trades foreman	3	450	1350	3 × 200 = 600		1,950
Ganger	1	400	400	200		600
Engineer	0.2	650	130	0.2 × 100 = 20	30	180
Assist. Engineer	0.2	400	80	0.2 × 100 = 20	30	130
Chainman	0.2	350	70	0.2 × 100 = 20		90
Planner	0.5	650	325	0.5 × 100 = 50	75	350
Quantity surveyor	0.5	650	325	0.5 × 100 = 50	75	350
Safety officer	0.1	600	60	0.1 × 100 = 10	15	85
Secretary	1	300	300			300
Typist	0.6	250	150			150
Cleaner	1	200	200			200
Total cost/week						6,735
Total cost/contract				10 months or	44 wks	296,340

* salary includes for employer's contribution to pensions and national insurance, CITB levy etc.

Table 13.10 Monthly site staff costs

Staff description	*Jun.*	*Jul.*	*Aug.*	*Sep.*	*Oct.*	*Nov.*	*Dec.*	*Jan.*	*Feb.*	*Mar.*
Contract manager	4113	4113	4113	4113	4113	4113	4113	4113	4113	4113
General foreman	6062	6062	6062	6062	6062	6062	6062	6062	6062	6062
Trades foreman	2814	2814	5629	11258	11258	11258	11258	11258	8442	8442
Ganger	2598	2598	2598	2598	2598	2598	2598	2598	2598	2598
Engineer	1559	1559	1559	779	779				779	779
Assist. engineer	1126	1126	1126	563	563				563	563
Chainman	779	779	779	390	390				390	390
Planner	1516	2273	2273	2273	1516	1516	1516	758	758	758
Quantity surveyor	758	758	1516	1516	1516	1516	1516	1516	2273	2273
Safety officer	368	268	368	368	368	368	368	368	368	368
Secretary	1299	1299	1299	1299	1299	1299	1299	1299	1299	1299
Typist				1082	1082	1082	1082	1082	1082	
Cleaner	866	866	866	866	866	866	866	866	866	866

assumed, for simplicity, that all the staff are paid on a monthly basis. Under each monthly column in Table 13.10 is the total cost of each person to the contract, based on 4.33 weeks per month.

Site accommodation (1.2.2). The assumption is that all the accommodation required for both the contractor's and client's requirements is costed here. Further, that after the initial set-up costs and dismantling costs, the accommodation provided is rented on a monthly basis over the period of the contract. This identifies the time-related and fixed charges. Table 13.11 represents a typical calculation excluding furniture.

Table 13.11 Site accommodation

Fixed charges		*Time related charges*		
Description	*Cost (£)*	*Description*	*Rental/ month (£)*	*Total cost (£)*
Prepare foundations	500	Staff offices	1000	10,000
Transport and delivery	400	Canteen/mess room	500	5,000
Crane hire	150	Materials store	300	3,000
Installation	200	Clerk of works office	200	2,000
Service connection	65	Washing facilities	300	3.000
Dismantling	200	Toilets	250	2,500
Total installation	*1,515*	Security office	200	2,000
Crane hire	150	*Total monthly cost*	*2,750*	
Transport and return	400	*Total contract cost*		*27,500*
Dispose of foundations	250			
Total removal	*800*	*Total accommodation costs*		*29,815*

Service and facilities (1.2.3). Table 13.12 demonstrates the breakdown of the various services typically considered when pricing this section. The costs per item have not been shown since they are extremely variable depending upon the particular site and authority. There are fixed and time-related charges. Some items such as electrical wiring and cleaning comprise both labour and materials.

Table 13.12 Services and facilities

Fixed charges		*Cost (£)*	*Time related charges*	*Cost (£)/mth*
Electricity	Connection charge		Electricity charges	
	Wiring		Site temporary electricity supply	
Telephone	Connection charge		Calls and line rental	
	Answering machine			
Water	Connection		Water charges	
Rates	Temporary accommodation			
Fire	Extinguishers		Extinguisher maintenance	
Office	Computers		Stationery and postage	
	Fax machine			
	Photocopiers			
Waste			Chutes	
			Skips	
Cleaning	Office if not included in 1.2.1		Roads	
			General site cleaning	
			Cleaning building on handover	
Protection			Protection of work in progress	
Small tools			Approx. 0.2% of contract value.	
			Can be in A43 instead	
Safety	Signs		Protective equipment	
			Training	
Security	Alarm installation		Staff if not included in 1.2.1	

The provision of services for the client's representatives can either be included here or separated out depending upon the contractual requirements. Some contractors include attendance allowances in this section, but with the increased use of subcontractors there is a growing tendency to add these to general site overheads, as many of the facilities provided are used by all.

Mechanical plant (1.2.7). This is for plant not included in the bills rates. There are two distinct categories to consider. First, tower cranes and hoists etc., which have a fixed charge for both installation and dismantling and a time-related charge when in use, and those such as dumpers, which are purely time related.

Temporary works (1.2.8). The provision of access roads and scaffolding are the main items included here. Items such as strengthening roads, temporary walkways and traffic control are not necessary on every site and are calculated as required. Signage can be priced here and would be a lump sum, but can be included in general overheads instead.

The cost of access roads varies considerably from site to site depending on their length, amount of traffic usage and the quality of the subsoil upon which they are laid. A detailed look at the materials available and design criteria are explored in 19.3. It would be considered a fixed charge, as it is a one-off event, hopefully requiring little maintenance, if designed properly. A monthly allowance can be included to take account of minor repairs and cleaning involving both labour and materials. The costs will include for preparing the subsoil, providing and laying the material and occasionally the provision of drainage.

The majority of scaffolding is carried out by a subcontractor who gives a price for both erection and dismantling, and then charges a hire rate for the intervening period. Minor modifications, such as the removal and replacement of handrails to allow the passage of goods into and out of a building or the removal of ladders at the end of the working day, may be carried out by the main contractor, providing liability issues are resolved.

If the contractor is carrying out the scaffolding, the material needed is hired from the supplier and then erected by the contractor. The weekly or monthly hire charge is based upon the length of scaffolding tubes, numbers of fittings and length of boards ordered. It will also include an extra for delivery and collection.

Dayworks (2.13.3). The procedure for calculating daywork is to record the amount of labour, materials and plant that has been used to carry out an operation. This record is signed by the architect or clerk of works to confirm that the details recorded are correct. At a later stage the client's quantity surveyor and the contractor can agree some other method of valuation and the daywork sheet discarded. If the daywork procedure is continued, the data is priced. This has to comply with the definitions prepared for building works as published by the Royal Institution of Chartered Surveyors (RICS).

The definition for building works of prime cost of daywork states that the component parts that make up a day's work are: labour, materials and plant. The contractor then adds for any incidental costs, overheads and profit at the tender stage. The effect of doing it then is that it creates competition for daywork.

The prime cost components are calculated as follows:

LABOUR

Labour is expressed as an hourly rate, but differs from the all-in rate labour rate as worked out in 13.8.2 in so far as many of the cost components are omitted, but taken into account in a percentage addition made by the estimator later. The hourly prime costs are calculated by establishing the annual labour prime cost and dividing by the number of working hours per annum. The annual prime cost comprises:

- the guaranteed minimum weekly earnings;
- differential payments for skill and extra responsibility;
- public holidays;
- employer's National Insurance contributions;
- the CITB levy;
- annual holidays and contributions to death benefits.

Table 13.13 demonstrates the calculation for general operatives and tradesmen. Note that the all-in rate calculated in Table 13.5 is £21.56 for craftsmen and £14.80 for general operatives compared with £14.01 and £10.64 respectively for the prime costs.

The percentage addition incorporates the following incidental costs, some or all of which will be added depending on the nature pf the daywork activity:

- head office charges;
- site staff and site supervision;
- additional costs of overtime;
- bonuses and incentive payments;
- subsistence allowance, fares and travelling;
- time lost due to inclement weather;
- sick pay;

Table 13.13 Labour prime cost rate

Prime cost of labour		*Hours*		
Working hours per week	40			
Working hours per year (40 × 52)		2080		
Annual holidays (days)	21	–168		
Public holidays (days)	8	–64		
Total working hours per annum		*1850*		
Wage calculation	*Craft*	*Gen. op*	*Craft*	*Gen. op.*
Guaranteed weekly wage	442	332		
Annual costs for working hours	46.2 weeks	20,420.40	15,338.40	
Annual costs public holiday	1.6 weeks	707.20	531.20	
Extra for skill	None			
Sub Total			*21,127.60*	*15,869.60*
NI contributions	13.8%	13.8%	2,915.61	2,190.00
CITB levy (based on LOSC)	1.5%	1.5%	316.91	238.04
Annual holidays & death benefits	11.05/hr	8.30/hr	1566.40	1394.40
Annual prime cost of labour			*25,926.52*	*19,692.04*
Hourly prime cost rate (÷1850)			£14.01	£10.64

- third party and employer's liability insurance;
- redundancy payments, tool allowances;
- use of scaffolding;
- provision of protective clothing, lighting, safety, health and welfare facilities and storage;
- variations to the basic rate of pay;
- profit.

The calculation of these depends upon the size and nature of the site and is beyond the needs of this text. In practice the percentage addition can exceed 100 per cent of the prime cost rate.

MATERIALS

The prime cost of materials is the invoice cost after deducting trade discounts in excess of 5 per cent and the cost of delivery if ex-works. The cost of unloading, storage and handling will be included in the labour calculation. The percentage addition is based upon those items listed above under labour percentage additions that are considered applicable. The percentage added will be normally in the range of 15 to 25 per cent, as it comprises only overheads and profit.

PLANT

The prime cost of plant excludes non-mechanical tools and scaffolding, the latter being included in the labour percentage addition. The cost of the plant operative is costed as a labour prime cost. There are two methods of assessing the value of plant:

- the RICS Schedule of basic plant charges. This is a regularly updated published list of the hourly hire rates for a variety of mechanical and non-mechanical plant. Extremely comprehensive, it includes the majority of plant likely to be encountered on a construction site. The schedules also include for fuel, oils, grease, maintenance, licences and insurance;
- current rates of hire. If the contractor has to hire plant, currently not available on site, he is entitled to be reimbursed at the rates charged by the hiring company irrespective of the rates stated in the RICS schedule.

The percentage addition is calculated based upon the applicable incidental costs listed under labour prime cost. Since in the daywork calculation the plant is only costed for the time it is in operation, the estimator needs to take account of the time lost due to inclement weather and standing time as well as overheads and profit. Depending upon the piece of plant this percentage addition could be anything from 20 to 50 per cent.

13.13 Subcontractors and suppliers quotations

Since in today's marketplace the majority of work is subcontracted either in the traditional way or as packages, it is essential that the selection of both subcontractors and the supplier of materials is carried out expertly. There are four types of subcontractors: nominated, named, domestic and labour-only.

13.13.1 Nominated subcontractors and suppliers

It should be noted that nominated subcontractors and suppliers are now used only with a small number of standard forms of contract.

Nominated subcontractors are those who have been selected by the contract architect to supply and fix materials as part of the contract. The contract architect will have usually gone out to tender and obtained a quotation for the work at an earlier stage. One of the main reasons is that the lead-in time for commencement of this work from receipt of order is such that if left to when the contract is awarded, it would cause delays to the construction process. Traditionally, specialist subcontractors such as lift installers and mechanical and electrical service engineers were nominated, but the practice is less common than it was. Clients may also insist upon a nomination if, for example, the work could be carried out by one of their subsidiary companies.

The contractor can make 'reasonable objection' to the nomination in writing, usually based on previous experience of working with the subcontractor such as their quality of work, inability to keep to the programme or unwillingness to work as a team player. If these can be demonstrated to be valid, the architect would be unwise to ignore these comments.

There are risks in nominating because if the subcontractor delays the project as a result of their performance on site, or if there are defects or delays in their design, the contractor can legitimately claim for an extension to the contract and would not have to pay any contractual penalties for late handover as a result. Further, the contractor has no incentive to ensure that the subcontractor completes the work on time and could manipulate the situation to use the delay to cover up his own inefficiencies elsewhere on the project. In today's current climate of cooperation, many contractors find the lack of control of both the design and productivity sometimes to be a hindrance in completing the contract on time. On the other hand, the contractor not being responsible for the work can still generally expect to receive overheads, attendance and profit applied to the prime cost sum. It is not always the case, but the risk relative to employing subcontractors direct is lessened.

Examples of why a supplier is nominated is the selection of an architecturally influenced material such as ceramic tiles that can only come from one source or a material where the lead-in time may be problematic. Finally, there are situations where the nominated supplier or subcontractor offers a near unique service, especially if this involves a design component.

13.13.2 Named subcontractors

This is used especially if the form of contract, such as the JCT Intermediate Building Contract 2011 (IC 2011), is selected where there is no provision in the contract for nominated subcontractors. Still chosen by the contract architect, unlike with nominated subcontractors, the contractor is required to take responsibility for the subcontractor's performance both in terms of delays and quality. The main contractor is also responsible for certifying payments.

The contract architect invites tenders from chosen subcontractors and after selection, if this occurs prior to the preparation of the main contractor's bid, the named subcontractor's tender is given to the contractor who is required to include it in the main contract and not as a PC sum. This means the contractors can add any percentage they like to cover attendance, overheads and profit. Unlike with domestic subcontractors, the main contractor is not in a position to renegotiate the price with named or nominated subcontractors if their bid is successful.

Another method is the contract architect gives a list of named subcontractors from which the main contractor can choose one to include in their bid. They may be allowed to suggest alternatives to the contract architect to be included on this list. In this case, in spite of the 'named' list, the contractor makes the contract with the subcontractor so these are classified as domestic subcontractors.

13.13.3 'Domestic' subcontractors

Whether using the JCT11 (IC 2011) or most other forms of contract, the majority of subcontracted work falls into this category. The main contractor is responsible for the final selection and engagement of the subcontractor and has full responsibility to ensure the work subcontracted is carried out properly and to programme. Theoretically in the conditions of most contracts, the contractor can only sublet work with the architect's written consent, but the clauses go on to say that this should not unreasonably be withheld. Such reasons are more likely to be based upon the client's ethical beliefs and requirements.

The main reasons for the main contractor to subcontract are as follows:

- the reallocation of risk by shifting all or part of the risk to another party;
- to overcome the problem of finding continuous employment for full-time employees when the work load fluctuates as this may be difficult or uneconomical, which in turn can give more flexibility in planning the works;
- to supplement the contractor's own labour force to overcome peak loads, thereby maintaining or accelerating the programme;
- to deal with specialist work, such as piling, structural steel, mechanical and electrical services.

It does mean that the management changes from managing one's own staff to that of coordinating packages. This is a different emphasis. It means there is potentially a loss of control especially if the subcontractor sublets part or all of the work. The main contractor may wish to prevent this by having a clause prohibiting this. There is also an argument that the amount of profit that can be made by the main contractor is limited as they can only make the percentage added to the subcontractor's quotation and cannot benefit from any productivity gains made by the subcontractor. The only other opportunity to earn more is renegotiating the subcontractor's tender on being awarded the main contract.

As with the client, contractors have their own preferred lists of subcontractors and suppliers upon whom they can rely for all the reasons as follows:

- their reputation for good quality workmanship;
- the quality of their management;
- completion on time;
- their experience on similar work;
- their financial state;
- the experience the contractor has had with the subcontractor on previous work.

The domestic subcontractors as outlined above are those that supply and fix all the necessary labour, materials and plant. This will usually involve providing some site accommodation such as toilets and washrooms and on large sites, and canteen facilities.

13.13.4 Labour-only and labour- and plant-only subcontractors

Labour-only and labour- and plant-only subcontractors provide, as the titles indicate, only labour or labour and plant, the materials being provided by the main contractor. Typical examples of the former are brickwork and concreting gangs and of the latter, a reinforcement steel subcontractor who may provide the cutting and bending equipment.

They are selected and engaged in the same way as domestic subcontractors, the main contractor having the responsibility to ensure quality and programme commitments are met. The contractor needs also to be satisfied that the labour-only subcontractor has adequate insurances and they are registered with Construction Industry Scheme (CIS) to satisfy the Inland Revenue (HRMC) and are not working on the black market.

The reasons for employing labour-only are similar to those outlined in domestic subcontractors (13.13.3), but there are other issues to be clarified before signing any contract. These include:

- Are any attendances required especially the use of plant if not provided as part of the contract?
- Are they or the main contractor responsible for unloading their materials and does this include moving from storage to the point of production?
- Who is responsible for setting out their work?
- Is the subcontractor providing their own supervision to ensure quality, safe working practices and keeping the work on programme?
- Are they to be paid weekly or after the monthly valuation of work done?
- Does the contractor or subcontractor carry the burden for retentions? If the latter, then at what point is it released? It would be unfair for a small contractor to have to wait 12 months after the main contract has finished before receiving any retention monies.
- Does the contractor collect the CITB levy from the subcontractor or do they pay it direct themselves? If the latter, the contractor needs to be satisfied that this is happening.

Construction Industry Scheme certificate (CIS)

The HMRC became concerned that as much of the labour employed in the construction industry was labour-only subcontracted there was a significant gap in the legislation that permitted operatives to work, be paid and then not pay tax. The initial solution was the 714 scheme, but this was replaced in August 1999 by the CIS.

Construction companies must only employ subcontractors who are registered under the scheme; for the subcontractors to get paid under the scheme they must hold either a registration card or have gross payment status. To obtain either of these the subcontractor has to register with the Inland Revenue. Gross payment status means the subcontractor is responsible for paying tax and national insurance contributions.

Those issued with registration cards are usually individuals or small gangs. In this case the contractor must make a deduction from all payments for labour of an amount to cover the subcontractor's tax and national insurance contribution (NIC) liability. Full details of the CIS scheme can be found on the government website: www.gov.uk/what-is-the-construction-industry-scheme.

13.14 Selection and invitation to tender

The procedure for selecting a subcontractor is very similar to the client selecting a contractor. It is increasing likely that the subcontractors selected to tender will either be on a preferred list, which means much of the checking procedures will have already been carried out, or have a supply chain agreement with the contractor (see Chapter 24). If neither of these exist or the contractor wishes to extend the preferred list, a series of questions need to be answered before giving the subcontractor the opportunity of tendering, as listed in section 13.13.3.

It is essential to obtain as accurate a price as possible from the subcontractor. This can only happen if the information submitted to them is as comprehensive as possible. The information they require is similar to what the main contractor needs for their tender plus issues specific to the subcontractor such as:

- basic details about the project such as location, names of employer and consultants;
- a general description of their work;
- other appropriate pages from the preliminaries including information about access and any other restrictions as stated such as working hours, noise levels etc.;
- where further information such as drawings and other relevant reports can be inspected;
- the relevant extracts from the bills of the measured work to be priced by the subcontractor with the relevant specifications and schedules;
- drawings needed to permit a tender to be produced;
- the type of contract they will enter into if successful and any amendments identified – this may well be a contract document developed by the contractor for their own use with subcontractors;
- what attendance facilities the contractor is to provide for their use. For example, temporary accommodation, scaffolding, unloading and storage of equipment, day-to-day setting out and provision of waste skips;
- the programme requirements and any apposite method statement;
- the date and time by which the quotation has to be returned;
- financial matters including the method of measurement and frequency of payments, the amount of retention and defects liability period, fluctuations if any, and the daywork schedule of rates;
- either a request for discounts offered or a requirement of a percentage discount payable to the contractor;
- the contractor's health and safety plan.

13.15 Opening of tenders

Whether or not these tenders should be as sealed bids depends upon circumstances. If it is for a local authority direct works department (they are in effect contractors) then this would be required and all bids would be opened at a given time under controlled supervision to demonstrate that no preferential treatment or fraudulent act occurs. However, in a general contractor's office, while it is in the contractor's own interest to ensure that the process is open and fair, the routine is usually less vigorous.

On receipt of the quotations there are two prime processes to carry out: to ensure one is comparing like with like; and then to select the most appropriate quotation. This is not just comparing each competing subcontractor's 'total' price. It is normal procedure to enter information of each submission on a form such as demonstrated in Table 13.15, so that a true comparison can be more easily made, as it is important that all the submissions are inspected carefully. However, before analysis, the figures are checked to see that the subcontractor has quoted for as required. Issues to look for are:

- Have all the items been quoted for? The subcontractor may consider that part of the bill section sent does not come within their remit and expertise. For example, the masonry section of the bill may have been sent in total and includes some dry-stone walling. One subcontractor specialising in brick and block work does not price this part whereas another may have access to specialist dry-stone wallers and has priced the section of the bill in its entirety.
- Have they made any specification changes? This is a common ploy as by offering an alternative it can create a competitive edge over others. The revised specification may in the end be a satisfactory alternative solution, but at this point in time it is essential to compare like with like. It does give the contractor the opportunity to renegotiate later as well as assist in value engineering the project.
- Have they added or subtracted any contract conditions? It is not unusual for subcontractors to have their own schedule of conditions and include these with their tender in spite of being asked to quote based on the contractor's conditions.
- While the contractor has specified the attendance allowances to be provided, in practice the subcontractor's needs may not be fully understood so the tender may include for extra provision.
- The letter of enquiry will have stipulated start and finish times for the subcontractor's work. It is necessary to check they have accepted these and not modified them in any way.

Having investigated the above and resolved any matters arising, the data can then be entered. As can be seen by the example in Table 13.14, Firm B appears at first to be the most competitive, but on analysis after the discount rates are taken into account, Firm A is the lowest. The breakdown of the various bill items could be useful once the contract has been awarded to act as a basis for negotiation. Why, for example, does firm B have a lower rate for items M50/1/A and D than the overall lowest net price of firm A? Alternatively, Firm A was more expensive than Firm B, before the discount was deducted. Is this negotiable?

13.16 Suppliers' quotations

On requesting quotations from a material supplier the list of requirements is stated to enable them to put together an informed quotation, although this is not as large as for the subcontractor. The kind of information needed includes:

Table 13.14 Comparison of subcontractors

Subcontractor comparison
Project: Jasper Winston offices
Subcontractor type: Carpet Tiles

		Firm A		*Firm B*		*Firm C*	
Bill Item	*Quantity (m²)*	*Rate (£)*	*Total (£)*	*Rate (£)*	*Total (£)*	*Rate (£)*	*Total (£)*
M50/1/A	450	24.50	11,025	24.25	10,912.5	23.75	10,687.5
M50/1/B	1000	23.75	23,750	23.90	23,900	25.00	25,000
M50/1/C	630	34.60	21,798	35.00	22,050	35.20	22,176
M50/1/D	350	29.25	10,237.5	27.60	9,660	29.00	10,150
	Total		*66,810.5*		*66,522.5*		*68,013.5*
	Discount	2.5%	1,670.26	1.5%	997.84	2.0%	1,360.27
			65,140.24		65,524.66		66,653.23
	Lowest Net		65,140.24				

- the name and location of the contract to which they have to deliver;
- any restrictions on delivery, such as time and parking;
- a description of the materials, and most important, the specification;
- the approximate total quantity needed for the contract;
- drawings of components if they are to manufacture;
- the schedule of components, for example doors and windows;
- the approximate frequency, timing and quantity of delivery – this can only be indicative as the contract start date will not necessarily be fixed yet, nor are all the designs complete;
- date by which to return the quotation;
- whether or not it is a fixed or fluctuating price;
- any discount terms being offered;
- whether they could offer a different material that meets the specification requirements, but is cheaper.

On receipt of the quotations they are checked in line with the summary shown in 13.9.2. Suppliers often have a standard set of conditions printed on the rear side of all their quotation stationery and this needs to be checked for any unacceptable conditions. Once this has been done the various quotations can be entered on a form for comparison and analysis, as shown in Table 13.15. For simplicity this table assumes all conditions of sale are the same and no discounts are offered from any of the suppliers.

For the purposes of pricing the bills it is assumed that the rates can be used in isolation, picking the cheapest from each supplier, unless the supplier has specifically stated that each price is subject to both materials being ordered. If awarded the contract, and the prices remain the same, the contractor is in a position to negotiate a lower price, not just because the order is now in the bag, but because it may be in both the supplier's and contractor's interest to obtain similar materials from the same source. For example, the total price for Firms A and B are similar, so would Firm A be prepared to lower their price for hardcore to match that of Firm B.

Table 13.15 Materials comparison sheet

Materials comparison sheet *Project: Jasper Winston Offices*				*Firm A*		*Firm B*		*Firm C*		*Rate selected*
Material	*Specification*	*Quantity*	*Unit*	*Rate*	*Total*	*Rate*	*Total*	*Rate*	*Total*	
Hardcore		2,358	tonne	18.3	43,151	18.1	42,680	18.8	44,330	18.1
Sand		566	tonne	14.8	8,377	14.9	8,433	14.9	8,433	14.8
			Total		*51,528*		*51,113*		*52,763*	
				Firm D		*Firm E*		*Firm F*		
Concrete	C.75: 20mm	1,324	m^3	69.1	91,488	70.1	92,812	68.9	91,224	68.9
	C25: 10mm	677	m^3	73.3	49,624	72.3	48,947	72.4	49,015	72.3
			Total		*141,112*		*141,759*		*140,239*	
				Firm G		*Firm H*		*Firm I*		
Masonry	Facing brick	90,000	1,000	249	22,410	249	22,410	249	22,410	249
	Eng. Brick	6,000	1,000	194	1,164	174	1,044	180	1,080	174
			Total		23,574		23,415		23,490	

Table 13.16 Method statement

Item	*Method*	*Quant.*	*Labour*	*Material*	*Plant*	*Rate*	*Duration*	*Comments*
Reduce piles to datum level	Compressed air breakers	$100m^3$	2 No. labs	–	Compressor and 2 No. breakers	4 man hours per m^3	12.5 days	Noise levels to be controlled

13.17 Method statements

Method statements are a written description of how specific activities or groups of activities are to be carried out. Increasingly, method statements are required by the client to demonstrate the contractor's knowledge of production, safety and quality. Contractors can be reluctant in open tender to submit their detailed method statements at this stage as if not successful the client could utilise their ideas with the successful bidder. As a result, the method statement offered to the client may be less detailed and more of a public-relations exercise, designed to demonstrate the contractor's competence and experience. On the other hand, for the experienced client it demonstrates the contractor's ability in complying with their needs.

There is a more fundamental reason for method statements: to establish the best method for the needs of the contract and also as part of value engineering. The method adopted may be determined by the duration given to complete the contract and is therefore linked to the programme time. It might also be a useful tool to demonstrate to the client that, for a small sum more, the contract could be completed much faster, giving the client the opportunity of increased revenue. Finally, it provides the basic information to enable the pre-tender programme to be produced.

Analysing the work carefully at this stage can resolve many later issues, as it could be that the method adopted alters the proposed sequence of construction requiring design information to be produced in a different sequence and it can be a great starting point for the management team when planning the construction of the contract.

The first stage is to analyse the operation in question. To do this the estimator and advisors need to look at the overall site conditions such as limitations of space, access, the footprint of the building, availability of labour, requirements for phased handover of the works and so on. They will consider the climate implications such as high winds that may mean that tower cranes are not used because of the amount of down time predicted. They will make note of how other operations will affect each other. Finally its size and scale are taken into account, as for example, large-scale operations would allow higher capacity equipment to be used than employed on smaller operations.

After all these considerations a summary of the decision can be produced and Table 13.16 is a typical way of presenting the information giving the planner the resource information needed to produce a plan. An alternative form of presentation, often used for the client's benefit, is to produce the same information in a narrative format rather than tabular.

13.18 Tender programme

The tender programme is required to establish whether or not the project can realistically be completed within the time prescribed by the client and to establish the method and sequence of operations from which other issues flow and calculate the amount of resources needed to complete the work. Up to a certain limit the amount of effort put into the production of plan and method statement relates to the accuracy and competitiveness of the tender price. Its production demonstrates the contractor's intention at the time of tender and the basis upon which the tender was priced. Some clients require a programme to be produced as part of the tender submission and this document can be used also for this purpose. How to produce a programme can be seen in Chapter 20.

The programme enables the estimator:

- to establish when subcontractors and suppliers are needed so that this can be incorporated into the invitation to tender documents;
- to act as a basis to establish when staff are required for the contract so that site overheads can be calculated (preliminaries);
- to establish when plant is required on site, giving the planner the opportunity to better resource its use;
- to identify building works likely to be affected by weather conditions, such as ground works in winter and high wind;
- to establish when temporary works such as scaffolding and falsework are required for the preliminaries pricing;
- to identify possible savings in time that may be offered to the client.

It can also show an allowance for public and peak annual holidays and is an ideal mechanism for calculating the cash flow predictions.

13.19 Cash flow

One of the important pieces of information the senior management team requires when deciding how much overheads and profit to add to the tender is how the cash flow is predicted to contribute to the business as a whole. It is normal to expect a negative cash flow at the commencement of the contract which then becomes positive towards the end of the project. They look at the overall cash flow of the business and determine whether there are any periods in the financial year when the business is in or approaching deficit so that, if required, a financial loan can be sought. Ideally, the workload of the business will be such that the positive cash flow contributions from contracts nearing their conclusion will compensate the deficit cash flows on recently started projects.

To produce the cash-flow predictions, the tender programme and bill costs are brought together to produce a financial graph showing how much income the contract generates each month and when the contractor will receive monies from the client. Added to this are the outgoings to the suppliers and subcontractors. Since the bills are produced in the way that they are, it is relatively simple to abstract the costs of each stage of the programme, as each bar line on the tender programme is similar to the structure of NRM2.

All that has to be done is to divide this sum approximately equally along each bar line in weekly or monthly increments, and total up each week's/month's expected contribution, not forgetting the preliminaries that are fixed or time related, and deduct the retention percentage, in this case 10 per cent, as shown in Table 13.17.

The entries for the payment out and the cash flow are not completed in the table. There are other considerations to take into account such as the date that monies are received, invoices paid and retentions withheld. A more detailed discussion of this subject is given in Chapter 16.

13.20 Overheads and profit

While the preliminaries cover the site overheads, there is still the matter of the regional and/or head office overheads. These overheads comprise:

Table 13.17 Monthly income

Month *Activity*	*1*	*2*	*3*	*4*	*5*	*6*	*7*	*8*	*9*	*10*	*11*	*12*	*13*	*14*
A	2.8	2.8												
B		4.3	4.3	4.3	4.3									
C		7.7	7.7	7.7	7.7	7.7	7.7							
D				5.5	5.5	5.5	5.5	5.5						
E		8.4	8.4	8.4	8.4	8.4	8.4	8.4						
F						1.8	1.8	1.8	1.8	1.8				
G						9.8	9.8	9.8	9.8	9.8	9.8			
H								3.3	3.3	3.3	3.3	3.3	3.3	3.3
J								6.7	6.7	6.7	6.7	6.7	6.7	6.7
K									5.1	5.1	5.1	5.1	5.1	5.1
L							7.5	7.5	7.5	7.5	7.5	7.5	7.5	7.5
M									12	12	12	12	12	12
etc. ↓												etc.	→	
Prelims – time related	2.2	3.8	5.6	6.1	6.1	6.8	7.3	7.0	7.8	7.8	7.6	5.4	5.4	5.4
Prelims – fixed	89													
Gross monthly income forecast	94	27	26	32	32	40	48	50 →	54	54	52	etc.		
Less retention 10%														
Net interim value	85	24	23	29	29	36	43	45	49	49 →	47	etc.		
Payments subcontract and suppliers														
Balance/cash flow														

- the salaries and costs of the directors and staff employed at the head/regional office – this figure includes for all the additional employment costs, such as employer's pension and NIC contributions, holidays, and the CITB levy;
- the rent, or depreciation, rates and maintenance of these offices and other facilities such as workshops, stores and playing fields;
- heating and lighting;
- the office furniture and furnishings;
- the office equipment including information technology provision and telephones;
- the various insurances cover that need to be provided;
- company cars, fuel, personal car allowances and travel;
- postage, stationery and printing;
- the provision of subsidised canteen facilities and welfare in general;
- the costs of advertising and public relations;
- entertainment expenses;
- auditors and other outside consultants;
- interest to be paid on retentions and other loans.

The amount of all these costs can be readily calculated to give an annual sum. This sum is paid for by adding the percentage to successful contracts bids.

To give a simple example of how overheads might be charged to demonstrate the principle (there are more sophisticated accountancy methods for this), assume the head office overheads are £100,000 p.a. and the annual turnover planned for in the company business plan is £2,000,000, then the percentage needed to be added to each contract to cover the overheads is:

$$\frac{\pounds 100{,}000 \times 100}{\pounds 2{,}000{,}000} = 5\%$$

This figure of 5 per cent assumes interest on retention money and working capital has already been taken care of when calculating the total office overhead.

When business is tight there is a temptation to reduce the percentage overhead to less than the amount calculated and to take zero profit to become more competitive, but this is a dangerous policy. What to do under these circumstances becomes a strategic decision, because the question has to be asked whether or not the economic situation of the time is going to continue and if so whether or not it is sensible to remain in this market place.

If the business is growing successfully and has outdone its budgeted turnover, a different situation arises. For example, if after say nine months, the business obtained equates to the annual turnover planned for, it may be that the overheads already added to the work won have now covered the annual overhead bill. If so, then it would not be necessary to add any overhead to any other tenders submitted for the remaining part of the year, resulting in being more competitive and hence probably achieving a higher turnover. Alternatively, the company could add the same percentage of overhead or somewhere in between. This is a strategic decision. However, if the business grows too fast, this will mean recruitment of more staff. They have to be trained and it takes time for them to absorb the culture of the organisation. Once they are 'full' members of the team, there would be some reluctance in parting with them, which means the work required the following year would have to be increased to keep them in employment.

Deciding upon the level of profit to add is a senior management decision. It is a complex decision, as many factors have to be considered, but in the end it boils down to the instincts and experience of the management. This is why, after all, they are in their senior position and if they get it wrong too often, receive their marching orders. Examples of the issues upon which they make their decision include:

- How seriously do they wish to win the contract? This is influenced by
  - the strength of the current order book;
  - the wish to break into a new market;
  - the knowledge that this is the first of many contracts with this client;
  - an existing client who they have built up a good working relationship with;
  - a prestigious project that will raise the profile of the company.
- The company has worked with the client and or the design team before and all the contracts have run smoothly and profitably.
- What is the other competition for the work? Experience will indicate how successful bidding against these companies has been in the past and what the difference in margins has been.
- Will winning the contract overstretch the resources of the company and what are the implications of this?
- What levels of profit have been obtained on recent successful similar bids as this gives an indication of the market and where to pitch the bid? In other words, how much will the market stand?
- How much and how has finance to be provided to fund the earlier part of the project because of negative cash flow.
- Have changes to the standard form of contract been made so as to shift the risk away from the client and towards the contractor and if a non-standard contract what risks are involved?
- Are there production risks involved because of lack of information upon which the tender was priced, or are there complex and difficult construction details, or novel design features?
- How tight is the programme to complete the building and what is the likely risk of overrun and consequent penalty implications?
- Having reviewed all the above what is the overall risk if the bid is successful?

13.21 Bringing the tender information together for final review

The purpose of this report is to provide the senior management with all the relevant information needed to enable them to make an informed commercial decision on whether or not to proceed with the tender submission and, if yes, then how much profit and overheads should be added. Some of the information will be produced in case it is required, but may not be used. The contents of the subtitles below inevitably overlap, but in general the information needed includes:

General contract information:

- a brief description of project and its location to give a general flavour of what the contract is all about supported by appropriate drawings;

- details of the client and consultants, which reminds senior management of past experiences if worked with before;
- the form of contract being used;
- specifications, ground investigation reports and any other specialist report.

Production information:

- an outline description of the methodology of construction used as the basis for the estimate;
- the programme duration and confirmation this complies with the client's requirements;
- any information that would be pertinent to offering a reduction in the time of the construction programme;
- any unresolved technical problems that might increase the financial risk;
- any major assumptions made in the estimate, for example, permission to over fly with the tower crane only being agreed in principal;
- site visit report.

Contractual matters:

- a list of special or unusual contract conditions imposed by the client and risks as this may affect judgement on profit figures;
- special terms or conditions stated by any of the subcontractors, especially those nominated;
- any unresolved contractual problems that might increase financial risk;
- any client's special requirements such as bonds and insurances.

Financial matters:

- the overall cost of the project before overheads and profit are added;
- a summary of the completed bills of quantities or tender summary – management is not interested in the detail, just the final figure and a breakdown of this into discrete but usable information e.g.:
  - own labour costs;
  - materials cost and bulk quantities and materials comparisons forms;
  - plant costs and comparisons;
  - domestic subcontractors' work costs and comparisons;
  - nominated subcontractors and suppliers;
  - other provisional sums;
  - overhead contribution based on company formula;
  - schedules of suppliers, subcontractors and plant and duration;
  - schedules of prime cost sums and attendances;
- a cash flow calculations sheet as described in 16.4;
- if a fixed price contract, how long the tender is to remain open for acceptance.

A typical financial summary spreadsheet is shown in Table 13.18. It is designed so that the estimator can immediately modify any changes management may wish to make at the final review meeting (13.22). The percentage analysis calculation column is based on the revised estimate figures, expressed as a percentage of the net total.

Table 13.18 Tender summary sheet

Tender summary	*Contract: Jasper Winston Offices*				
Description		*Estimate (£s)*	*Adjust-ments*	*Revised estimate*	*% analysis*
Own work	Labour	103,542		103,542	7.12
	Plant	25,678	−1.0%	23,110	1.60
	Materials	323,456	−1.5 %	318,604	22.09
Domestic subs	Net	784,230	−2.5%	764,624	53.02
Nominated subs	Net	25,600		25,600	1.78
Nominated suppliers	Net	3,456		3,456	0.24
Total		*1,265,962*		*1,238,926*	*85.85*
Provisional sums					
Preliminaries					
Employer's requirements	Insurances etc.	12,659		12,659	0.88
Contractors requirements	Staff	48,432		48,432	3.37
	Accommodation	5,600		5,600	0.39
	Services and Facilities	8,098	−0.5%	8,056	0.58
	Plant	134,654	−1.0%	121,189	8.41
	Temporary works	6,800		6,800	0.47
Statutory authorities	Water-connect	380		380	0.03
	Electricity-connect	250		250	0.02
Sub total		*214,280*		*203,366*	*14.15*
Net total		*1,480,242*		*1,442,292*	
Overheads	5%	74,121		72,115	
Profit	2.5% pre overheads	37,061		36,057	
Prime costs/attendances		7,000		7,000	
Estimated dayworks		20,000		20,000	
Tender value		1,618,424		1,577,464	

The adjustment column is based upon the estimator's feel for the commercial market about how much they believe the quotations received from others could be renegotiated if the company wins the contract. The percentage columns are useful in the final review meeting as an indicator of the proportion of the work set against the total price before overheads and profit are added. This gives information to management to assist in assessing the risk and whether the proportions are consistent with previous successful bids. The amount of suggested overheads and profit could be entered here to give management an indicative price, but may be omitted and entered at the final review meeting. The dayworks and attendances are separated in this case to permit management to make a separate decision on the amount of profit and overheads to add, which may differ from the main part of the tender.

Estimators comments:

- the quality of information upon which the estimate was made and any assumptions as a result;

- the estimator's view on the current market place and conditions;
- the estimator's view on the likely profitability of the contract that is often a function of how buildable the works are;
- who the other competitors are, if known;
- the likelihood of more work from the client especially if the one under review is the first phase;
- recommendations of any conditions the contractor should lay down when submitting the tender;
- date and time of submission.

13.22 Final review meeting

This meeting is to make the final decision on the tender submission price, which includes the overheads and profit. The key players in this are at the very least the following:

- the senior manager, usually the managing director, who has the overall picture of the financial state of the business and is therefore commercially aware of what is in the best interests of the company in terms of the need to obtain the work, and the margins to be delivered;
- the estimator, having completed the process and being the most familiar with the tender documentation and price;
- the buyer, or in some organisations the QS, who is there because of their commercial awareness of the current market place – very important when reconsidering the possibility of reducing some of the estimator's prices for subcontractors and suppliers;
- the planner, whose particular contribution is concerned with answering resource questions and the implications of altering the sequence of work or reducing the overall timescale;
- the contracts manager, or failing that, it is important that a 'practical' input is available to discuss methodologies that could be used in the event of changes being made and to cast an eye over the chosen method used for the basis of the tender.

Most companies have a standard agenda for this meeting, reflecting the topics discussed in 13.21. In the background is the overarching consideration of what the client will use as their criteria of selection which could be dependent upon the nature of their business. Typical issues might be as follows:

- Is price the overwhelming consideration? It is more likely to be if a public body and may be in their constitution as a standing order. There may be some history of the client that demonstrates that price is not always the key criterion.
- The client's business may well indicate the sensitivity for completion on time or any advantages gained by an accelerated programme, even though the latter may cost more. Examples would include retail developments when completion is keyed into the Christmas selling and sports stadiums being built primarily in the closed season. Any acceleration, which limits the loss of sales, could be advantageous.

- Whether or not the client is experienced can affect how much effort and time is spent on presenting such information as methods and issues of environment, quality and safety.
- Equally if the client has a particular interest, say environment, it would be appropriate to design the contents of the final document and/or presentation appropriately.
- If an oral presentation is to be made, the contractor needs to consider the format and content prior to putting the information together for rehearsal.
- Finally, the client will always be interested in the amount of money that has to be spent on the project. On the one hand this could be presented as a total sum or alternatively as a predicted monthly outlay. However, the latter may be seen as overstepping the mark and cause irritation. Equally excessive front loading (13.23) or apparent high fixed costs in the preliminaries could have a similar effect.

The purpose of the meeting is:

- to make the final decision as to whether to submit the tender. Normally this decision would have been made earlier on receipt of the tender documents (13.4), but issues may have arisen subsequently and this would be resolved at this meeting;
- to assure management that this is the best way, so alternatives need to be evaluated, as the tender will have been produced based upon a selected method of construction. Another method may be in the client's interest if the effect is to be able to complete the contract earlier;
- to reduce (usually) the domestic subcontractors and supplier's prices where appropriate. The reason for the opportunity to reduce prices is based on the 'bird in the hand' principle. Being asked for a quotation is very different from being told by the successful contractor the work is secured. The estimator and especially the buyer have a good commercial feel for what the market will bear. Based upon current trends, the amount of work available, and the relationships with the suppliers, and subcontractors, it is not unusual to be able to reduce the sums by between 1 and 3 per cent. Possible reductions may have already been indicated by the estimator as shown in Table 13.18, but the buyer will be even more sensitive to the current market place and is able to fine tune these figures;
- to consider any other modifications to the contract, especially if the client has deleted clauses that transfer more risk to the contractor;
- to consider any offer from the contractor, as a result of the deliberations at this meeting, of an improved service than that which the client asked for. This can include improved functional performance of the building, reduced costs by modifying the design and a shorter construction programme;
- to reassess both in the light of the information brought to their attention in Table 13.18, where a figure for both overheads and profit has been added by the estimator, and the considerations outlined in 13.20;
- to make the final decision on the costs. The advantage of the use of computer-aided software is that the modifications requested by management can be entered at the meeting and an instant answer produced. This gives management the opportunity to try out a variety of different options to assist in making the final decision.

13.23 Final adjustments and submitting the tender documents

As a result of the review meeting the tender documents are modified accordingly. If not asked for it may be desirable not to show the full extent of overheads and profits. As bills are usually priced using computer software, it is not difficult to add a percentage to all bill items, thereby reducing or eliminating the shown profit figures. This also can be carried out for a proportion of the preliminary items and in the quotations given by the domestic subcontractors. However, if this is done too much it will become clear to the client's QS that the bill items costed do not reflect the market place.

There are a variety of ways that the bills can be modified to benefit the contractor financially, but at the same time not increase the overall tender price. When the contract commences there is a period of approximately two months before the contractor receives money from the client for work done. This, coupled with retentions, causes a negative cash flow during the early part of the contract resulting in the contractor having to financially support the project at the beginning. To compensate for this the contractor sometimes utilises a technique called 'front loading'. This is where the estimator inflates some of the larger items at the beginning of the bill and to compensate, reduces some of those which will be carried out towards the end of the contract. While there is a general acceptance that this occurs, there is a limitation to how much this can be done before the client's QS would raise questions. If the contractor becomes insolvent during the contract and there has been excessive front loading, then the client would have to pay out more to have the rest of the contract completed by another contractor. Unless site management is aware of this strategy and the amount of the adjustment, they can be lulled into a false sense of security, believing they are profitable at the beginning of the contract when they are not. The opposite of front loading is that of 'back loading' which is only relevant in times of high inflation and where fluctuation clauses are included. Here, items to be completed at the end of the contract are inflated at the expense of the earlier ones, resulting in the contractor being paid proportionately more. Both of these practices raise some ethical questions.

Another technique used is to identify certain items in the bills that are currently shown as small quantities, but it is suspected are either a measurement error, or are likely to increase as a result of variations, sometimes referred to as 'item spotting'. If the price for these items is inflated, the effect on the overall tender price is minimal, but if the measure is subsequently increased then the contractor can make substantial gains. An example of this would be in the excavation of foundations, when unexpected problems are revealed, such as pockets of peat, and much more has to be excavated and carted away as a result. However, to do this conflicts with the current philosophy of openness between all parties and so has much wider implications.

One cannot underestimate the need to submit the final documents in a well-polished form. Presentation has become increasingly important and reflects the image of the company. If an oral presentation is required to support the bid, then all the participants have to be well coached and rehearsed. The content of the accompanying letter that will outline any modifications or suggestions needs to be carefully composed to communicate clearly, but written in such a way as not to cause offence.

It is normal for the documents to be sent in a sealed envelope, known as a sealed bid, so that they can be kept under lock and key until all are opened at the same time under strictly controlled conditions to ensure that honesty prevails. It is not unusual for the bid to be delivered by hand either by courier or a company representative. Sometimes this is because the contractor has been on a very tight schedule, or because it leaves an opportunity to make last-minute adjustments if new information arises.

After the results of the bids have been announced, and the successful bidder notified, it is increasingly common practice for the employer to notify all competitors of each other's bids. The timing of the disclosure of this information must remain confidential until after the contract has been signed otherwise the lowest bidder could try to negotiate higher prices. This needs only to be a list of the various tender prices, as some contractors may not wish their competitors to know what their bid was. If this is not published, then it is important for the estimator's grapevine to come into play to discover the value of the other bids. This applies whether successful or not as this data gives an indication of the current market place and the company's relative position in it, which is useful knowledge for current and future tenders (see Chapter 14).

13.24 Vetting of tenders

The final stage in this process is with the client. They will usually open all the tenders together and list the contractor's bid prices. This is done in a controlled manner by the client, or representatives, and would normally include the quantity surveyors responsible for putting together the bills and contract documents. It should be done as soon as possible after the deadline given for submission of the bids. Precisely how this panel is constituted is not prescriptive. What matters is that it is, and is seen to be, a fair and honourable process. It is normal for two members of the panel to countersign each tender submission. Any unsolicited bids must be rejected and any late bids returned to the sender unopened.

All except the lowest three should be advised immediately their bid has been unsuccessful. The documents of the lowest of these remaining bids are then vetted and checked for genuine errors and confusion. The most common occurrences are:

- mathematical errors made when adding up the columns of numbers, decimal point errors and multiplying quantities with rates;
- obvious errors such as pricing a material in say kilograms rather than tonnes, or clearly pricing for a lower or higher specification than is required;
- prices for similar items at different parts of the bills that are considerably different in price;
- extreme cases of 'front loading', 'back loading' and 'item spotting' – in all of these cases, the contractor cannot be compelled to change them, but the quantity surveyor can advise the client not to accept the bid.

The contractor should be advised of these errors and given the opportunity to revise their bid. The contractor then needs to consider whether or not they wish the revised bid to remain on the table or withdraw it. It could be that these amendments raise the tender price above that of the next lowest, in which case this one is subjected to the same rigorous inspection and so on until a final decision can be made as to which one is selected to execute the contract.

If the lowest tender exceeds the client's budget price and a maximum sum has been indicated that must not be exceeded, negotiating will take place. If accepted, the contractor would usually accept the risk with the following caveat, 'the price is for the scheme as drawn now'. There may well be other clauses and exclusions included to protect the contractor. They will try to spread the risk to the suppliers and subcontractors. For example, if the price calculated by the contractor was 5 per cent above the guaranteed maximum price, the subcontractors would be asked to reduce their price by a similar margin.

In the event that the client had valued the job at say £42 million and the contractor believes the cost to be £44 million, the contract might be let at £44 million on the condition that anything saved between £42 and £44 million would be handed back to the client and anything below £42 million would be divided say 25 per cent to the client and 75 per cent to the contractor.

13.25 Risk and uncertainty in estimating

Construction is by its very nature an industry with a significant amount of risk. The estimating process carries its fair share and, while not intending to become involved in the mathematical and statistical analysis of risk evaluation, the reader should be made aware of the issues involved. This is a summary of the issues the estimator takes into account when pricing a contract and management when reviewing the tender prior to submission. The factors built in by either the client or designer that can impact on the estimating process are:

- the accuracy and availability of the tender documentation necessitating making assumptions that may have inherent error;
- a very tight project programme that is likely to have an overrun as most programmes have some optimism built in and any delays that occur are more difficult to reclaim which can result in penalty clauses being invoked;
- fixed-price contracts that have to take into account most increases that may occur in the future – the longer the duration of the contract, the more difficult these are to predict thereby increasing the financial risk;
- provisional sums, which are on the one hand an advantage in so far as the client takes the financial risk; however, the contractor has less control on performance which may have an impact on the overall performance of the project;
- complex projects, which are more difficult to price and quantify in terms of duration;
- monthly interim valuations that have to be agreed by the client and contractor – over-zealous quantity surveyors representing the client's interest can try to undervalue the work to be seen to be protecting the client's interests. Even if the valuation is only a small percentage lower than the true value it will have a knock-on effect on the contractor's cash flow;
- any variations to the works that are measured separately. If it is appropriate to use the bills rates then there is no inherent risk – unfortunately it is not as simple as this and there is a risk attached especially if a large number of variations are instructed;
- the way architects, engineers and clerks of works interpret the specification – this can cause problems that can be costly and cause delays. For example a contractor,

with the agreement of the consultant engineer, decided to precast the reinforced concrete stairs on site in lieu of in-situ to permit ready access between floors during the construction. The precast stairs were rejected on the grounds that the blowholes formed in the surface, while complying with the in-situ specification, did not meet that specified for precast.

The following risks are inherent in the tendering process itself:

- While taking account of the amount of time predicted to be lost as a result of inclement weather, it is notoriously difficult to assess.
- In the euphoria of wanting to successfully obtain the contract and meet the client's needs there is a danger of making optimistic assumptions when programming.
- The time available for carrying out the estimate is limited so there is always the risk that not enough time is allocated to thinking through all the implications of the construction processes and the wrong method of works is selected as the basis for the tender.
- Deciding upon the levels of overheads and profit to be added is the greatest risk of all, which is why senior management takes the decision.
- Predicting the increased costs of plant, materials and labour throughout the duration of the project is difficult, especially if a long contract. The risk is reduced in the event of having a fluctuation price contract.
- If during the course of the project shortages of labour, plant and materials occur, the price of obtaining them will rise.
- In the end, other than when insufficient overheads and profits have been added, the contract's success or failure is usually a reflection of the quality of the management.

It is interesting that many of the risks can be reduced significantly if all parties work with a spirit of cooperation rather than confrontation.

13.26 Estimating in construction management and management contracting

Much of construction management and management contracting is about dividing the construction work into work packages and employing subcontractors or work-package contractors to carry out the work. The result of this is that the B of Q type of contract is not used to the same extent, because the subcontractors carry out the estimating function for that part of the work.

Before submitting their tender, the company will go through a similar process to that described in the previous sections. They will provide and cover the costs of many of the preliminaries items and others will be covered in the package estimate. This distinction needs to be clarified.

Their tender submission will include the following:

- acceptance of the client's conditions and requirements;
- their fee to cover the pre-construction phase;
- their fee to cover the construction phase;
- the organisation structure for the project and key personnel demonstrating their ability to carry out the work;

- their overall strategy for the project including the overall programme, method statements, safety policy, environmental policy and recruitment policy;
- the work-package estimates;
- estimates for any other work not covered in the work-packages.

Accuracy depends on when in the RIBA Plan of Work the management contractor is brought in. For example, if in the early stages then the latter packages such as finishings may not be available for pricing. A gap analysis needs to be carried out to find out what information is missing; for example, who makes the holes in the structure for services etc., who is responsible for making adaptions to the scaffolding, and who is responsible for a design element – the main team or the subcontractors. In the latter case these issues need to be agreed with the design team before putting in one's bid. The mechanical and electrical services information available is often a grey area and may need clarification. Where information is not available it may be necessary to resort to approximate estimating techniques, flagging up the likely percentage of error. Where there is uncertainty it is usual to put in a contingency sum to cover the risk involved.

13.27 Computerised estimating

Computer-aided estimating has come of age and is accepted by most medium and large contractors. Software packages are also available for the smaller organisations. The process described in this section covers traditional methods of procurement where bills of quantities are still in use. However, the general principles apply to all procurement where both detailed estimates and those based upon approximate quantities are used.

13.27.1 Entering the bills of quantities

The first stage of the process is to enter the B of Q into the system. This can be executed in three ways:

- re-typing the bill into the system, but this is time consuming, tedious and prone to errors that need checking for;
- putting the B of Q through optical scanners to prepare an electronic B of Q. Depending upon the type used there can be a need for checking as errors occur as some types of scanners misinterpret the information;
- using an electronic version provided by the client's quantity surveyor. This saves everybody a lot of time, providing the contractor and quantity surveyor's software are compatible. Most of the technical problems have now been overcome, but the use of them is a two-way process and both receiver and sender have to understand the needs and methods of working with each other. If the contractors do not ask for them, there is little point in the quantity surveyors producing them. There is, however, increasing use of this method and it is preferred by many.

13.27.2 Linking the bills to the estimating library or data base

The quality and quantity of data in the estimating library, linked to the ease and speed of access, is the key to the success of any computer-aided system. Like with any filing system the speed of access is a function of how the data is stored, filed and referenced. A typical structure is a resource-based model in which data is primarily divided and stored in two categories, that which is changing and needs regular updating, such as the costs of labour, materials and plant and that which is relatively static such as production rates of output. This type of system is most appropriate for unit rate estimating.

The advantage of such a system is that when an item is changed it automatically updates the other linked data. For example, a change in labour or concrete costs affects many other bills items costs and without this facility it would take a very long time to make the amendments.

There are two fundamental ways of relating the bills to the data stored in the estimating library. First, by storing the data in the library so that it matches typical bills items. They may not always match perfectly so some manual adjustments may be necessary. Second, to collect and keep the most up-to-date prices for the more frequently used resources. These would include all-in labour rates, gang sizes and outputs, different mixes for concrete and mortar, typical plant costs and outputs, and materials costs. Although the library will store many inclusive rates, the estimator has to build up more bill item rates from the basic data stored.

In both cases, when developing a library, the estimator needs to be assured that the computer-aided estimate system is capable of supporting the approach to be taken and before entering the data is fully cognisant of the way it works, otherwise the full benefits will not be realised.

It is possible to purchase a library of data with most systems and then modify or adapt the data according to the contractor's view on costs and production. The advantage of these packages is that they provide a filing system that estimators can adopt without designing their own. However, before purchasing such data it is important to investigate the package to ensure that the way the information is calculated coincides with one's own. Databases are also available using the Internet.

Any system adopted must have the ability to include and calculate variables that are either pertinent to the contract or the company experiences. Typical examples of these include:

- allowances for waste on materials that vary depending upon material, the complexity of the contract and the history of the contractor in controlling waste;
- bulking in aggregates, especially sand as a result of excess water;
- conversion factors – for example, aggregates and sand are purchased by the tonne and are measured in m^3 or m^2. Bricks are obtained in 1000s and normally measured in m^2;
- materials suppliers offering different discounts often with conditions attached – the system needs to be able to deduct these from the calculation;
- subsystems that permit analysis of the quotations from suppliers and subcontractors;
- mechanisms that identify all outstanding quotations from suppliers and subcontractors;
- cost codes that are the same as being used throughout the company;
- monetary conversions – useful for contractors working overseas;
- mark-ups being added to each individual bill item rather than as a percentage at the end, as required by some companies;

- fixed-price contracts that are of such duration that inflationary factors are likely to come into play – thus an assessment of the likely increases has to be made and added.

The estimating library is continually modified from various sources, namely:

- performance data such as plant outputs in different conditions and labour rates for executing different tasks with a variety of materials of prescribed dimensions;
- costs that require regular updating such as labour rates, CITB levy and NIC;
- material suppliers' costs;
- subcontractors' costs.

The estimating library can then be used to provide information for producing the preliminaries and bill item rates or, if designed to do so, approximate quantities rates. The estimate can then be compiled from the suppliers' and subcontractors' quotations, the bill build-up rates and the preliminaries to provide the document required for the final review meeting (13.22).

At the final review meeting, any modifications the management require can be entered and immediate feedback given on the effects of these decisions, which gives the opportunity to inspect various scenarios before making the final decision on the tender price.

Bibliography

Brook, M. (2016) *Estimating and Tendering for Construction Work*, 5th edn. London: Routledge.

Buchan, R.D. Fleming, E. and Grant, F.E.K. (2003) *Estimating for Builders and Surveyors*, 2nd edn. Oxford: Butterworth Heinemann.

Cartlidge, D. (2015) *Estimator's Pocket Book*. London: Routledge.

Chartered Institute of Building (1997) *Code of Estimating Practice*, 6th edn. London: Longman.

Chartered Institute of Building (2009) *Code of Estimating Practice*, 7th edn. Chichester: Wiley-Blackwell.

Co-ordinating Committee for Project Information (1987) *Co-ordinated Project Information for Building Works: A Guide with Examples*. London: CCPI.

Egan, J. (1998) *Rethinking Construction*. London: Construction Taskforce: HMSO.

EMAP Glenigan Cost Information Services (2001) *Griffiths Complete Building Price Book*. Bournemouth: Glenigan Cost Information Services.

Harris, F. and McCaffer, R. (2013) *Modern Construction Management*, 7th edn. Oxford: Blackwell Science.

JCT11 Intermediate Building Contract (2011). London: Sweet & Maxwell.

JCT11 Standard Building Contract with Quantities (2011). London: Sweet & Maxwell.

Johnson, V.B. (2005) *Laxton's Building Price Book*. London: Butterworth & Heinemann.

Latham, M. (1994) *Constructing the Team*. London: HMSO.

Sher, W. (1996) *Computer-aided Estimating: A Guide to Good Practice* London: Longman.

Skoyles, E.R., Jr. (1987) *Waste Prevention on Site*. London: Mitchell.

Smith, R.C. (1999) *Estimating and Tendering for Building Work*. London: Longman.

Spon's Architects' and Builders' Price Book (1986) London: E & FN Spon.

Standard Method of Measurement of Building Works (1998) 7th edn. London: RICS.

The Aqua Group (2007) *Procurement, Tendering and Contract Administration*. Oxford: Blackwell Publishing.

14 Building strategy

14.1 Introduction

In capitalist society most business is obtained by tendering a price to carry out work or sell a product or service in competition with others, the lowest being the winner. An alternative approach is to auction off an item, starting with an initial figure and allowing competitors to outbid each other until only one is left paying the highest price. The flip side of this is the concept of bargaining when a price is offered and the purchaser sets about negotiating the proposed price downwards. In all of these cases the purchaser and/or the seller needs a bidding strategy for the process. In essence they need to establish a range from the lowest to the highest price that is acceptable to them. Bidding is the relationship between the levels of profit and overheads required, the accuracy of the estimate and the number of competitors.

It is necessary to develop a bidding strategy because of the uncertainty associated with tendering for work. This is made more difficult because of the low mark-ups added by the contractor, resulting in an increased risk of winning the contract and losing money at the same time. Ideally, the contractor wishes to produce the highest possible tender price that will obtain the work, but which will also cover overheads and contribute towards profits. To achieve this the contractor relies heavily on historical data provided from previous bids and analysis of the competitors for the same work.

The purposes of any bidding strategy are:

- to determine whether or not to tender for the work as there is no point in wasting time and effort pursuing an already lost cause;
- to increase the chances of winning the contract(s);
- to obtain an adequate return relevant to the risk taken;
- to maximise the profits;
- if bidding as a loss leader, to minimise expected losses;
- to minimise losses if the market place is very tight and too low a price has been submitted;
- to provide an even workload throughout the financial year.

Before looking at procedures of producing a strategy it should be noted that there is still considerable scepticism by much of the UK construction industry in using a mathematically based system in lieu of working from instincts and experience. Also, many of the models produced do not allow for these subjective judgements to be included which is a further concern. Indeed some would argue that the significant losses made by companies in recent years have not been as a result of bidding, but because of the errors made during the estimating process in misunderstanding the implications of the problems of the construction work. If the estimating quotation is not accurate, with margins being as low as they are, the proposition that strategic bidding is of value collapses.

So why should bidding models be utilised since the industry has apparently successfully developed without them? It is argued that it is not so much whether they should be used, but rather what are the principles behind them, which will clarify the thinking processes of those making the decisions and by doing so supplement and improve their subjective judgements. It would be imprudent to suggest that substituting a mathematical model should ignore the expertise and commercial acumen gained. Currently many of these models are found in the research arena of academia rather than in the boardroom. However, there will be a divergence as the models are seen to be relevant and the estimating process becomes more accurate or margins increase.

14.2 Comparisons with competitors

In 1956 Friedman originated one of the most accepted methods of comparing competitors. The process is relatively straightforward, but relies on having competed against the same contractor on several occasions. Each of the specific competitor's bids is expressed as a percentage compared against the estimated cost (not the price tendered by one's own company), i.e. net before overheads and profits are added. This information is then presented as a frequency distribution or a histogram form as shown in Figure 14.1. In this example competitor A has bid 27 times against the company. For the purposes of this demonstration the bids have been grouped in bands of 2.5 per cent. There could be situations where there is a negative showing on the histogram if the competitor's is lower than the company's.

Since the comparison of the competitor's bid is the total tender price compared with the net cost, it can be seen, for example, that if the company had added 5 per cent to its estimate, it would have lost three of the contracts (11.1 per cent) and won

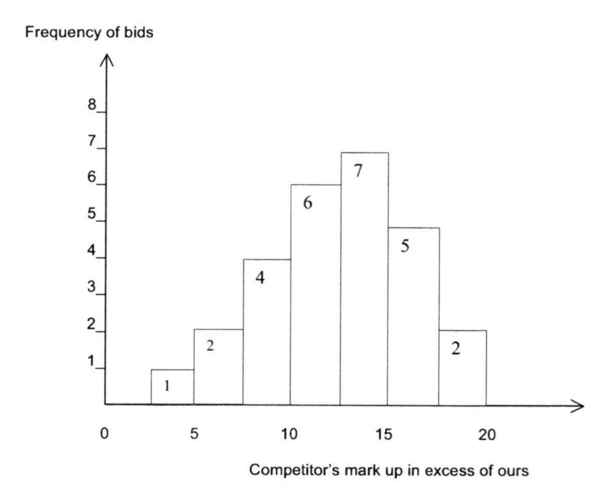

Figure 14.1 Frequency distribution of competitor A

24 (88.9 per cent), and if 10 per cent had been added then 13 (48.1 per cent) would have failed and 14 won (51.9 per cent). The probability of winning the work in both these two cases is therefore 0.89, and 0.52 respectively.

Using the data shown in Figure 14.1, a cumulative frequency curve, Figure 14.2, can be plotted showing on the vertical axis the probability of beating the competitor and on the horizontal axis the percentage mark-up. If zero mark-up is added then the probability the company will win the work is 100 per cent and if 20 per cent is added the probability of beating competitor A is zero. A word of caution though, this form of analysis relies on having competed enough times to have sufficient number of comparisons to be statistically valid.

The same process can be adopted for all competitors added together to assess where one fits into the market as a whole. However, various alternative models have been proposed since Friedman and there has been considerable debate as to which is the most accurate. Interested readers might read chapter 10 of Harris and McCaffer (2013).

14.3 The accuracy of the estimate

As mentioned before, much of the validity of this approach depends upon the accuracy of tenders because of low mark-ups. To apply such techniques requires the estimator to know their range of accuracy so that comparative information can be logically applied. Different estimators within the company and in competitive companies will all vary to a certain extent, but generally all of them aim to produce the real cost of executing the works.

As indicated in Chapter 13, the estimator has to make various assumptions when preparing the estimate, all of which have a margin of error associated. These include such issues as:

- that the all-in labour calculation reflects reality (Table 13.4), when variable assumptions were made about inclement weather, the bonus to be earned, overtime worked and levels of sickness;

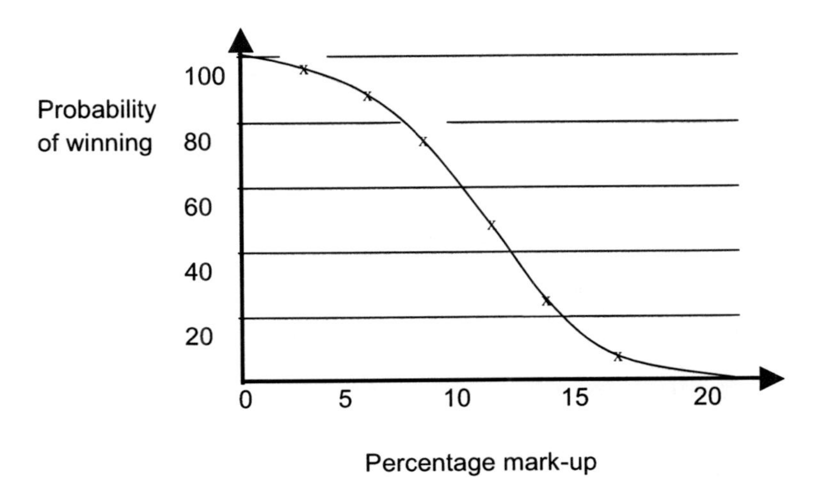

Figure 14.2 Probability of outbidding competitor A curve

- the selection of the productivity rate to be used in calculating the bills rate (see 13.10);
- the percentage waste allowed for materials (see 13.9);
- that the programme will be adhered to, which will have an impact on the preliminaries;
- inflation costs on materials, labour and plant.

The result of these assumptions is that the estimator will produce a likely cost estimate range of the true costs of say ± *N* per cent, the value of *N* depending upon the accuracy of the estimator. If the senior management assumes that the estimate cost is correct and adds a percentage to cover for overheads and profit, then one of two outcomes can be expected, depending upon which end of the range is the estimated cost. If at –*N* then the contract will probably be won making either a loss or minimum contribution to overheads and if +*N*, the tender price will be too high and the bid unsuccessful. Figure 14.3 illustrates this point. In the first case the estimator accuracy is ± 5 per cent. When a mark-up figure of 7.5 per cent is added, then if successful the contract will return from 2.5 per cent to 12.5 per cent if lucky. In the second case where the estimator accuracy is only ± 10 per cent, the return will range between –2.5 per cent and + 17.5 per cent.

Studies carried out using simulation programmes have indicated that if the estimator estimates more accurately, then the achieved profit margin will be increased, which confirms the belief that the more confident senior management is about the estimate, the more likely they are to pitch their mark-up at the right level, especially if they take into account knowledge of their position in the market and against the competitors for the project being tendered for.

More accuracy can only be obtained if estimators have accurate data at their disposal. This emphasises the need to keep up-to-date records and to have appropriate feedback from the site of the actual costs, causes of delays etc. This can only occur if the site has a good post-contract cost control system in place (see Chapter 16) during the construction phase and from feedback after the final account has been settled.

14.4 The number of competitors

What is clear is that the more competition, the less likely the odds are on being successful. It is therefore in the interest of the contractor to select contracts with a limited number of competitors. The reason for this is that research suggests that

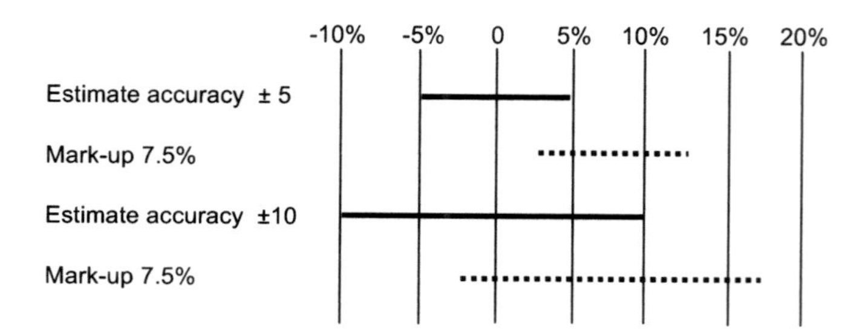

Figure 14.3 Effect of estimate accuracy on likely outcome of bid

compared to the mean bid of all the competitors, the lowest bid reduces as there are more bidders, perhaps due to stiffer competition.

14.5 Summary

There are clear steps that can be taken in developing a bidding strategy. First, establish the base data by:

- investigating all previous contract tendered for and tabulate the names of the competitors on each contract and their tender price including one's own company;
- producing a comparison of each major competitor's price against one's own;
- constructing a probability curve for each major competitor;
- constructing a probability curve for all major competitors;
- continually updating these after the result of every tender submission.

Second, for the specific bid:

- identify the competition;
- construct a graph superimposing all the competitors' probability curves and determine the probability of being successful based on this information;
- make a judgement as to what mark-up should be added to stand a good chance of being successful with the bid, or if this is insufficient to satisfy the company requirements either decline the tender or submit the tender with the mark-up needed to remain profitable and in business.

Finally, it should be noted that the market place is changing. Many contracts are now not awarded just on price, but rely upon negotiation when the contractor is selected on other issues, such as design, reputation, previous work for the client, meeting deadlines and so on.

Bibliography

Cook, A.E. (1991) *Construction Tendering: Theory and Practice*. London: Batsford.

Friedman, L.A. (1956) 'Competitive Bidding Strategy'. *Operations Research*, 4, 104–112.

Harris, F. and McCaffer, R. (2013) *Modern Construction Management*, 7th edn. Oxford: Blackwell Science.

Park, W.R. and Chapin, W.B. (1992) *Construction Bidding: Pricing for Profit*, 2nd edn. Chichester: Wiley.

15 Purchasing

15.1 Introduction

It is not intended to give a detailed analysis of the purchasing function as much has been considered in other chapters such as Chapter 13, but rather address some of the key issues affecting this role in an organisation. Purchasing is sometimes referred to as procurement, but this can be confusing in the construction industry; procurement is also applied to the process of selecting a contractor to carry out the work. It is also referred to as the buying function.

15.2 Ethics

It is important to put the subject in context and consider the ethical implications of purchasing. Morality and ethics affect every branch of the organisation, but the purchaser in a construction company has a high spending power and must pay particular attention to ethical concerns. There are two main considerations: exploitation of labour, especially in third-world countries, and environmental concerns. Both have serious implications for sustainability summarised in the Natural Step's fourth system condition 'that in society people are not subject to conditions that systematically undermine their capacity to meet their needs'.

Typical examples of the problems the purchaser should consider include:

Exploitation of third world labour

Labour is consistently paid low wages, which do not permit those working to have sufficient finance to live satisfactorily, or surplus money to either invest in their business, such as agriculture, or to purchase material goods. In the case of agriculture, if farmers are able to invest in their businesses by, for example, drilling for a water well to yield better and more crops, then they generate more finance that in turn allows them to purchase goods. Once labour is able to purchase goods, then somebody needs to manufacture them. The result is improved standards of living all round and a more sustainable and equitable society worldwide. This is a simplistic view on very complex issues, but was highlighted in 1980 in the Independent Commission on International Development Issues publication *North–South*. Only by raising standards of a given country can that county begin to make strides towards development. This is a very complex issue as, while it is naturally considered to be unacceptable by those in the developed world, if a society is poor it cannot afford education for its children, so what should the children do instead? This is why there are many countries that employ child labour. Only by raising the standards of that country can education be made more widely available to children. The greater the finance available, the higher the age to which education can be offered.

Exploitation of labour in developed countries

It is often forgotten that exploitation still happens in the developed world. The employment of illegal immigrants is an obvious example but many people, especially women, are exploited by being paid to carry out work in their home at rates of pay well below the national minimum wage usually paid on piece-work rates, i.e. based upon output rather than an hourly wage. This is difficult for a purchaser to identify, because it is hidden from view.

Exploitation of the environment in the manufacturer of materials

There are relatively strict regulations within the developed world concerned with environmental pollution during extraction and manufacture of materials and components. However, many of our materials are sourced from elsewhere in the world, where legislation either does not exist, or if in place, is less stringent. In effect this means Western countries can export their environmental problems. Typical examples are the obtaining of timber from non-renewable sources and the extreme amounts of uncontrolled pollution caused in the extraction of metals such as waste rock and slurries, not to mention the pollution of watercourses. This is all well documented as occurring, but the purchaser may have some difficulty in ensuring the material bought actually comes from a less polluting source. Timber from sustainable sources has over the years become easier to trace.

Unnecessary transportation

Linked to environmental issues is the unnecessary transport of materials which should be avoided if possible. As a simple example, a design decision specifying Skye marble extracted approximately 150 miles north of Glasgow was sent south by road to Manchester over 350 miles from the quarry to be fixed on the face of precast concrete panels then transported back north to Glasgow to a building being constructed there.

Unscrupulous suppliers

There are recorded examples of suppliers and those in the service industry who have sufficient financial muscle to lower their prices for a period long enough to put the competition out of business, enabling them to monopolise the market. Others have been accused of stealing ideas from others having the finance, then bringing the product to the market place.

Gifts and favours

This is a highly contentious issue, but interestingly what is acceptable varies from generation to generation, company to company and society to society. It is about peer review, i.e. what is acceptable within the culture and organisation. It is easy to take the moral high ground and state categorically that all gifts and favours are a bribe so none should be accepted, but it is argued here that this is a decision either the senior management makes on behalf of the company or one an individual takes. A late-morning meeting that overruns, where it is suggested to continue over lunch begs the question, who should pay? Christmas presents vary from a diary, calendars, to bottles or crates of wine or spirits. Firms have hospitality suites at important sporting or entertainment venues that are offered as 'a thank you' for the business received. In extreme cases an all-expenses paid holiday including members of the family can be offered. At what point do these become a bribe? It is a problem because many of the smaller gifts, especially those offered at Christmas, have become a tradition and generally accepted. The question left to the reader is what would you feel about the company if they had regularly given you a bottle of whisky and then for no reason it does not arrive this year? In the interest of openness, some companies will keep a register of gifts received.

Conflicts of interest

These are such situations where the purchaser is buying products from a relation, a company they own shares in or from close friends. The transactions may be honourable and honest and in the best interest of the company, but can raise suspicion from others. In these cases, it may be sensible to keep such suppliers, but the deals need to be very transparent and clearly vetted by senior members of the organisation, so that everybody can see that the deals are above board.

Tax evasion by supplier or subcontractor

In spite of continuing efforts by the UK Treasury, there is still a significant number of people, many of whom are employed in the construction industry, who are either evading the payment of taxes or who are also being paid social security benefits of one sort or another. This is often referred to as the black economy. It is only right and proper the company purchasing these services has systems in place that deny access to employment for these individuals. Equally, if a supplier is offering materials for cash, thereby avoiding the payment of VAT, this should also be resisted.

Unsustainable business practices

When prices are clearly well below the current market value, this will almost certainly be because of one or a combination of any of the above practices. Besides having an ethical dimension they could potentially fail to deliver, resulting in delays and higher expenditure.

15.3 Fair price

With these issues as a background to the purchasing process, obtaining goods and services is not just about obtaining the lowest price. While this is important, it is equally imperative that a fair price is sought. This is because without a fair price the supplier cannot run a sustainable business and will either go out of business or have to take short cuts to survive.

A fair price must permit the supplier to:

- not be forced to carry out any of the issues identified in section 15.2;
- have sufficient income to pay fair wages, which means that this improves the likelihood of maintaining a stable labour force;
- be able to invest in new plant and machinery and keep abreast with modern developments; this makes the organisation more competitive and enables the business to survive in the competitive market place;
- depending on the nature of the business, be able to invest in research and development to ensure their products keep up with technical development;
- invest in people within their organisation – this involves development and training so that personnel are up to date with appropriate legislation, new developments and management skills;
- use materials of the appropriate quality to the specification requested; and
- produce a safe working environment.

15.4 Procedures for procuring suppliers and subcontractors

This depends upon the size and scale of the procurement. If a subcontractor is employed to erect a steel frame, the contractual arrangements and documentation would be very different from obtaining a plumber to repair lead around the junction between a domestic chimney and a pitched roof.

However, the purchasing process will involve some or all of the following stages:

- identifying the requirements at the estimating stage;
- determining the specification requirements;
- determining the delivery requirements;
- obtaining quotations from various potential suppliers some of which may be preferred suppliers;
- obtaining outline design proposals;
- receiving and evaluating the quotations;
- on winning the contract, rechecking the specification, delivery needs and renegotiating the prices;
- issuing contractual documentation;
- tracking the supply to ensure the product or subcontractor arrives as required – this in the case of the former may mean making visits to the factory to assess progress;
- monitoring performance.

It is important to understand, especially in the case of suppliers of materials and components, how their process works. Failure to understand leaves one vulnerable to being misled or unable to sensibly negotiate both prices and variations to the order.

15.5 Preferred suppliers and subcontractors

It makes sense to engage suppliers and subcontractors who on past record and performance are reliable. Hence monitoring of performance is crucial. It is important to consider what percentage of the supplier's business the order is. Too high a percentage and problems can occur as they will lack the flexibility to redirect production from another client to you.

New blood has to be introduced from time to time, otherwise there is a danger of complacency and the possibility of missing new opportunities from fresh inputs into the process. This can be done from time to time, partially as an experiment, but not before the new organisation has been carefully scrutinised. This is carried out in a systematic manner, by interviewing and reviewing past performance with other organisations. In all cases it is important to set down performance standards against which the organisation can be compared and monitored. For example (although in some cases setting absolute standards would be difficult):

- delivery on time;
- quality of material/workmanship;
- response if things go wrong;
- general cooperation with the main contractor and other subcontractors;
- attention to safety issues and standards agreed;
- attitude to claims.

15.6 Supply chain management

This subject is explored in more depth in Chapter 24. For the purpose of these comments, the design stage of suppliers or services, such as mechanical and electrical, is ignored.

Supply chain management is a natural progression from having preferred suppliers and subcontractors. Its success relies on a relationship of trust between parties. A supplier needs to fully understand the requirements of the contractor and equally vice versa. In other words it becomes a partnership.

The smaller supplier may not have the wherewithal to provide the necessary controls for such issues as total quality management (see 25.3) the contractor needs, so the contractor has to assist them in this. They may also need financial investment to purchase the most up-to-date machine tools and other equipment. It will usually mean guaranteeing minimum quantities of orders each year so they can invest. The supplier/subcontractor may also need to amend their internal controls systems to be compatible with the main contractor.

Bibliography

Brandt, W. (1980) *North–South: A Programme for Survival.* Report by Independent Commission on International Development Issues. London: Pan Books.

Fewing, P. (2008) *Ethics for the Built Environment.* London: Routledge.

Mirsky, R. and Schaufelberger, J. (2014) *Professional Ethics for the Construction Industry.* London: Routledge.

Murray, M. and Dainty, A. (2008) *Corporate Social Responsibility in the Construction Industry.* London: Routledge.

16 Post-contract cost control

16.1 Introduction

For the purposes of this text, post-contract cost control is defined as the control of costs after the contract has been awarded to carry out the building work. The emphasis is to be on the cost control from the contractor's perspective, but the client, sub-contractors and suppliers are also looking to control their costs. Mention will be made from time to time about the client's cost-control mechanisms, as, while similar, the emphasis can be different especially since the client is usually looking for a much longer period for their returns.

16.2 Pareto's Law

Pareto's Law states that: ' In any series of elements to be controlled, a selected small fraction of items in terms of number of elements will always account for a large fraction in terms of effect.'

Conversely, the majority of items are only of minor significance in their overall effect. Simply put, 20 per cent of your decisions will affect 80 per cent of your business. This is similar in principle to 'management by exception'. A typical example to demonstrate this is approximately 80 per cent of all construction work is carried out by only 20 per cent of the companies, the remaining 80 per cent carrying out 20 per cent of the work. In other words it is unproductive to spend an equal amount of time controlling each element, in this case costs, but rather concentrate on controlling the key or major elements only. Figure 16.1 illustrates this point.

The significance of this in controlling construction costs is that there are potentially a large number of cost centres that can be created. The number of items in B of Q clearly demonstrate this. If the cost-control system developed and adopted becomes too detailed then the cost of controlling costs becomes disproportionate to the benefit derived. Therefore, rather than design a system of cost control that considers every item, it is better to concentrate on the key elements.

16.3 The basic needs of a cost-control system for the contractor and client

Any cost-control system used by the contractor must ensure payment for work done by having automatic triggers built in. It should be able to monitor costs as the work progresses at least monthly and indicate the amount of losses or gains being made,

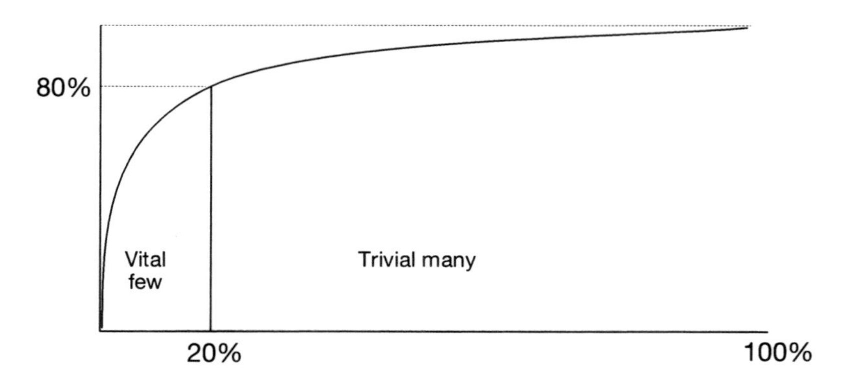

Figure 16.1 Pareto's Law

anticipating any cash flow problems ahead. There should be limitations in the system that prohibit any changes to the work in excess of a given amount without the authority of senior management. Similarly, certain personnel may have restrictions on the amount they can spend without recourse to a senior person. It should be transparent and understood by all those whom it affects, whether in-house, suppliers or subcontractors.

The client needs a system that does not permit any alteration to the tender cost of the contract without authorisation by either the client or representatives. The client or representatives should agree any changes made that affect the tender sum above a certain limit before the instruction is given to the contractor. Any likely overspend from the budget agreed should be highlighted as soon as possible to enable the employer to either have time to finance the excesses or rein back expenditure to the initial budget. The system should be such that the client can monitor, on a monthly basis, the current financial state of the project. It should be readily understood by the client, representatives and the contractor.

16.4 Cash flow

Cash flow is the transfer of money into and out of a company. As indicated previously, the developer needs to know what financial demands are going to be made during all stages of the development. Equally, contractors need to know about their predicted cash flow so as be able to cover any deficit in funding. Remember that the contractor aims to cut this to a minimum as the margins on a contract are small and any interest paid on money borrowed will eat into this.

In practice, on a traditional contract using the B of Q the timescale is as follows. To save time the contractor's QS will produce an interim valuation of the work that has been carried out by the contractor during the month. A meeting then takes place with the client's Private Quantity Surveyor (PQS) on site at the start of the following month and this valuation will be agreed. It normally takes one week for the PQS to submit this valuation to the architect. Usually the architect takes two weeks before issuing a certificate and, depending upon the clause in the contract, the employer is obliged to pay the contractor the agreed amount within 14 days. This can sometimes be 28 days. Generally, then, from completion of the month's work the contractor is paid within four to five weeks.

It used to be that when the contractor received this money, the subcontractors would then be paid. However, this practice on major works has largely died out, as the subcontractor is entitled under their contract with the main contractor to receive their payment after a fixed period of time, often 28days. In practice this means that by and large the main contractor receives payment at the same time as it is necessary to pay the subcontractor. It is important that this latter payment is made, as the subcontractor has to pay the staff and operatives carrying out the work, many of whom will be paid hourly, which means they are paid the week following the one they have worked. Withholding this payment could land them in a financially difficult position.

The tightness of the timescale also reinforces the previous discussion on the importance of the decision on whether or not to tender, by the main contractor, in establishing the financial strength and reliability of the client's payment record.

16.5 S-curve

An S-curve is a graph plotting the cumulative income against time. The timescale is normally in monthly increments to coincide with the way contractors are paid by the client normally in contracting.

The S-curve in Figure 16.2 demonstrates what happens when the client pays the contractor. At the end of each month, the contractor submits an interim valuation for the sum of work carried out. However, these monthly valuations will not be paid for another 4 to 5 weeks, shown by the horizontal line on the steps. The vertical line on the steps represents the interim payment from the client. If the payment is made within the month then it can be seen that the main contractor will not have a cash-flow problem However, if the contractor front loads the tender by a small amount then, as can be seen in Figure 16.3, the contractor receives the money from the client

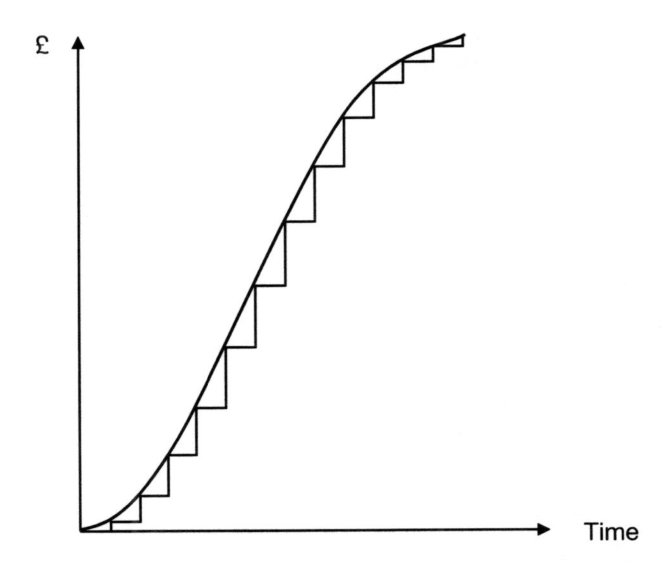

Figure 16.2 S-curve

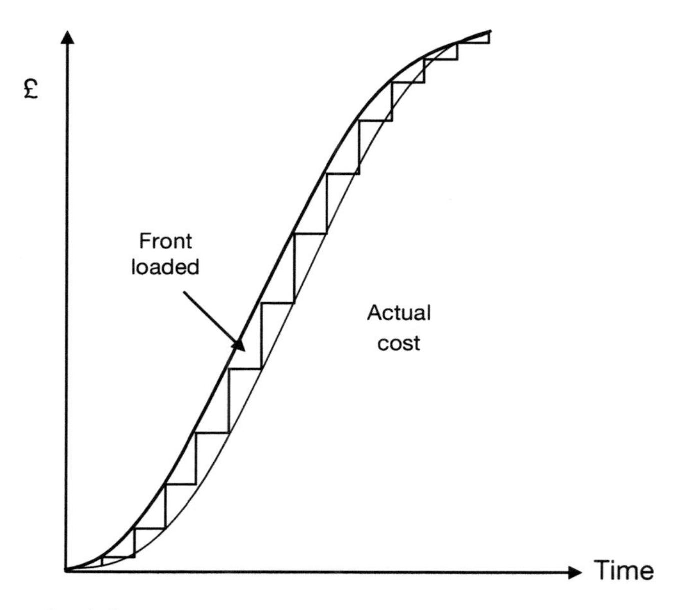

Figure 16.3 Front-loaded S-curve

well before it is necessary to pay the subcontractors. The cash-flow deficit probability is then limited to the back end of the contract.

16.6 Further refinements

To carry out a full and detailed calculation, there are many more factors that have to be taken into consideration. Adapted from Harris and McCaffer (2001), Table 16.1 shows these factors and indicates how the calculations are derived. These would be carried out at monthly intervals throughout the duration of the contract in a similar way to that used in calculating discount cash flow (see 6.5.3).

16.7 Monitoring using the S-curve

Simply by monitoring the actual performance with the predetermined budget, it can be seen what is happening and the results. The graph below on the left of Figure 16.4 represents the budget forecast of expenditure and the work actually carried out at a point in time on the project. This graph can be produced for all the project costs or sections of it such as materials. In this case the trend is positive. The graph on the right represents the budget forecast as before, but in this case the actual value of work carried out is less. Any continuing trend in this direction will result in increased financial problems for the company.

16.8 Saw-tooth diagram

Another means of demonstrating the cash flow is by using a saw-tooth diagram. In many ways this is a better way of demonstrating the position so that it can be seen

Table 16.1 Factors affecting cash flow calculations

	Description	*Comments*
1	Cumulative value	This is the value of work to be carried out each month, usually based upon the activities from the construction programme
2	Cumulative value less retention	As 1 above less the percentage retention specified in the contract
3	Cumulative payment received after certification	This will be the same figure as in 2 above, but delayed by a month plus
4	Cumulative retention payment	This will be paid usually 12 months after the handover of the building
5	Cumulative cost	This is the value (column) less the profits and overheads on the assumption this is the same for every item of work done
6	Cumulative direct labour costs	These payments are in effect made as and when the work is done Note comments in 9 below.
7	Cumulative materials costs	Suppliers are usually paid 28 days after delivery, so this cost goes in the following month after receipt of goods
8	Cumulative plant costs	As 7 above
9	Cumulative subcontractor costs	Normally 28 days after work is completed paid on a monthly basis, so shown in the following month to when work was valued Note that in certain cases of labour-only subcontracts this can be as 6 above
10	Cumulative cash out 6 + 7 + 8 + 9	Direct labour + Materials + Plant + Subcontractors
11	Cumulative cash flow 3 + 4 – 10	This line becomes the data to demonstrate the cash flow situation month by month
12	Fixed preliminaries	This is the amount paid up front by the client
13	Term related preliminaries	The monthly outgoings in running the contract
14	Contribution to overheads	This is the difference between 1 and 2 above

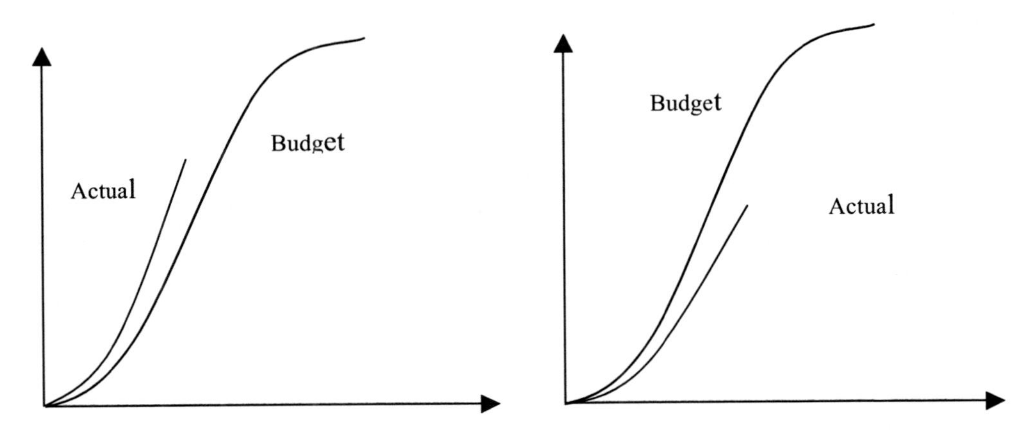

Figure 16.4 Monitoring trends

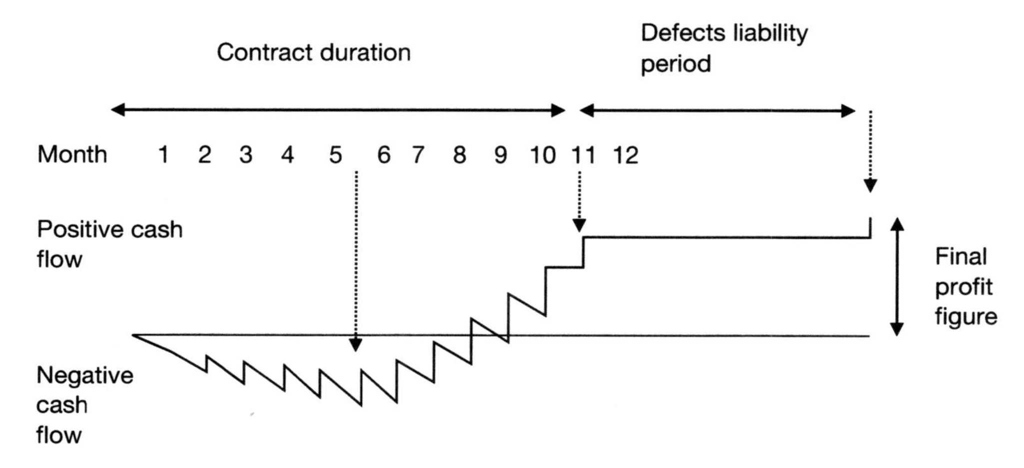

Figure 16.5 Saw-tooth diagram

what financing is required month by month. The cash flow is shown in a linear form, plotting negative and positive cash flows against time.

At the start of the contract as shown in Figure 16.5, the interim payment for the first month's work is received at the end of the second month (approximately) as shown by the vertical line and so on throughout the duration of the project. For simplicity, it is assumed the expenditure is constant throughout each month, although in practice this may not be the case.

In this example, the lowest point of negative cash flow occurs at month six. This is the maximum cash requirement the contractor has to provide to fund the project and it would be important to know at the tender stage. If during the course of the contract the position worsened because of delays, then this would be continually looked at to see what the implications are.

At month 12, the defects liability period commences, at the end of which the retention monies are paid. The final profit released is shown at the end of the defects liability period. It is possible to superimpose the actual position in terms of valuation or cost on this diagram.

16.9 Bad practices

There are a number of questionable practices that have arisen over the last few decades that assist the contractor and subcontractor to maintain a good cash flow. Since Latham (1994), and subsequently Egan (1998), there has been a trend and pressure to eliminate such practices and be more open with the client and those involved in the design and construction processes. This trend must be continued for the health of the industry, especially with the development of supply chain management and partnering initiatives.

Practices to be avoided include:

- front loading: not changing the total tender price, but increasing the items constructed early and reducing those at the end to balance the overall price;

- back loading: at times of high inflation, loading the latter items to be constructed at the expense of the earlier ones;
- over measurement: for example agreeing that a floor on a multi-storey reinforced concrete building is seven-eighths complete rather than the three-quarters in reality;
- delaying payments to subcontractors and suppliers until receiving payment from the client – this may incur a loss of discount for not paying within the contracted period for payment, but may still be profitable;
- entering into dispute with the subcontractor over variations to delay payment;
- disputing the account to delay payment.

The client's representatives can employ similar practices, such as:

- entering into dispute with the contractor over variations for work done;
- disputing the account sent in;
- delaying passing information to the architect for the issuing of interim certificates;
- delaying issuing the interim certificates.

The ombudsman in 2016 stated that a well-known company in the UK seriously breached the industry's code of conduct, by disputing accounts and delaying payments to its suppliers.

Bibliography

Egan, J. (1998) *Rethinking Construction*. London: Construction Taskforce: HMSO.

Harris, F. and McCaffer, R. (2001) *Modern Construction Management*. 5th edn. Oxford: Blackwell Science.

Heinze, K. (1996) Cost Management of Capital Projects. New York: Marcel Dekker.

Latham, M. (1994) *Constructing the Team*. London: HMSO.

Potts, K. (1995) *Major Construction Works: Contractual and Financial Management*. London: Longman.

17 Interim valuations, claims and variations

17.1 Introduction to interim valuations

Interim valuations are for work completed during each month of the contract and are a standard feature of construction forms of contract. What is written into the terms of the contract will determine who is responsible for measuring the work done, but for most building contracts it will be the Quantity Surveyor (QS) who carries out this role. It is normally the responsibility of the client's QS to measure and value the work, but to expedite the situation it would be unusual if the contractor's QS was not involved. Indeed, it would be normal for the contractor's QS to have already measured the work and then just agree it, with some auditing of course, by the client's QS.

Not only do these valuations act as the basis for the contractor and supply chain to receive money for work done, they also act as a major source of data for any monthly cost-control system in place, as this is the actual money received and not a theoretical sum.

The valuation for the interim certificate is the gross agreed valuation of work less any amount deducted for retention and materials on and off the site.

17.2 Sums subject to retention

The amount of retention will have been agreed between the parties as part of their contract, and is usually 5 to 10 per cent. If the client has a strong belief the contractor will complete the works on time and to the quality specified with minimal remedial work required after the completed building has been handed over, then the percentage will be in the lower range. The sums normally included in this calculation are:

- the total value of the work properly executed by the contractor including any agreed variations;
- the total value of the materials and goods delivered to site for incorporation into the building, providing they are protected against weather and other damage, and that they are not prematurely delivered – the contractor cannot obtain all the materials for the whole contract, have them delivered and expect to be paid for them, so there must be a phased delivery unless there are good reasons why this cannot happen (see section 17.4);
- the total value of goods off the site, but that are to be included in the building. These are known as 'the listed items' (see section 17.5);
- the total amount due to nominated subcontractors which is also subject to a retention;

- the profit that the contractor makes on the work executed by nominated subcontractors;
- fluctuations if the contract allows for this.

17.3 Sums not subject to retention

While the majority of items have the retention percentage applied to them, there are some payments the contractor has to make that are exempt from this. These are generally costs the contractor has to make that would be unreasonable for them to bear and can include:

- any fees or charges made to local authorities or statutory bodies;
- where the contractor has been instructed by the architect to open up for inspection any work covered up, or to test any material or goods;
- if instructed by the architect, any costs incurred in infringing any copyright or patents;
- if instructed by the architect, any added insurance costs;
- any justifiable costs not covered for elsewhere in the contract;
- any costs in respect of any restoration, replacement or repair of loss or damage and including the removal and disposal of debris;
- final payment due to the nominated subcontractor after their agreed defects liability period;
- any increase of costs due to fluctuations on national insurance contributions, levies or taxes;
- monies payable to nominated subcontractors that are not subject to retention.

17.4 Materials on site

These are all materials delivered and stored on the site that the contractor intends, within the foreseeable future, to become part of the finished building. Theoretically when paid for they become the property of the client although there are sometimes conflicts if the supplier's conditions of sale stipulate that they remain the property of the supplier until such time as the contractor pays for them.

The materials once paid for must not be taken off site without written permission; this also applies to domestic subcontractors once materials are paid for. When this payment occurs, a transfer of title to the client from both contractor and subcontractor is made, providing this can occur legally, depending upon ownership at the time.

17.5 Off-site materials

These are usually components, plant and equipment such as air-conditioners, boilers etc., that have been made and are awaiting installation. It could also include other materials such as bricks and blocks especially if specials have been made that are specific to the contract. The conditions to be met include the following:

- The item(s) have to be listed by the client and supplied to the contractor in the annex to the contract bills.
- The contractor has to demonstrate that the components etc. are for the building.

17.8.1 Site establishment costs

These are to do with the time-related costs such as the accommodation, servicing of the same in terms of labour for cleaning and electricity, heating, insurances etc., and any plant that is standing idle as the result of a delay. It also includes site supervisory staff and non-productive labour, including attendance to subcontractors. The claim will normally be based upon hire invoices and bills for the accommodation, plant, and for the heating, insurance etc. The staff and labour costs include the total cost to the contractor of any benefits such as pensions, holidays and health insurance.

17.8.2 Head office overheads

This is a difficult item to value and then to agree with the client who could well argue that those employed in the head office can in part be employed doing other activities in lieu of supporting the contract.

The simplest way, known as the 'Hudson formula', is to take the percentage added in the bills, divide by the contract period and multiply by the length of the delay. The weakness in this approach is that it relates to the tender, which is based on value, rather than actual costs. Equally, it assumes that the overheads for the contract are constant throughout its running period. There are a variety of other ways of doing the calculation, some more popular than others.

17.8.3 Visiting head office staff

This is the additional time spent by staff visiting the site. It can be argued that this is taken care of in the general head office overheads and this is valid. However, the contra argument is that the staff still have to visit and their workload may increase as a result of all the claims being made by the subcontractors as a result of the delay. It can cause conflict between the contractor and the client and is probably best, on balance, not followed through.

17.8.4 Uneconomical working

Any disruption on site causes problems that can be exhibited in a variety of ways. If the disruption is severe, the morale of the workforce takes a dive with resulting loss of productivity, and possible industrial relations problems. However, it is difficult to quantify the cost of the disruption. It is not just the loss of production that costs. If extra work of modest proportions is added to the contract it may mean bringing back a piece of plant, operatives having to go through the learning curve again and extra costs being involved in administering the cost of procurement etc. It would under these circumstances be unfair to use the current bills item for this purpose.

17.8.5 Uneconomical procurement

This will not happen very often, but there are circumstances that arise where it might be applicable, especially where large deliveries are involved. The conditions of sale of the supplier may be such that if delays over a certain length of time accrue, storage costs could be charged. Alternatively, it may be necessary for the deliveries to be continued, resulting in storage problems on site and double handling with possible damage.

17.8.6 Loss of profit

Since profit is usually a percentage based on the cost of the works, it could be argued that the contractor will receive the profit as contracted for in any case. The other argument is that delays restrict the ability to earn more profit on potential opportunities. This is a difficult one to quantify and tactically it may be better to ignore it unless the disruption and consequent delay is of significant magnitude to have an impact on the business as a whole.

14.8.7 Acceleration

In spite of the delay, the client may still require the project to complete by the original date. A good example of this is in retail. If the project was due to be handed over in time for the retailer to capitalise on the Christmas period, the costs of the delay may well outstrip paying the contractor extra to accelerate production to maintain this date. The client offering financial inducements normally achieves this, or by the contractor being paid for the extra costs incurred in meeting the client's revised requirements, although there are some legal risks in this. Alternatively, the client can persuade the architect to deny extensions or the contractor will take responsibility for meeting the original deadline, without passing on extra costs, as a marketing tool for future work and enhancing their reputation.

17.8.8 Third party settlements

In today's market place, when a project is delayed because so much of the work is subcontracted there is an immediate impact on them. This means that the financial liabilities traditionally borne by the contractor are passed down to the subcontractors. The problem arises in that the negotiation between the contractor and the client over extra costs can take a considerable amount of time. In the meantime the subcontractors want their money and may well be financially distressed if settlement is not made in a reasonable time. In these cases the contractor may decide to settle with the subcontractor before agreeing terms with the client.

17.8.9 Inflation

If the contract includes for fluctuations then the comments in this section do not apply, as inflation will automatically be taken into account in the sums. However, if the contract is a non-fluctuating cost project, and the client causes delays, in principle, the contractor is entitled to be reimbursed for increased costs due to inflation. The calculation would be based upon the start times of the activities as compared with the original programme. Again, whether or not the contractor wishes to invoke this depends upon the extent of the delay.

17.8.10 Financing other cost centres

When borrowing rates from the banks and financial institutions are high, then the costs of financing delays reflect this. If the delays occur at a time when the contract is at a stage when the contractor is running at a planned negative cash flow (section 16.4) then the impact is even greater. However, even with a positive cash flow it is still costing the contractor money, as any shortfall of profit cannot be re-invested at this now higher interest rate.

17.8.11 Financing retentions

The contractor when tendering for a contract produces a cost flow analysis indicating the shortfalls in cash flow. A cash flow for the whole company operation will also be produced which might, over the period of the financial year, eliminate the need to finance any negative cash flow as the profits from other contracts cover this deficit. The contractor will have made arrangements accordingly. If a contract is delayed there is a similar delay in the payment of the retention monies, which may either cause negative flow, and hence the need for financing, or lower the positive cash flow, reducing the amount of potential interest earned from investment. This would in both cases be a legitimate claim.

17.8.12 Interest for non-payment

When payments from the client are paid late the contractor has a right to simple interest.

17.8.13 Cost of producing a claim

Clearly to put together a claim costs money in terms of the man-hours needed to collect the information and make the case. In practice, claims of this nature are made for breach of contract against the client. However, as before it is a matter for management to decide what the long-term implications are to the relationships with the client and the architect.

17.9 Introduction to variations

Variations to the works tendered for are almost certain to occur for some reason or other. Ideally, a building would be designed so that this would not happen. This would remove the flexibility, useful to a client, especially since there may be a long gap from the time of the feasibility study to the day the contractor arrives on site. In the meantime the client's requirements can change.

Typical reasons for variations are:

- unexpected ground conditions resulting in extra excavation or foundation design changes;
- a change of requirement by the client;
- a design error – considering the complexity of a modern building it is almost inevitable that errors will occur.

The calculation for variations is beset with potential points of disagreement between the client, the contractor and subcontractors. It can involve considerably large sums of money. The more thorough the company's cost-control documentation, the easier is it to provide evidence to demonstrate the validity of the claim.

The JCT11 contracts clearly lay down a series of conditions that contribute to a variation and lay down the rules for claiming variations. The reader should refer to these documents for further details. If the claim can be measured using a bill item and the variation agreed, there will be no dispute. However, if the claim is of a higher magnitude, how the calculation is considered is an issue. It should be transparent and honest.

17.10 Example calculations

An item in the bills is to excavate a basement 3.0 metres deep over an area of 25m × 40m. The calculation to establish the bills rate could be as follows:

Stage 1: calculate the cost of excavation

Table 17.1 Excavation costs

Bills Quantity 3000m³ *Labour*	*Rate*	*Plant*	*Rate*	*Total*
		Excavator	£30/hr (inc. driver)	£30
Ganger	£8.8/hr			£8.8
2 No.	£8.3/hr			£16.6
			Total cost	£55.4
Output rate from excavator is 20m³/hr				
Cost of excavation per m³ is £55.4/20				£2.77

Stage 2: calculate the cost of removing the excavated soils and cart away to landfill site

The 3000m³ has to be carted away using 8m³ capacity lorries. It takes 24 minutes to load the lorries and the round trip journey to the landfill site is 30 minutes including tipping time. This will require three lorries for the following reasons. Ideally the excavator needs to work continuously to keep the production on the site moving forward. Table 17.2 demonstrates the cycles of excavator and the wagons required to make this happen. It means that each wagon will have to wait for 12 minutes on return from the landfill site. This is not a major problem as it would be necessary in practice to build in some slack to account for variable traffic, delays at the tip and natural breaks for the drivers. Increasing or decreasing the capacity of the excavator can also adjust the balance between the excavator and the wagons, but it must be remembered that the size of the excavator will have been chosen to suit the programme requirements. On the other hand, when value engineering the project, an increase in the excavation duration may be one of the ways of achieving a cost reduction, in which case the balance between excavator and lorries can be reconsidered.

The cost of the transportation per metre cube is therefore based on the decision to utilise three wagons, which is: 3 lorries costing £60 in total remove 20m³ in one hour, therefore the cost per m³ is £60/20 = £3.00.

Table 17.2 Ratio of lorries to excavators

		24mins	24mins	24mins	24mins	24mins	24mins
Excavator	$20m^3$/hr						
Lorry 1	$8m^3$ capacity	Load	Transport		Load	Transport	
Lorry 2	$8m^3$ capacity		Load	Transport		Load	Transport
Lorry 3	$8m^3$capacity			Load	Transport		Load

Stage 3: is to calculate the bill item

Table 17.3 Bills item calculation

Description	*Quantity*	*Cost/unit measure*	*Cost*
Excavation	$3000m^3$	£2.77	£8,310
Transport to tip	$3000m^3$	£3.00	£9,000
		Total cost	£17,310
Site overheads	10%		£1,731
Profit and overheads	8%		£1,385
		Gross cost	£20,426
		Cost/m^3 (divide by 3000)	£6.80

Now assume that the excavation depth has to be increased by 2 metres as a result of discovering a layer of peat covering a major proportion of the site area, but unfortunately missed by the site investigation. The calculation may now change because the excavator previously selected cannot excavate to this depth. In this case the calculation will be as follows:

Stage 1: calculate the cost of excavation

Table 17.4 Excavation costs

Revised quantity 5000m³ *Labour*	*Rate*	*Plant*	*Rate*	*Total*
		Excavator	£40/hr (inc. driver)	£40
Ganger	£8.8/hr			£8.8
2 No.	£8.3/hr			£16.6
			Total cost	£65.4
Output rate from Excavator is $32m^3$/hr				
Cost of excavation per $m^3$is £65.4/32 =			£2.04	

Note that the cost of excavation in this case is cheaper than in the bills by £0.73/m^3.

Stage 2: calculate the cost of removing the excavated soils and cart away to landfill site.

The same size lorries are employed taking the same time to transport and tip the soil, but the volume has increased to 5000m³.

Table 17.5 Ratio of lorries to excavator

		15mins	*15mins*	*15mins*	*15mins*	*15mins*	*15mins*
Excavator	$32m^3/hr$						
Lorry 1	$8m^3$ capacity	Load	Transport		Load	Transport	
Lorry 2	$8m^3$ capacity		Load	Transport		Load	Transport
Lorry 3	$8m^3$ capacity			Load	Transport		Load

In this example the transportation works out that still only three wagons are required, but the waiting time has been eliminated on return. If this number of lorries is adopted, there could be hold ups in the excavation from time to time due to the possible delays to the lorries indicated before, so it would become a management decision to decide whether to have an extra lorry or not. The cost of the transportation per metre cube using the three lorries: £60 in total to remove 32m³ in one hour is £60/32 equating to £1.875.

Stage 3: is to calculate the revised rate

Table 17.6 Revised bills rate calculation

Description	*Quantity*	*Cost/unit measure*	*Cost*
Excavation	$5000m^3$	£2.04	£10,200
Transport to tip	$5000m^3$	£1.875	£9,375
		Total cost	£19,575
Site overheads	10%		£1,957
Profit and overheads	8%		£1,565
		Gross cost	£2,3097
		Cost/m³ (divide by 5000)	£4.62

Note that the rate in this case is less than that in the bills – that means that pro rata the client is paying less per m³ although in real terms is paying the difference between that in the bills of £20,426 and the revised price of £23,097.

Had the travel time been say 48 minutes, then the situation would have been different again as in the first instance the lorries usage would have been 100 per cent and in the latter, less, as shown in Tables 17.7 and 17.8.

Table 17.7 Ratio of lorries to 20m³ excavator

		24mins	*24mins*	*24mins*	*24mins*	*24mins*	*24mins*
Excavator	$20m^3$/hr						
Lorry 1	$8m^3$capacity	Load	Transport		Load	Transport	
Lorry 2	$8m^3$capacity		Load	Transport		Load	Transport
Lorry 3	$8m^3$ capacity			Load	Transport		Load

Table 17.8 Ratio of lorries to 32m³ excavator

		15 mins	*15 mins*	*15 mins*	*15 mins*	*15 mins*	*15 mins*
Excavator	$32m^3$/hr						
Lorry 1	$8m^3$ capacity	Load	Transport				Load
Lorry 2	$8m^3$ capacity		Load	Transport			
Lorry 3	$8m^3$ capacity			Load	Transport		
Lorry 4	$8m^3$ capacity				Load	Transport	
Lorry 5	$8m^3$ capacity					Load	Transport

For the $3000m^3$ excavation the comparative costs would be as before £6.80. In the case of the $5000m^3$ excavation, the cost of the excavation is the same at £2.04/m^3, but the cost of the transportation has increased to 5 lorries costing £100 in total to remove $32m^3$ in one hour making a cost per m^3 of £100/32 equating to £3.13.

Table 17.9 demonstrated the revised calculation and it can be seen that while more expensive than shown in Table 17.6 it is still cheaper per m^3 than the original bills item. What is demonstrated in these two examples is that as the parameters vary, there will be an effect on the cost of the variation. Another solution would be to reduce the number of wagons on the $5000m^3$ excavation and let the excavator wait the three minutes to allow the wagons to return.

Mention was made earlier in this section to using this information for value engineering, and altering the size of the excavator can reduce the cost of the operation.

Table 17.9 Revised cost for 32m³ excavator

Description	*Quantity*	*cost/unit measure*	*Cost*
Excavation	$5000m^3$	£2.04	£10,200
Transport to tip	$5000m^3$	£3.13	£15,650
		Total cost	£25,850
Site overheads	10%		£2,585
Profit and overheads	8%		£2,068
		Gross cost	£30,503
		Cost/m^3 (divide by 5000)	£6.10

From the examples above it can be seen that in the case of the shorter haulage distance and the longer one, the use of the larger excavator reduces the unit cost of excavation and carting to tip rate. If the excavation remained at the initial volume ($3000m^3$) it would be cheaper to use the larger excavator. The effect also in this case would be to reduce the time for the excavation from 3000/20 = 150 hours to 3000/32 = 94 hours, i.e. from four weeks to just over 2.5 weeks.

Bibliography

Egan, J. (1998) *Rethinking Construction*. London: Construction Taskforce: HMSO.
Latham, M. (1994) *Constructing the Team*. London: HMSO.
JCT11 Standard Form of Building Contract with Quantities (2011). London: Sweet & Maxwell.
JCT11 Design and Build Contract (2011). London: Sweet & Maxwell.
Ndekugri, I. and Rycroft, M. (2000) *The JCT98 Building Contract: Law and Administration*. London: Arnold.

18 Cost systems

18.1 Introduction

The purpose of a cost-control system is to ensure that costs are controlled so that the contract is completed within budget and lessons are learned through appropriate feedback. It is important to note that this means investigating not just the causes of loss-making activities, but also those that produce a high rate of return – the latter because in today's market environment the price of similar works in the future might be reduced, making the tender more competitive. This chapter looks at contractor cost-control systems rather than that of the client.

Traditionally the industry used to record costs and compare with the budget (see 18.2). The problem with this approach is that it provides historical information when it may be too late to do anything about it on the current contract, although it does provide feedback after analysis for the future. However, where there is repetitive work the information can be analysed and used for the rest of the contract. For example, it was established that the foundation slab of the first of four 16-storey blocks of flats had lost a considerable amount of money compared to the budget. On analysis it was concluded that the main cause was the method adopted. By this time, the second block's foundations were too far advanced to change, but a revised method was adopted for blocks three and four.

Ideally what is needed is a-cost control system that is cost forecasting rather than cost recording.

18.2 Budget cost

The budget cost can be derived from a variety of sources depending upon the form of contract and accompanying documentation. For example, with a traditional B of Q, the cost of each item is derived from the build-up of cost elements of plant, materials and labour, so budgets can easily be produced for either a unit measure, for example metre run of skirting, to the cost of a larger element such as a complete storey of reinforced concrete including walls, columns, beams and floors.

Another way of establishing budgets is to cost the activities on the programme then assess at the time of the cost analysis what percentage of the activity has been completed. The advantage of this approach is that there are programme-management packages that will produce costs based upon the programme constructed.

At the moment with the industry using a high percentage of subcontractors, the budgets are based upon the subcontractor's price quoted for the contract; the work done by them can be calculated as a percentage of their work completed and the materials consumed in the process.

18.3 Actual cost

For directly employed labour this can be quite a complicated process depending upon the system in use. The following considerations need to be taken into account:

- employees being paid each week for the hours they have worked and a bonus based on their productivity;
- the bonus usually paid one week later than their wage as it takes a week to calculate;
- the wage also including guaranteed hours and bonus due to say inclement weather;
- employer's contribution to national insurance;
- the contribution to pensions in some cases;
- craft employees paid a different rate to general operatives and skill rates differing;
- wages that may include travelling cost;
- plus rates in lieu of bonus payments;
- holiday pay.

The budget labour figure from the B of Q includes for other items itemised in 13.8 used to build up the tender rate and also takes account of the summer and winter working hours to provide a wage rate to be used in the estimate.

Reconciliation between these two sets of information and figures is not as easy as it seems. This means that either the actual cost has to be adjusted starting with the wage sheet information, deducting the bonus paid and adding the previous wage sheets bonus, and then adding a figure that accurately reflects the other cost items not shown on the wage sheet, or the budget figure is amended by deducting all the cost items that do not appear on the wage sheet. In the latter case, these costs have to be reflected somewhere else in the overall balance of costs.

It is probably simpler to adopt the former approach. It is necessary to calculate the total number of hours both the general operatives and the craftsmen are paid, rather than the number hours they have worked, shown on the left hand side of Table 18.1, and calculate the total wage each type of labour has received in total as shown on the right hand side.

By dividing the total of the wages columns by the total of the hours column an hourly rate can be calculated for both general operatives and craftsmen.

When dealing with subcontractors actual costs, the matter is of course much simpler as one is paying for the work that has been completed and this is as per their quotation to the contractor. The monies that contribute to the contractor's account would include any attendance allowance and or profit and overheads associated with their work.

18.4 Direct labour costing systems

18.4.1 Historical

These systems are based on the accountancy principle of recording everything in detail and producing accounts later that demonstrate what has happened in the past. It is

Table 18.1 Weekly labour costs

General operatives (GO)			
Breakdown of hours		*Breakdown of wages*	
Normal working hours		Flat rate wages	
Overtime hours worked		Plus rates	
Non-productive overtime		Bonus (previous week)	
Travelling time		Tool money	
Maintenance time		Expenses	
Inclement weather time		Lodging allowance	
Guaranteed make up time		National insurance contributions	
		Holiday pay	
		Pension contributions	
Total	GO hours		GO wages
Craftsmen (CM)			
	ditto	TM hours	TM wages

GO + CM Hours should equate to the total hours on wage sheet
GO + CM wages should equate to the total monies paid out taking account of the bonus adjustment and any costs not actually paid directly to the operatives.

a very slow and laborious process. It used to be assumed that one cost surveyor was needed to calculate the bonus and costs for one hundred operatives and the operation would take a full working week. The reason for discussing it here is that with the advances made in information technology, possible developments in software and the use of pointers able to measure dimensions from computer-aided drawings, there is no reason why this approach should not be resurrected, providing the opportunity for some manual intervention is permitted. If this was developed then the operation could be completed within one or two days and management would have much more up-to-date information to make decisions upon. At the moment though, it is far too costly to implement.

The first stage is to collect information about the work that has been carried out by each person or gang of operatives and how long they have taken to carry out the work including any non-productive time. This is provided by the supervisor daily or weekly on a time sheet. This information is then abstracted and entered on an allocation sheet (Tables 18.2 and 18.4).

On the left hand side of the page as shown in Table 18.2 the details of the operatives and how much time they have spent working on the contract during the week is entered. It can also be used for the bonus calculation. The times allocated against the gang are those the operatives have actually worked, based on the system used by the contractor recording entry to and leaving the site such as clock cards. In practice these may be different from the time sheets provided by, or on behalf of the gang (Table 18.3), where it is not unusual for the hours to be rounded up to the nearest half and ignoring the fact that the operatives may have been quartered. Quartering is when the operative has clocked in more than 3 minutes past each quarter. The operative will be paid as if having commenced at the start of the next quarter period.

Table 18.2 Actual hours worked

Allocation sheet *Week Commencing:*								
Operative	*Works No.*	*M*	*T*	*W*	*Th*	*F*	*Total*	*Bonus*
Jones	1345	9	9	8.75	9	8	43.75	£75.00
Smith	1243	8.5	9	8.75	8.75	8	43.00	£74.00
							86.75	£149.00

Bonus calculation
Hours earned = 124
Actual hours = 86.75
Hours gained =37.25 @ £4/hr = £149

£149/86.75 = £1.72 per hour

A typical daily sheet for Monday is shown in Table 18.3. In this example the two second fix joiners have allocated a total of 18 hours, yet according to the times booked into and out of the site as shown in Table 18.2 they have only been recorded for a total of 17.5 hours.

The right hand side of the allocation sheet, shown in Table 18.4, is used to allocate the work and is taken from the time sheets. This has to take into account the actual hours worked rather than those submitted on the time sheets. It is usual practice to make these minor adjustments in either the main non-productive items or, if none are identified, in those productive items with the largest amounts of time allocated. This adjustment is made on 'await instructions' shown in italics in the Monday column of Table 18.4.

The total hours for the gang when added up should be reconciled with the total hours worked by the gang as shown on Table 18.4, i.e. 86.75 hours.

Table 18.3 Daily time sheet

Daily worksheet	*Names*	*Works No.*	*Trade*
Date:	Jones Smith	1345 1243	Joiner Joiner
Operation	*Quantity*	*Hours worked*	
Productive			
Fix doors	10 No.	7 hours	
Fix skirting	22m run	4 hours	
Fix architrave	16m run	3 hours	
Non-productive			
Await instructions		2 hours	
Replace broken door		2 hours	
	Total	18 hours	

Table 18.4 Work allocation

Hours *Description*	*Quantity*	*Rate*	*Earned*	*M*	*T*	*W*	*Th*	*F*	*Total*
Productive work									
Fix doors	26	1.5hr/No.	39	7	5	3	5.5	3	23.5
Fix skirtings	93	3m/hr	31	4	2	5	4	4.5	19.5
Fix architraves	120	3m/hr	40	3	7	5	5	6	26
Fix window cills	6	2hr/No.	12		2.5	3	3.25		8.75
Fix notice board	1	2hr/No.	2					2	2
Non-productive									
Await instructions			0	***1.5***				0.5	2
Replace broken door		0	2					2	
Clean up			0		1.5				1.5
Re-fix skirting			0			1.5			1.5
			124	17.5	18	17.5	17.75	16	*86.75*

The quantity and descriptions of the work done, usually separated into productive and non-productive work, is then entered. The column 'Rate' is the bonus rate agreed between the company and the operatives and from this the bonus hours earned for each of the described activities can be calculated, giving a total, in this example, of 124 hours. The bonus calculation is relatively simple as is shown in Table 18.2 where the hours worked are deducted from the hours earned and this is then multiplied by whatever rate has been agreed between the parties. In this case £4 using a 50 per cent scheme (see 29.12.2) sect giving a total bonus for the gang of £149 that is pro-rated for each member of the gang depending upon the hours worked. In this calculation, the non-productive items are given no earnings at all. This can be a cause of dispute, since it can be logically argued that, if for example in the case of the 1.5 hours 'awaiting instructions', it is management that is preventing the operatives from earning bonus they should be at the very least given the time (1.5 hours) as a contribution to the bonus or be paid a bonus pro-rata to the bonus paid.

The quantity and hours spent fixing, in this case skirting, are then transferred from all the appropriate allocation sheets to a cost ledger sheet as shown in Table 18.5 so that a weekly total of this item can be ascertained, as well as a running total for the contract. The weekly rate is as calculated from Table 18.1, and thus the actual cost (weekly rate × hours) can be deduced. The budget cost comes from the bills and with this the loss or gain for the week along with a cumulative position can be calculated for the contract for this particular task, that of fixing skirting.

This information can then be transferred to a weekly account for all the second fix joinery as shown in Table 18.6 and to a weekly summary for all the work carried out on the contract that week, Table 18.7, and finally to a monthly statement Table 18.8.

The weekly summary is used by the site management as a control tool, the monthly being sent to the regional or head office to advise on how the site is progressing and along with the same information from all the other sites, the overall financial position of work in progress.

Table 18.5 Cost ledger fix skirting

Cost ledger *Second fix joinery* *Sub set*								
Fix skirting		Labour Budget rate £X per metre (derived from the bill)						
Date	*Quantity*	*Hours*	*Weekly rate*	*Actual cost*	*Budget cost*	*Loss*	*Gain*	*Cumu-lative*
Wk com.	metre		£	£	£	£	£	£

Table 18.6 Weekly summary for second fix joinery

Cost ledger – Weekly account *Second fix joinery*				
Description	*Actual cost* £	*Budget cost* £	*Gain* £	*Loss* £
Doors				
Skirtings				
Architraves				
Notice boards				
Cills				
Window ironmongery				
Door ironmongery				
etc.				
		Total		
		Profit/Loss		

Table 18.7 Weekly summary for the contract

Cost ledger – Weekly summary					
	Description	*Actual cost* £	*Budget cost* £	*Gain* £	*Loss* £
Superstructure					
Cladding					
First fix joinery					
Second fix joinery					
Partitions					
etc.					
		Total			
		Profit/Loss			

Table 18.8 Monthly summary for contract

Cost ledger – Monthly summary				
Description	*Actual cost* £	*Budget* £	*Gain* £	*Loss* £
Superstructure				
Cladding				
First fix joinery				
Second fix joinery				
Partitions				
etc.				
		Total		
		Profit/Loss		

18.4.2 Forecasting/predicting

These are systems that rapidly anticipate future problems, especially on matters concerning cash flow. Figure 18.1 shows a network diagram of a small project which converts into a bar line as shown in Figure 18.2. How to produce a programme can be seen in Chapter 20.3. The number in the top left of each circle is the earliest time an activity can finish, in the right the latest it can start and still complete the programme on time and the number in the bottom is used to identify an activity. The programme management package used to produce these programmes needs this, so for example activity A is defined as 10–30, B as 10–20 and so on.

The bar line programme can be produced to demonstrate the budget income expected in terms of labour, materials and plant (if so devised) for each activity. This can be further refined by, on the assumption that the earning is constant over the duration of the activity, a cumulative total cost for all activities and an appropriate S-curve drawn.

Using the programme, Table 18.9 demonstrates the breakdown of the costs for each activity into labour, materials and plant. The costs can either include or not the retention percentage. It is based on the network in Figure 18.1.

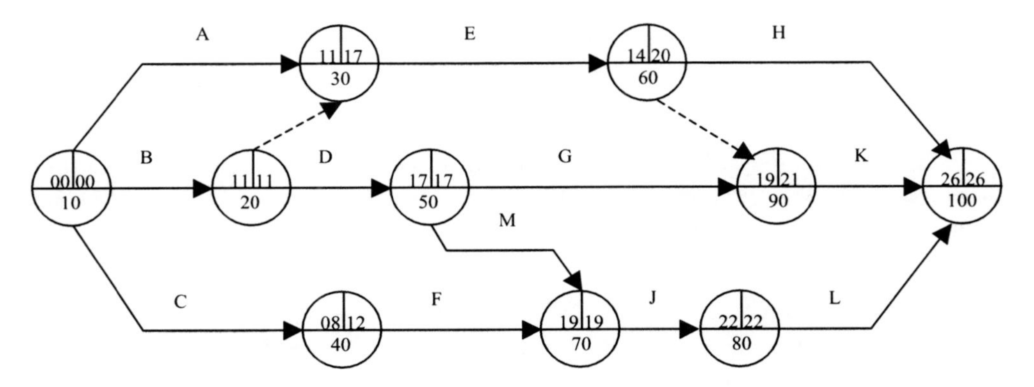

Figure 18.1 Network

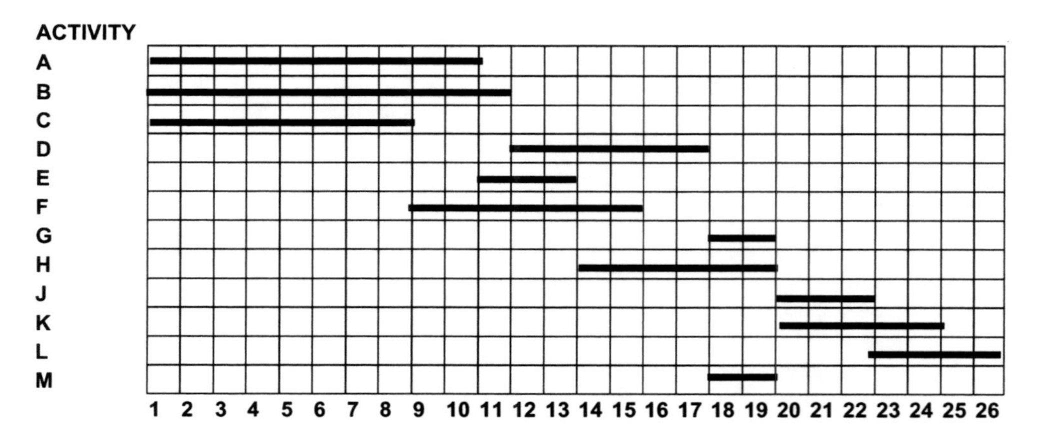

Figure 18.2 Bar line chart

Table 18.9 Breakdown of costs per activity

Activity Description	*Activity Comp No.*	*Duration*	*Predecessor*	*Gang size*	*Labour cost (£K)*	*Material cost (£K)*	*Plant cost (£K)*	*Total cost (£K)*
A	10–30	10	Start	4	40	37	8	85
B	10–20	11	Start	3	31	30	2	63
C	10–40	8	Start	5	37	30	0	67
D	20–50	6	10–20	6	36	25	0	61
	20–30	dummy	10–20	0	0	0	0	0
E	30–60	3	10–30, 20–30	3	8	9	1	18
F	40–70	7	10–40	5	33	38	7	78
G	50–90	2	20–50	6	11	10	0	21
	60–90	dummy	30–60	0	0	0	0	0
H	60–100	6	30–60	3	15	15	0	30
J	70–80	3	50–70, 40–70	5	15	18	4	37
K	90–100	5	60–90, 50–90	6	28	25	6	59
L	80–100	4	70–80	4	16	17	0	33
M	50–70	2	20–50	3	5	4	0	9

Taking the labour cost and assuming it is used constantly through each activity, Table 18.10 demonstrates the amount of money the contract would generate during each of the first 15 weeks to pay the labour bill if the programme is adhered to. The sums are rounded to the first decimal place.

18.4.3 Package costing

Now that so much work is let out in packages a further method of controlling costs has evolved that is used particularly in management contacting and construction management contracts where the management team has a clear responsibility for

Table 18.10 Breakdown on weekly labour budget costs

Act.	*£K*	*£K*	*£K*	*£K*	*£K*	*£K*	*£K*	*£K*	*£K*	*£K*	*£K*	*£K*	*£K*	*£K*	*£K*
A	4.0	4.0	4.0	4.0	4.0	4.0	4.0	4.0	4.0	4.0					
B	2.8	2.8	2.8	2.8	2.8	2.8	2.8	2.8	2.8	2.8	2.8				
C	4.6	4.6	4.6	4.6	4.6	4.6	4.6	4.6							
F												6.0	6.0	6.0	6.0
E											2.7	2.7	2.7		
F									4.7	4.7	4.7	4.7	4.7	4.7	4.7
G															
H														2.5	2.5
J															
K															
L															
M															
Total	*11.4*	*11.4*	*11.4*	*11.4*	*11.4*	*11.4*	*11.4*	*11.4*	*11.5*	*11.5*	*10.3*	*13.4*	*13.4*	*13.2*	*13.2*
Weeks	1	2	3	4	5	6	7	8	9	10	11	12	13	14	15

ensuring the building is built to budget by controlling both design cost implications and construction costs.

The client's quantity surveyor produces a cost plan of the design that is modified and agreed by all parties. This then acts as the cost-control document. The management contractor then allocates these costs to the various activities to be provided by suppliers and subcontractors. These are referred to as packages. When the managing contractor goes out to obtain quotations, hopefully the tender prices equate to those in the cost plan. If not, then negotiation will take place to try to narrow or close the gap. If the price in the plan cannot be met, then there are various options:

- the deficit can be balanced against other packages that may come in under budget;
- there can be a modification to the specification to bring it in line;
- a new method can be value engineered.

As the project progresses, any increases in the cost of a package caused by variations or claims are immediately noted and the effect analysed so that the three options above can be considered. As all this information is put on a spreadsheet, the cost impact can be seen immediately so that action can be taken.

Bibliography

Griffith, A. and Watson, P. (2004) *Construction Management*. Basingstoke: Palgrave Macmillan.

Harris, F. and McCaffer, R. (2013) *Modern Construction Management*, 7th edn. Oxford: Blackwell.

Pilcher, R. (1992) *Principles of Construction Management*, 3rd edn. New York: McGraw-Hill.

Part IV

Operations management

19 Site organisation

19.1 Introduction

It is often said that if you get the blinding level correct at the onset of the construction of the building, then the rest of the building will be built correctly. So it is with site organisation. Get the basics in place and the contract stands a much better chance of being managed well. The problem is that often, given the time available, especially in traditional types of procurement, from the contract being awarded to the contractor being on site there is only a limited time to think this process through properly. It is important that due attention is given to this during the estimating process. In other forms of procurement where the contractor or project manager is brought in during the design process there is further opportunity to get it right as part of the continuous thinking process.

The prime areas of concern are those of:

- site boundaries
- access roads, on and off the site
- provision of services
- accommodation for contracting staff, subcontractors and client's representatives
- material storage and handling
- waste disposal
- site logistics
- location of fixed plant and selection of heavy equipment
- hoardings
- communications
- security
- managing the neighbours
- environmental protection and pollution control.

In all of the above, the most important document is the site plan, showing existing services, the site boundary and the footprint of the building. Other information needed includes elevation drawings, floor plans and other drawings showing the way the building is to be constructed, quantities of bulk materials, component schedules, client's requirements such as staff accommodation and phased hand-overs, restrictions to work, maps showing roads to and around the site, and the programme for the contract. The majority, if not all, should be available in the preliminaries section of the contract documents (see 12.4,12.5 and 13.12). Finally a site visit is a must to obtain a real understanding of the site and surrounding area, road congestion and so on.

19.2 Site boundaries

Where the site boundaries are and the position of existing buildings relative to the site of operations need to be established. Any encroachment onto a third party's land can lead to dispute and extra costs. Since the ruling on *Woollerton & Wilson v Richard Costain* (1970), tower cranes over-flying neighbours properties can be considered an act of trespass unless the deeds permit it. If not, then permission has to be sought. It is not unusual for the owner of the affected land to request financial compensation in return. To avoid this or the delays and cost incurred in challenging the claim, alternative methods of construction might have to be considered, or repositioning the crane or selecting a different type, such as one with a luffing jib.

The condition of buildings in close proximity to the works needs to be inspected and recorded before work commences, as contractors need to protect themselves against spurious claims and only accept any redress sought for further deterioration caused by the construction works. Equally, steps need to be taken to ensure that trees close to the border are not damaged, especially if they have a tree preservation order placed upon them.

19.3 Access roads

Properly positioned and appropriately constructed access roads are essential to the successful running of the contract as they are an integral part of the production process. They are the artery for the flow of materials during the construction process and if they fail the consequences to the site programme can be significant. They should be considered as early as possible in the process and certainly at the tender stage as there can be considerable financial implications. For example, when building a five-mile stretch of motorway, do you put a temporary road the full length to ensure access throughout its length or do you use existing roadways to access the various structures being built such as bridges and other parts of the site? On being awarded the contract the issue should be revisited, as the information upon which decisions were made at the tender stage may have to be amended as more detail becomes available.

Access roads are not just confined to the site, but also involve those to be used in bringing materials and people to the site. When constructing the Rakewood Viaduct on the M62, large heavy steel plate girders had to be taken along a narrow winding road to access the site. In places this road was not strong enough to cope with these heavy vehicles and had to be strengthened. If any out-sized long, high, wide or heavy component has to be brought to the site a route has to be determined. The roads around the site may have restrictions in terms of parking and off-loading, which can have both cost and production implications. Depending upon the site activity, it may be necessary to provide wheel-washing facilities to prevent vehicles leaving the site dirtying the public highway. The public, especially those on foot, need to be protected from being either splashed or getting their clothes and footwear dirty when walking past the site.

Consideration should be given to the needs of the personnel working on the contract. It is useful to establish what public transport is available and, for those travelling by car, the parking facilities provided in the local area and on site. If car-parking areas are to be provided as part of the finished project, consideration should be given to constructing them earlier rather than later. Some sites in urban location provide secure bike docking bays for local workers. These can also be used for site accommodation and clean materials storage if the area is large enough.

Other construction work in the close vicinity can have an effect on access to the site. For example, the football stadium in Coventry was awarded to Laing O'Rourke, but the adjoining road works to Nuttalls. Besides delaying normal users of the road, diversions put in place could impact on deliveries to the stadium unless consultation between the two companies took place.

The police need to be consulted to discover whether others have made any arrangements during the period of construction that would result in access being difficult or impossible. For example, marathon runs, festivals, parades, demonstrations and marches.

19.3.1 Design and location of temporary roads

There are several considerations to be taken into account when planning, designing and locating access roads on site:

- One-way flow is preferred to two-way, because the road can be made narrower if passing places at key unloading points are provided. All vehicles enter at the same place and can be checked to see that the load is as stated, before being directed to the correct part of the site.
- It was always argued there should be an exit and entrance to the site so if either became blocked the direction of flow could be reversed. However, in the current security-conscious atmosphere having only one gate into and out of the site improves security.
- The entrance to the site needs to be positioned so as to minimise the interruption to the general flow of traffic on the main highways.
- The route should be as short and direct as possible as this is cheaper to provide. Factors affecting this are the likelihood of the road being dug up to permit the positioning of underground services or any overhead obstruction such as a temporary electrical or telephone supply. The phasing of construction can also influence the decision, for example if certain areas of the site have to be handed over completely before the rest of the contract is finished.
- They should be designed so that water drains naturally through the thickness of the material or by being laid to falls permits rainwater to run off. The latter could in extreme circumstances mean the provision of some form of drainage. In any case excess water must be removed or it can combine with soil deposited by vehicular traffic making the road impassable. It should be noted with the exception of 'tipper' lorries used for disposing of excavation spoil, the majority of vehicles coming onto site are articulated and they have more difficulty in moving over muddy roads so temporary roads have to be kept relatively clean.
- The design must be sufficient to support the point loads caused from the trailers of articulated vehicles, if left on site without the prime mover (the lorry part). On large contracts, concrete may be mixed on site, in which case the hard standings on which the aggregates and sand are stored should be of sufficient strength to support the high point load caused when the vehicle tips its load.
- If the roads to be constructed for the contract are conveniently positioned to coincide with the construction programme, the road foundation could be used instead. Obvious examples of this are on low-rise residential estates.

- From a safety point of view it may be necessary to provide temporary lighting particularly during autumn and winter.
- If tower cranes are being utilised, the temporary road should run within the lifting radius of the crane (19.9) so that components can be lifted directly into the building or into storage areas.
- Storage areas need to be located adjacent to the access road so goods can be unloaded and stored safely.
- If the site is on either side of a public road and it is necessary to move plant across, it may be necessary to provide traffic control either manually or with automatic traffic lights.

Access to the site is not always by road. On rare occasions there is a rail line connected to the main network running into the site such as at the naval base Marchport at Portsmouth where the navy loaded much of the task force ships on their way to the Falklands. After the war it was extended, enabling the contractor to bring in certain bulk materials by rail. Containers can also be used to deliver materials over longer distances provided the supplier and the site are within approximately an hour of their respective Freightliner depots. The majority of the structure and cladding of Gartnavel General District Hospital in Glasgow was made from precast concrete. It was manufactured on the north side of Manchester and on average 40 tons was shipped by this means overnight to arrive on site at commencement of work the following morning. This was approximately 8 per cent cheaper than using roads.

In Hong Kong barges brought in the aggregate and sand used for some of the large structures constructed on the waterfront. In one case, because of the restricted site space, the mixer set up was also constructed on a moored barge and the mixed concrete brought ashore on conveyors and distributed with concrete pumps.

19.3.2 Materials of construction

There are a variety of materials that can be considered for the construction of temporary roads. Much depends upon the frequency of traffic on the road, the type of soil on the site, the climate and the availability of materials. In extremely hot climates, the natural soil may be perfectly adequate for moving vehicles. There is a problem only when there is torrential rain that may occasionally occur, or more predictably, such as during the monsoon season. Providing construction only takes place during the dry months, this may be a perfectly adequate solution. Often when it is raining the precipitation is so great construction ceases in any case.

Concrete, sometimes reinforced, while expensive, is ideal where the passage of vehicles is excessive. This would be used where the distance is small such as access from the public road when the footprint of the building covers most, if not all, of the site as well as around a mixer set up. Hardcore and quarry bottoms (material from the quarry of variable size and shape down to dust that cannot be used for structural purposes) are commonly used for temporary roads.

In areas where there is significant demolition of brick buildings, broken brick can be used. This is a diminishing market in the UK due to the lack of housing replacement, compared with the slum clearance programme of the 1960s, but also because bricks, especially facing bricks, are being recycled.

Geo-textiles are also used to strengthen the road base, providing the edges can be restrained. Fabric such as Terram have the properties of allowing the passage of surface water through it, but prohibits the vertical movement of the soil beneath. Hardcore is then placed on top to complete the roadway.

In severe soft and weak ground conditions, large slabs of expanded polystyrene have been used as a means of cushioning the load, but this would be an exceptional solution used on only very rare occasions. Timber railway sleepers were used a lot in the 1960s as a result of the cuts made to the rail network after the Beeching Report (1961), but today sleepers are usually made from pre-stressed concrete that is unsuitable for access roads. However, there are still parts of the world where they are readily available. Finally, in emergencies the army use metal roadways. These are in roll-form housed on the back of the vehicle. The leading edge comes over the cab and is laid by driving the lorry forward over it, thereby unrolling the metal sheeting.

19.4 Provision of services

Usually the construction process requires water, electricity, and telephone and sewerage services. Other services such as gas and cable services for television may be required for the completed building(s). The services needed for construction can come either from existing services running close or adjacent to the site, be brought in by the statutory authorities, or are provided by the contractor. Those requiring excavation during installation will cause some disruption, but if this is planned for it should be minimal, being early in the contract programme.

During construction water is required for the temporary offices, general use and for the wet construction processes such as the production of mortar, plaster and concrete. In the case of the latter if mixed on the site, the quantities involved can be considerable and if the rate of supply to the site is inadequate, it may be necessary to have storage facilities on site to guarantee the volume and speed of flow required. On sites where there is no provision, it will be necessary to bring water in tankers or bowsers and store it on site.

Waste water and sewerage disposal can be accomplished by connecting to the existing mains or the provision of portable lavatory and washing facilities. On a large site the volume to be disposed of is considerable, so access to the mains is highly desirable and may well determine the positioning of this facility, providing adequate fall can be achieved and the existing system is capable of coping with the extra load. De-watering systems can generate very large amounts of water.

Electricity is required for the offices, to power plant and equipment, and hand tools. For large fixed machines such as tower cranes, this can require a 400-volt three-phase supply that has to be brought in specially by the statutory authority. Generators will have to be provided where local power sources are not available. It should be noted that because of safety, the supply used on the site should be 110 volts single-phase only. For site buildings and fixed lighting a 230-volt single-phase can be used (see IEE Regulations BS 7671).

The sophistication of communication methods is changing rapidly and covers a wide range of options from the provision of land telephone lines, mobile phones and broadband. Precisely what is required depends upon the size of the contract, the types of management systems in use, whether or not these are site-based or linked to the head office, and the number of personnel working in the site offices.

19.5 Accommodation for contracting staff, subcontractors and clients representatives

There are two prime requirements to be satisfied: first, to satisfy the minimum construction regulation requirements; and second, to be able to function efficiently as an organisation in managing the contract. The Construction (Health, Safety and Welfare) Regulations 1996 Regulation 22 cites the welfare facilities to be provided on a construction site. It includes regulations on the provision of sanitary conveniences, washing facilities, drinking water, accommodation for clothing, facilities for changing and the facilities for rest, to be used for meals, boiling water and a place to go in the event of inclement weather causing a cessation of work.

The location on the site of the accommodation may be limited by the amount of space available and alternative solutions have to be considered. Typical points are:

- Ideally the operatives' accommodation for eating and changing should be as close to the workplace as possible to minimise the loss of productivity due to travelling between the two. For example if it takes 5 minutes to get to and from each, at the beginning and end of the day, two tea breaks and lunch, the total loss would be 40 minutes.
- On restricted sites, contractors can look for accommodation close to the contract. Although there will be no view of the site, all the services are connected, the space within may be greater than can be provided on site and the costs may be less than providing on-site accommodation.
- Many of the hired units can be stacked which saves on site space. Providing permission is granted, accommodation can be constructed over the public footpath, but steps must be taken to ensure the safety of the public. The main issue being the supports holding up the temporary building.
- Due to security issues, it is normal to have a security gate and office, for the signing in and out of visitors, at the entrance to the site. If the contractor is employing site operatives directly, there are various means for 'clocking' operatives' attendance such as electronic finger print scanners and this facility is usually in the near vicinity to also oversee potential theft scenarios.
- The cost of installation is a significant proportion of the overall cost of providing accommodation so once erected it should not have to be moved unless there is no other alternative. Sometimes the site is so restricted that this has to occur. Site staff could use, for example, an underground car park that would usually have minimal services, leaving a relatively free area from continuing construction work.
- Often there are size and shape limitations to the on-site office areas, but bearing in mind that the team may be together for some time, it is worth while taking account of the needs of the various functions and the frequency at which communications occur between each. Clearly, the higher the frequency, the closer they should be to each other.
- Accommodation may have to be provided for the client's staff such as resident engineers, clerks of works and, on large prestigious contracts, entertainment and public relation facilities.

Each function also has different needs. Without going into all the functions listed, but to give a flavour:

- Planners require a lot of wall space to post their programmes.
- Site managers spend much of their time meeting others so can either have a discrete office for themselves with an adjoining conference/meeting room or combine the two areas.
- Site engineers require lots of desk space to spread out several drawings at a time and need to access the site quickly and not bring mud on their boots through the whole of the accommodation complex.
- Site managers sometimes prefer to have a view of the site, but this is not of high priority in carrying out their function and is very much a personal decision.
- For major projects, a large meeting room for health and safety inductions and progress meetings equipped with a digital projector or large interactive touch-screen board is required.

The size and type of desk, chairs and filing capacity have to be looked at. This may become a determining factor in the amount of space the office user requires.

19.6 Material storage and handling

There is a conflict between current thinking of delivering materials 'just-in-time' (see 23.6) and the traditional way contractors have been paid for materials delivered to site within the period of the monthly valuation. The trend is towards the former, but it is unlikely the delivery of all materials will achieve this objective and materials will still need to be stored on site either at the place of work or in a designated storage area. Even when 'just-in-time' delivery is reached there will still be occasions when goods will have to be stored, be it temporarily, because of inclement weather such as wind causing tower cranes to stop work, and the occasional machine breakdown.

There are advantages and disadvantages of the traditional approach. If the material is there, material control is easier and it is readily available for use. However, the longer it is on site the more likely it is to be damaged or deteriorate, there is an increased risk of theft, and it may have to be double handled because it is in the wrong place or in a centralised secure holding point, with further transportation and the probability of more damage, resulting in increased costs. There is also a tendency to lose control of the stock especially because more tends to be taken than is actually needed.

An external area may be required to use for mock-up wall/floor/roof construction and/or finishes.

19.6.1 Methods of storage

This depends upon the value of the material, its vulnerability to damage and weathering, and where and when it is required in the construction process. Ideally materials should be delivered on to the site when needed and placed at the side of or by the operative who is fashioning or fixing it. There are many considerations concerned with material storage and handling. There is an important relationship between the supplier and the contractor in understanding and agreeing the way materials are handled at the factory and how best handled at site. For example bricks are packaged in lots of approximately 400 held together using metal or plastic strapping, stacked in such a way that there are two horizontal parallel holes set at centres that permit the packs to be lifted and transported using a fork lift truck. Cranes attached to the lorry and tower cranes have similar lifting devices that permit the loads to be off-loaded from the vehicle without the need for breaking down the loads.

A completed building comprises a very wide range of different materials and components all requiring specific attention to ensure they are not wasted. To give a flavour, Table 19.1 shows suggested categorisation of materials to demonstrate different handling and storage issues.

Issues concerned with storage and handling include the following, all with the aim to reduce waste:

- When unloading certain materials it is imperative to have the proper equipment to ensure safe and secure lifting, such as using lifting beams so that vertical lifting can be assured without structurally damaging the component.
- When transporting materials appropriate transportation methods and plant should be used to ensure no damage occurs.
- Storage areas must be clean and level enough to permit proper storage.
- Appropriately designed support structures, racking and spacers (dunnage) should be provided.
- All goods sensitive to damage from different climatic conditions should be protected by covers or housed in a secure cabin.
- Certain goods such as cement have a limited shelf life and need to be stored in such a way as to permit the earliest delivered material to be used first. Other goods left in store for a long time can change colour or start to look dirty. For example the edges and part of the face of vertically stacked cladding components are exposed to the elements, as the units stacked in front do not fully cover the face. Those at the rear of the stack can become discoloured with pollution. When they are fixed into the building they look different as a result. Sometimes a good thunderstorm will cleanse the panels when fixed in the building.
- Certain materials are delivered in bulk and adequate and appropriate storage has to be provided. This may mean silos in the case of cement or constructing bays to segregate the different aggregates to stop cross pollution.
- Certain materials such as diesel oil are potential environmental polluters if not stored properly. Steps must be taken to ensure that in the event of a leak, the

Table 19.1 Materials categories

	Examples
Valuable (small items)	Door and window furniture
Consumables	Nails, tie wire, brushes, nuts and bolts, fixings
Short shelf life	Cement
Medium shelf life	Paint, untreated timber, reinforcement steel
Bulk	Bricks, blocks, aggregates, sand, structural steel, dry mortar
Environmental hazards	Fuel, oils
Easily damaged	Plasterboard, polystyrene
Components	Windows, doors, cladding panels

material can be contained without risk of pollution of the ground and water-courses.

- Consideration should also be given to combating theft and vandalism.

19.6.2 Location of storage areas

Valuable items, small consumables and materials with a short shelf life need to be securely stored so that they can be issued when required. The size and amount of storage provision has to be calculated taking account of the delivery schedule. The location of the storage hut needs to be close to the workplace to reduce the travelling time of operatives collecting the material and positioned by the access road to facilitate deliveries.

Where tower cranes are in use, all significant materials within the lifting capacity of the crane should be stored within the radius of the crane in clearly designated clean and properly equipped compounds, made secure if necessary. These do not have to be adjacent to the access road, but the nearer they are, the less time it takes for the crane to off-load from the delivery vehicles. There is a strong case for providing a central secure storage area from which goods are distributed in amounts needed for up to three days production. This helps to control waste.

19.7 Waste disposal

The subject of waste management is covered in Chapter 22. There will always be waste because of the very nature of the work carried out on a construction site. Previously, waste was either buried on the site or taken away in skips and disposed of in landfill sites. In recent years as a result of increasing environmental awareness, legislation and the costs of tipping waste, attention has been drawn to minimising waste and recycling. To carry out the latter it is generally accepted that the most appropriate place to segregate waste is on the construction site.

The various elements to consider when preparing for waste management on site include the following:

- How much and what type of waste is to be planned for?
- Where does the segregation take place?
- How are materials collected and transported to the segregation point? This can involve the provision of chutes to take materials from the upper storeys of a building.
- How are the materials stored on site before disposal?
- What can be recycled and by whom? Does the company have a list of preferred organisations that will collect such waste?
- Are the waste quantity targets being met?

19.8 Site logistics

The quantities of materials delivered to a construction site are immense and should be planned for. Once the contract programme has been produced, the quantity and frequency of deliveries can be assessed. This then has to be equated against the amount of space available for storage, the number of vehicles the site can accommodate at any one time and the restrictions on parking on the public roads adjacent to the site. From this information programmes can be produced scheduling the timing of the deliveries, and suppliers and subcontractors advised accordingly. In the case of frequent and regular deliveries of structural elements or cladding components it may be necessary to locate a holding yard for vehicles to smooth out variances due to traffic conditions such as rush hours. The same might apply for deliveries coming long distances.

19.9 The location and selection of fixed plant and excavating machinery

While the selection of construction plant is part of the estimating process, it is also an important issue when setting up the site as it takes up space, and excavating machinery has to be stored at the end of the day where it is immobilised and for maintenance, which on large civil engineering projects can take up a considerable area and services: hence its positioning in this chapter. It would take up too much space to consider all plant so this discourse has been limited to the selection of the most commonly used plant.

Spending time on deciding upon the correct positioning of fixed plant on a site is well worth doing as the costs of relocating can be prohibitive. The main fixed plant on a site are the concrete mixers, concrete pumps, hoists and tower cranes. In all of these cases an electricity supply will be needed.

With the development of ready-mixed concrete it is rare to see a mixer set up anywhere other than on large civil engineering contracts and only then when the costs of bringing in large quantities of concrete some distance make it more economical to mix on site. The positioning of such set ups depends upon the type of contract. For example, a motorway contract will probably wish to have it sited near to middle distance providing there are good public access roads, whereas on tall structures where a tower crane is used, it will be more appropriate to have the set up within the radius of the crane's jib. Generally, the mixers need to be positioned as close as possible to the main uses of concrete on the contract to reduce transportation time. However, the closer it is to the site entrance the less likely the delivery wagons will get stuck on the access roads.

The use of static concrete pumps with stationary placing booms has increased in use especially for the construction of high-rise reinforced concrete buildings in inner city sites where space is at a premium. Their positioning is in part a function of the space available for the supply of the concrete using ready mix concrete lorries.

Hoists can either transport personnel or goods. In either case the positioning is determined by two criteria: access at the base and the efficient movement of materials on each floor level. Ideally on most contracts the latter is resolved, when using only one hoist, by positioning the hoist midway along the floor so that goods have the shortest distances to travel. Using two hoists, this may be at the one-quarter and three-quarter positions. However, it may not be possible to achieve this because of other building works, such as the construction of a podium area.

The selection of a tower crane is determined by the contract programme requirements, and will be there for a long time and lifting a variety of different materials and components of different weights and shapes, probably covering the full footprint of the building. It has to be decided what these components are, what they weigh, the frequency of the lift of each and the heaviest lift at the end of the jib. The key factors then for consideration include:

- Does the crane need to be fixed or on tracks? This is a function of the length of the building and the number of lifts. For example, if the footprint of the building is a slender rectangle and the total number of lifts is within the capacity of the crane, then one on tracks might be the solution. In both cases it will require substantial foundations.
- If a fixed crane is to be used, then is it free standing, fixed to the exterior of the building for support or constructed inside, often in the lift shaft? The advantage of the latter usually means a shorter jib, but the disadvantage is no work can be carried out on the lift installation while the crane is in position. Those fixed to the outside or within the building may require structural amendments to the building to take the extra forces exerted by the crane when lifting.
- The crane has to be high enough above the height of the building plus personnel, any obstructions such as handrails and construction equipment, to accommodate the lifting beams and accompanying chains and the depth of any component lifted (including formwork for in situ concrete).
- If more than one crane is being used, the jibs must be at different heights. Anti-collision systems are available to reduce the likelihood of the collision of the cables of adjoining cranes.
- If working in highly congested areas the use of a luffing-jib crane may be the solution, especially to overcome trespass issues. These have a short counter jib and variable working radius that permit work in extremely narrow spaces.
- Ideally the radius of the crane needs to be such that it can lift any load that needs to be raised, to all parts of the building. One of the simplest ways of establishing this is, once the weights of components and items have been determined, to use a sheet of Perspex scaled to match that of the site plan and drawn with circles of different crane radii. This can be moved over the site plan until an optimum position is found (see Figure 19.1). If fourth dimension BIM is being used, then this will equally solve the issue.
- It is often common on urban sites to have the crane sited within the new building's central stair core to ensure maximum site coverage.
- Once this position has been established, access roads and storage areas can be sited relative to the crane.
- Is the speed of the lift (hook speed) sufficient to cope with the maximum number of lifts required per day?
- Will the positioning of the crane make the erecting and dismantling processes easy and save costs?
- Cost comparisons between different solutions can then be made before final selection.
- How reliable is the machine and how available are maintenance and spares?
- If hired, what is included such as the driver and maintenance, oil and grease?

There are many factors that affect the selection of a piece of excavation plant:

- The starting point is the contract programme. How much time has been allowed to carry out the activity?

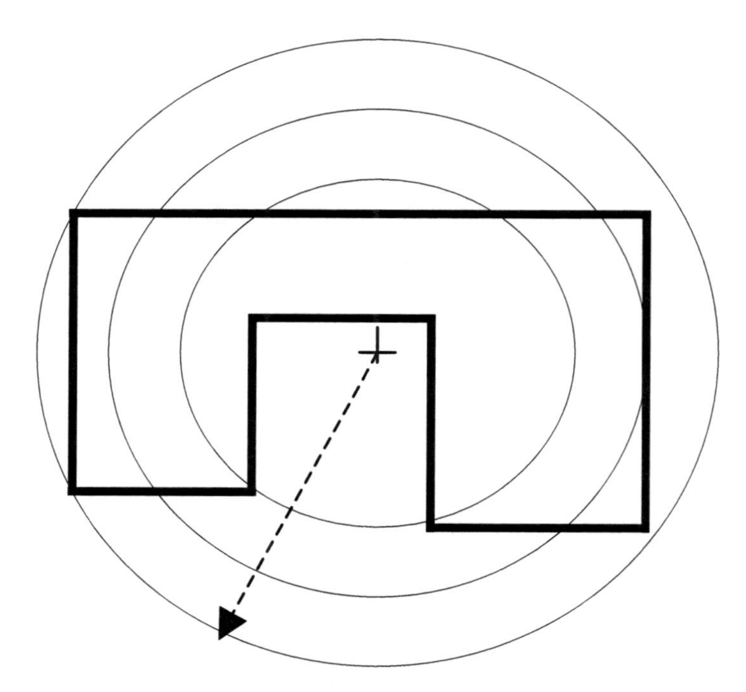

Figure 19.1 Positioning of tower cranes

- What is the composition and volume of the material to be excavated? This will determine the speed at which excavation can take place and what teeth may be required on the bucket.
- What type of excavation is it, such as a trench, bulk excavation, a general reduction in levels, or a shaft? In the case of a trench, the probable machine selected would be a back actor, for bulk a face shovel or back actor, in general reduction a scraper or bulldozer, and a grab for a shaft.
- Is there room for more than one machine?
- Are tracked or wheeled machines appropriate? Tracked vehicles move across the ground faster, but are restricted when the ground is wet and slippery.
- Can the machine rotate on its chassis which is a factor when discharging its load onto a lorry.
- From this information output per time period and the capacity of the machine can be calculated.
- Fuel consumption affects the cost of running and the carbon footprint of the site.
- Cost comparisons between different solutions can then be made before final selection.
- How reliable is the machine and how available are maintenance and spares?
- If hired, what is included such as the driver and maintenance, oil and grease?

With major excavations the number of lorries required to remove the soil elsewhere on the site or to a tip needs to be calculated. The distance from the site to the tip and the type of traffic to be negotiated will have an impact on the time it takes for each lorry and hence the number required. Remember the cost of the lorry is usually significantly less than the cost of the excavation plant.

19.10 Hoardings

Hoardings are the physical barrier between the public and the site activity. Their function is primarily to stop people wandering onto the site and potentially hurting themselves while also stopping the site activity from spilling over into the public domain. They also provide a first line of defence for security purposes, and if solid, assist in preventing the passage of noise and dust to adjoining properties. They can be used as a marketing tool for the main contractor and, if appropriate, the developer. If there is any danger that objects can fall on the general public, the hoardings have to be designed to give protection. In busy cities this can mean constructing a protected tunnel over the pavement using scaffolding or structural steel. On low-rise housing sites it is often uneconomical to provide hoardings, which means attention to the safety of the public and security become the key issues.

Whatever material is selected to construct the hoardings it should look neat and tidy, otherwise it sends out the wrong signals to the public and prospective clients. The materials usually chosen are chain-link fencing, plywood attached to timber posts, prefabricated corrugated metal units, prefabricated wire mesh units and the use of existing structures. Chain-link and prefabricated mesh units permit passers-by to see inside the site so that they can report any suspicious activity out of working hours, but also permit the thief to see what is around. On the other hand, plywood and prefabricated corrugated metal sheets do precisely the opposite. The material to be used may have already been determined by the client in the preliminaries. If plywood is used it will have to be painted either in the company's livery colours or that prescribed by the developer. There have been occasions in the past when the contractor has enlisted the help of the local art college, provided all the paint and allowed each student to create a 'painting' on a plywood sheet. This can generate good publicity and the final results become a local talking point.

The hoardings are normally used to place the company's logo or name at intervals along its length and some contractors provide either windows or observation platforms for the general public to observe the construction process. Some developers require an artist's impression of the final building to be attached to the boundary, which is a good idea as the public is interested in knowing how the building is going to affect their lives and what it is going to look like.

19.11 Communications

Good communication is vital for the effective running of a project and should be considered at a very early stage. Communication on site falls into several different categories:

- *Advising on how to get to the site*. This will include signs on the access roads to the site to direct plant and materials deliveries, or rerouting by diverting site traffic away from sensitive areas, and, on the site, clearly showing the entry points. Maps can be distributed to potential visitors that show parking arrangements, nearest tube and train stations and bus routes. All of these will save them time when accessing or delivering to the site.

- *Communicating with the general public.* Typical examples are apologies for any inconvenience caused, pictures and purpose of the completed building, observation points and the names of organisations involved in the process. These include the developer, architect, surveyors, engineers and main contractor.
- *Communication with the client and their representatives.* Systems, supported by appropriate equipment and software, need to be in place with agreed protocols. This will involve the distribution and control of drawings, matters contractual and points of contact. Similar issues need to be addressed when dealing with subcontractors and suppliers.
- *Communicating to staff on site and in the head office.* Systems and the provision of radio or telephone handsets are essential.
- *Communicating with trade unions.* Where there is union representation on the site, lines of communication should be established and all employees should be aware of the correct procedures.
- *Electronic communication.* More and more information is now communicated electronically so having sufficient computer capacity to store and send large amounts of data is essential, supported with the appropriate network system. There is also an increasing use of mobile technologies and applications such as tablets.
- *Health and safety.* Clear signs should be used to make any official visitor or site employee aware of mandatory requirements, such as hard hats, goggles, footwear, evacuation routes, the whereabouts of dangerous substances, restricted areas and so on.

19.12 Security

Security is required to reduce and hopefully prevent theft, trespass and protest, acts of vandalism and the risk from terrorists. There has always been theft from construction sites and children have found them to be exciting playgrounds. Protests, wanton vandalism and the possible threat of terrorist action have increasingly become issues that industry has had to take cognisance of.

19.12.1 Theft

Those thieving can be categorised as professional criminals, one's own staff and, especially on housing sites, the general public. Which plant and materials are the most vulnerable is a complex issue as the proceeds from organised crime are dependent on black-market forces. They can target large moveable plant such as excavators or materials in short supply. Many years ago there was a significant world shortage of copper as a result of sanctions against Zimbabwe, then Rhodesia, one of the world's major producers. Building houses without copper pipes was difficult as at the time there was no satisfactory substitute. A local builder took reasonable precautions, but the thieves broke in and stole all the fittings. When eventually the builder re-stocked, he parked and demobilised all his plant on all four sides of the hut. The thieves removed the roof.

Some of the company's own employees may resort to theft, for their own use or friends. They may sell information to organised criminals. Having everybody searched on leaving the site is expensive, the alternative being random searches, but in both cases there is the dilemma of how much this is a de-motivator, especially if the management staff is exempted. It is important to make it as difficult as possible by having systems in place that actively control the release and return of portable equipment and materials, so as to prevent these occurrences from taking place.

There are two types of theft by members of the public: those who pass by, see something they like and take it; or the residents living on the estate before building work is completed. It is not unusual for some of these to almost take it as their right to purloin materials. Cases have been cited where residents have built their own garage, garden walls and in one case started up the excavator and dumper truck to transport topsoil to their garden. To overcome this problem a contractor posted notices to all the residents offering a reward if a successful prosecution resulted from their reporting of an incident. The loss of materials dropped dramatically.

There is a free advice service offered by the crime prevention officer, but the quality of this advice depends upon the local force and the emphasis placed upon it. In some cases it could be a police constable and in others a chief inspector with considerably more experience. Other forces such as the Greater Manchester Police, have very effective architectural liaison officers who are primarily concerned with designing crime out of buildings and estates, but can also assist in the effectiveness of site layouts. Advice on the positioning of storage areas, lighting and alarm systems is also part of their remit.

Solutions include the provision of lock-ups, appropriate alarms and lighting, security patrols outside working hours, security gates and guards on duty at access points during the working day and painting equipment with forensic paint, which is a paint with a composition unique to the purchaser.

19.12.2 Trespassers, vandals, protesters and terrorists

All unauthorised persons that access the site are trespassers. However, the effect that each of the categories can have is different. Those intent on theft have been discussed in 19.12.1. Others may only be using the site to sleep overnight, to take drugs, or some may be children having fun exploring and playing games. In all of these cases, the intention is not wittingly to cause damage although they often do. However, if they have an accident while on the site, the contractor has a common duty of care and becomes responsible for any injury, unless it can be shown that all reasonable attempts have been taken to prevent it. So for example removing ladders as a means of access up the scaffolding would be a reasonable course of action, whereas it would not be if left in position where a child could climb and fall from a height.

Vandalism takes many forms on site from graffiti, deliberately damaging completed work and unused materials, to arson. Vandals may start up and drive plant to cause damage or as a form of joy riding with damage resulting. Besides the costs involved and loss of morale of personnel having to make good, certain acts may make parts of the site unsafe for operatives the following day. In all these cases the location of the site will have an effect on the likelihood of it happening.

On the other hand, some protestors will travel long distances and be prepared for a drawn-out campaign against the development. Normally these sites are predictable either because of potential threat from 'eco-warriors' and the like, or because of a build-up in the local press reporting on protests about the development long before construction takes place. The problem facing the contractor in the former case, is that they may well be encamped before the contract has been awarded, so the eviction process and the final securing of the site may be a lengthy affair with the need to go through the courts and the need to be sensitive to the opinion of the wider audience such as occurred at Manchester Airport's second runway and the Newbury by-pass. This makes pre-preparation impossible, unless considered at the development and design stages.

In the past, terrorism has been confined mainly to existing buildings, structures or events such as the bombing of the Grand Hotel, Brighton (1984) of government ministers attending the Conservative Party conference and the London bombings (2005), but in recent years it has been realised that any building being constructed, especially a prestigious one, is vulnerable and could be targeted. While the majority of the construction workforce are law-abiding citizens, the transient nature of the labour force means that special vigilance needs to be taken, not just in the security of the site, but also in the vetting processes during recruitment.

19.12.3 Security of information

Some of the information about the building may be sensitive. Drawings provided by the Home Office for the construction details of a prison would fall into this category. Protection of all information required for the management and construction of the building also needs protecting against loss by any cause and should be considered as part of the business continuity plan (see 30.9.2) as without it the construction work cannot be properly controlled without ensuing delays and extra costs.

19.12.4 Personal security

In certain locations the personal safety of staff, especially female, may be at risk travelling to and from the site. In these circumstances steps should be taken to ensure they are protected against possible harassment, verbal or physical attack.

19.12.5 Considerate Constructors Scheme

The Considerate Constructors Scheme is an independent non-profit-making organisation founded in 1997 by the construction industry to improve its image. It is a voluntary scheme to which construction companies and suppliers sign up and agree to abide by a Code of Conduct that is designed to encourage best practice beyond that laid down by statutory requirements. The Scheme is concerned with any area of construction activity that may have a direct or indirect impact on the image of the industry as a whole.

There are five areas identified:

- *Care about appearance*: constructors should ensure sites appear professional and well managed. This includes the external appearance of the site such as the hoardings, the way materials are stored, the condition of the plant, the appearance of employees and general cleanliness and tidiness.
- *Respect for the community*: constructors should give utmost consideration to their impact on neighbours and the public. This includes keeping others advised about

any disruption, and respecting and minimising the inconvenience caused by vibration, noise, dust and mud, staff parking, deliveries, and working on adjoining roads. Where possible the constructor should contribute to the local community and economy through training and employing local staff and purchasing policy.
- *Protect the environment*: constructors should protect and enhance the environment by identifying, managing and promoting environmental issues. This means looking for sustainable solutions, minimising waste, reducing the carbon footprint, protecting, and if possible enhancing, the local landscape, wildlife, flora and water courses.
- *Secure everyone's safety*: systems should be in place to protect employees, visitors and the general public and to ensure initiatives are enacted to continually improve safety, including embedding attitudes and behaviour concerning safety matters.
- *Value their workforce*: constructors should provide a supportive and caring working environment. This includes providing a workplace where all are respected and supported, treated fairly and given the opportunity for personal development and training. High standards of welfare provision should be provided and care given to employees' well-being and health.

The Scheme has developed an online resource, the Best Practice Hub, that gives tips for best practice and case studies that members can access free of charge. This information shows initiatives already in place and those currently being developed throughout the UK construction industry. They can be accessed to provide best practice ideas for the five categories above and assist in the development of Key Performance Indicators.

Bibliography

Forster, G. (2014) *Construction Site Studies, Production, Administration and Personnel*, 2nd edn. London: Longman.

Illingworth, J.R. (2000) *Construction Methods and Planning*, 2nd edn. London: E & FN Spon.

20 Contract planning

20.1 Introduction

The act of planning is not confined to industry and commerce. It is part of everyday life as can be seen every weekend in millions of UK households in the preparation of the Sunday dinner of meat, two vegetables, gravy and a pudding. The meat, vegetables and gravy all take different times to cook and yet all have to be ready at the same time. There is a need to check the progress of the cooking so that times can be adjusted, and on completion of eating the main course it is expected that the pudding will be ready. The times allowed for each of the operations will either be found in a recipe or have been established from experience. Planning the construction of a building follows the same basic principles, but because of the complexity, requires more sophisticated systems to support the process. Progress checks will also be carried out on several occasions commencing at the development stage, through procurement and the various stages of the construction process.

There is sometimes confusion in the interpretation of the terms planning and programme. Planning is the process of determining the sequence of events or activities that need to occur to complete the project. A programme is the diagrammatic demonstration of the act of planning.

It cannot be stressed enough that a programme must be realistic and therefore attainable, otherwise it is of no use and will fall into disrepute. Used properly it can be a means of engaging the people involved. A managing director of a small civil engineering company, whose work comprised mainly laying long lengths of drainage, approached the author, who was teaching on a part-time Chartered Institute of Building examination programme. The MD confessed he thought planning to be a waste of time and was only doing it to satisfy the Institute's examination requirements. After a few weeks, he had changed his mind and asked if planning could be introduced into his company. After discussion, a weekly programme of the works was produced and after the first week, he returned somewhat disillusioned, as only a part of the target had been accomplished. An analysis of the causes for the deficiency of progress ensued and identified some reasons so they could be anticipated in the future. Within a few weeks he was achieving his targets, but not only that, the foreman was now ringing him up and chasing him to ensure deliveries arrived on time otherwise he would not be able to meet the plan.

A good project manager will be able to look at a set of drawings and from their own experience sketch out an overall programme for the contract using perhaps only 10 to 20 key activities, but identifying each activity's completion date. The planner will then flesh out the programme incorporating further sub-activities. Experienced managers argue that the advantage of their programme is that it focuses the rest of the team on key completion dates they have to achieve and that too many activities can distract from this requirement.

20.2 Stages of planning

Planning takes place throughout the overall process, with the stage and details depending upon the size and complexity of the project:

- the development phase
- the tender stage
- the pre-construction stage
- the construction stage within which there can be various different levels of planning.

Table 20.1, adapted from Griffith *et al.* (2000) *Management Systems for Construction*, gives an overview of when formal planning is likely to take place. The most common planning techniques used are developed in 20.4 and 20.5.

20.2.1 Development planning

When the developer is calculating the financial viability of a potential project (Chapter 6), a key factor is the duration of the project design and the construction phases, as during these periods no income is being generated and interest has to be paid on the loans taken out to pay for these activities. Further, in retail development, the potential income generation is directly related to the time of year. For example, the run up to Christmas is generally the most profitable part of the year. Who carries out

Table 20.1 Stages of planning

	Development programme	*Pretender programme*	*Master or Contract programme*	*Medium (3 months) programme*	*Short-term (1 month) programme*	*Weekly programme*
Small project						
Complex			•	•		•
Not complex			•			•
Medium-sized project						
Long duration	•	•	•	•	•	•
Short duration		•	•		•	•
Complex	•	•	•	•	•	•
Not complex		•	•		•	•
Large project						
Long duration	•	•	•	•	•	•
Short duration	•	•	•			•
Complex	•	•	•	•	•	•
Not complex	•	•	•		•	•

Source: Adapted from Griffith *et al.* 2000.

the planning depends upon the type of procurement being adopted for the project. In a traditional form the quantity surveyor is the most likely person, and in management contracting the project management team would provide this service. The programme would be compiled from limited information, but based on the programmer's experience and the general information available, it is possible to produce a reasonably accurate timescale to complete the project. If this demonstrates the project as conceived cannot be constructed within the prescribed timescale, modifications to the design can be enacted, the project delayed or abandoned.

20.2.2 Pre-tender planning

This is to assist the tendering process. It is essential the estimator knows whether or not the contract can be built within the prescribed period stated by the client, because if not, then management may consider the project too great a risk and decide not to proceed with the tender. The timing of activities within the programme is required when asking subcontractors and suppliers to quote, since they need to know what is expected of them before submitting their price. The method on which the programme is based is important as it provides the basis of calculating the activities and preliminaries. It is used at the final tender review meeting to assist in providing information about the cash flow of the project, the amount of risk involved in meeting the programme requirements and any offers that can be made to complete the project earlier than the client has asked for.

The prime inputs to the process are from the design team, who produce the drawings, specifications, B of Q and contract details, and from the contractor who has the production management expertise and productivity data. This enables the contractor to produce method statements that assist in the production of the programme, site layouts, the programme and the completed tender documents.

The programme will not necessarily be the final outcome as many things can change between this process happening and the contract being awarded. The design may change and new methods adopted in the light of having more time for consideration.

20.2.3 Master or contract planning

This is sometimes referred to as pre-contract planning as it is produced when the contract has been awarded and prior to the work starting on site. At this point the personnel to run the contract have been selected and may have more developed views on how the works should be executed. This is because they have more information than at the pre-tender stage and can bring different ideas to the table.

They have at their disposal the tender documentation, additional project details from the design team, their own expertise, more accurate data from suppliers and subcontractors about delivery capabilities and the time they need to complete their part of the work, and the information provided from their site visit. In conjunction with this, schedules of when information is required and key dates for when resources are needed can be established so orders can be placed with suppliers, subcontractors and for the package contracts. At the same time, delivery schedules of materials and components can be processed. Method statements used for the planning process (see 13.7) can be adapted for Health and Safety purposes. In doing all of this, the contractor has the opportunity to develop good relationships with all the parties concerned with the construction of the project. Failure to do so could store up trouble for the future because of breakdowns in communication.

20.2.4 Contract planning

Throughout the duration of the contract, the planner will continually monitor progress and update the programme as changes in design and delays occur. As indicated in Table 2.1 the length of the contract and its complexity will determine the frequency of producing shorter-term programmes. On long and complex projects, programmes will be produced for three-month periods and updated every two months, and for areas of work that are either complex or critical, programmes of one month and one week will be produced. While there may be valid contractual reasons for the delays permitting extensions to the overall programme, the reality is that many contractors will strive to finish the contract within the original timescale if at all possible, to satisfy the client's needs. There is little point in finishing an Olympic facility after the games are over. There may well be cost implications in doing this, which will have to be resolved with the client.

20.3 Planning and producing a programme

Whatever technique is employed there remain certain fundamentals. The planner has to understand the sequence of operations and their interdependency, compare different methods for accomplishing the tasks in a safe manner, be able to establish the duration of an activity and resource it efficiently. The techniques employed are a function of the complexity of the project. Most important is to remember that the programme is a means of communication to others. Many of the techniques used for planning purposes are almost impossible to interpret by the layman and have to be converted into a readily understandable format, the most usual being a simple bar line (see 20.4 and 20.5.5).

20.3.1 Calculation of the durations

The planner will establish the duration of the activities from a variety of sources: in the case of the subcontractor, from their tender document that gives the duration they expect to be on site and amount of notice required before commencement; alternatively, they willl use their own experience based upon years of observing similar activities on other contracts, or calculation using standard production outputs and measuring the quantities of materials for each of the defined activities. The latter is not done to the accuracy of a quantity surveyor when producing bills of quantities, but is accomplished as quickly as possible either by taking the overall dimensions or scaling off from the drawing if not available. Quantities can also be taken from the B of Q, but the planner would normally only take off the main quantities and ignore the detail.

Table 20.2 shows a typical calculation for the brickwork from ground floor to first-floor level of a detached house, assuming an external skin of facing brick, an internal skin of block work with 50mm thick cavity insulation as follows.

In other words, in round figures, it would take one gang one week to complete the first lift of brickwork to this house. Utilising two gangs could reduce this to half a week, but there is a limit to the number of gangs that could be used in such a limited

Table 20.2 Activity duration calculation

Item	*Quantity*	*Rate*	*Duration*
Overall area, less openings	73m^2		
Labour rate, 2 bricklayers, 1 labourer		*0.5 gang hours/m^2	
Quantity × rate			36.5 gang hours

*Note the rate includes all the labour costs associated with producing one square metre of wall. It includes for the inner and external skin, the insulation and is inclusive of the laying operation, fixing the insulation and wall ties, fetching and carrying materials, and mixing the mortar.

work area as they could get in each other's way and reduce productivity. One of the planner's skills is to understand this. How many gangs are selected to do the work is determined by the necessity or otherwise to complete the activity faster. For example, there is little point in excavating every foundation on a housing estate within a couple of weeks when the overall contract period is one year as many of the excavations would fill with water and the sides collapse as they weathered, requiring remedial action to be taken.

20.4 Bar charts and linked bar charts

With the exception of line of balance, the majority of programmes will be shown in the format of a bar chart even though many of them will have been underpinned by other planning techniques. They are sometimes referred to as Gantt charts after Henry L. Gantt (1861–1919) who developed them in the 1910s for major infrastructure projects in the USA such as the Hoover Dam and interstate highway.

These types of presentations used to be produced by the planners using their experience and knowledge. They would calculate how long it would take to carry out each activity, work out the labour required, decide upon the sequence of events, consider the time delay necessary before an overlapping activity could start, and then sketch out the programme. They would then establish the amount of labour required in each week and readjust the programme to ensure there was a relatively constant use of the different types of labour. In those days over half the labour was directly employed by the main contractor. They would take the subcontract activity duration from the subcontractor and insert it. This method gave an overall idea of how the contract would run, but was fraught with difficulty because of the lack of serious logical thought underpinning the operation. However, for small contracts where the data are well known, they can work very effectively and can be used to monitor progress with some certainty.

Figure 20.1 shows a simple bar chart. The activities are listed as near as possible in sequential order, the first activity being at the top. The bar lines are drawn to scale, the duration having been ascertained as discussed in 20.3.1. Where activities cannot start before another activity is completed they can be linked together as shown. Where this linkage is used the chart is referred to as a linked bar chart.

Shading in the bar lines each week proportional to the amount of work done is used to monitor progress. Drawing a vertical line, shown in Figure 20.1 by a broken line, at the current date demonstrates whether or not an activity is behind, on schedule

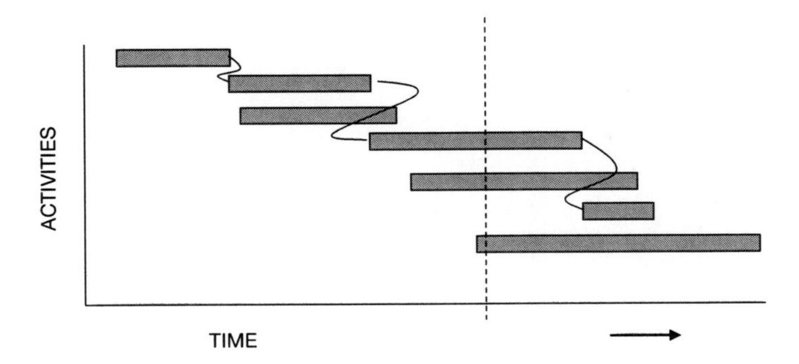

Figure 20.1 Linked bar chart

or ahead of programme. In practice, simply moving a length of string along the date line can do this. Software packages will provide a similar solution.

A traditionally constructed bar line is limited by the fact that it is not easy to reschedule if something needs to be changed as it takes many man hours to carry it out. The number of activities that can be used is also limited. It will normally be from 30 to 60, although sometimes they are produced with as many as 100.

20.5 Networks

Networks were first developed in the United States in the 1950s when the U.S. Navy Special Projects Office and their consultants devised a new planning scheme for special weapons systems. The outcome was the Program Evaluation Review Technique now known as PERT. Since the duration of many of the activities needed in designing new systems were unknown, this system was particularly concerned with assessing the probability of how long they might take. In construction, activities are more readily defined so their duration is predictable, so PERT was found not to be appropriate in the industry. However, a new activity-orientated system was developed for the industry called the Critical Path Method (CPM) or sometimes Critical Path Analysis (CPA).

These systems were first introduced into the UK construction industry during the 1960s, taking advantage of the computer technology being bought by the large construction companies. Unfortunately, these computers were relatively slow, the input was laborious and often the output was a network diagram that stretched around the walls of the planner's office and was incomprehensible to all but a few and of little use as a control document to those on site. Now that computers are powerful and fast, large quantities of data can be stored, manipulated and modified at an instant to produce many different outputs, such as the programme and resource implications.

The concept behind the production of networks is establishing the logic of a sequence of events. With bar charts there is a tendency to start with the first activity and decide which activities follow on. However, this is not the case with networks. It is a golden rule to remember that 'Which activity comes next?' is irrelevant in producing networks. To ensure the logic is correct the question must always be 'What activity or activities have to be completed before this activity can commence?'

There are two main ways of producing a network. These are called arrow diagrams and precedence diagrams. Each has advantages and disadvantages but produce the same outcome so the choice is a personal one.

20.5.1 Arrow diagrams

An activity can be a combination of smaller activities such as a 'reinforced concrete frame', which includes several operations such as form kickers, fix reinforcement, fix and strike formwork and pour concrete. What is included depends upon the detail it is necessary to go into. An arrow is used to represent an activity or operation, each of which must have a clearly defined start and finish as shown in Figure 20.2.

The length of the arrow is not related to the duration of the activity. Indeed, it is not unusual to find long arrows representing rather short durations and vice versa. At the two ends of the arrow is a point in time. These are referred to as nodes or events and are represented by circles, as shown in Figure 20.3.

Figure 20.4 demonstrates a simple example of the construction of a network. In this case activity B commences when activity A has been completed.

Not all activities are dependent upon the completion of just one other. In Figure 20.5 activities B and C can only commence when both activities A and D have been completed.

Figure 20.2 Arrows

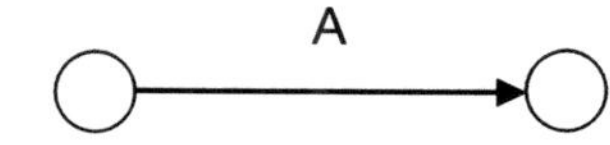

Figure 20.3 Arrows with nodes

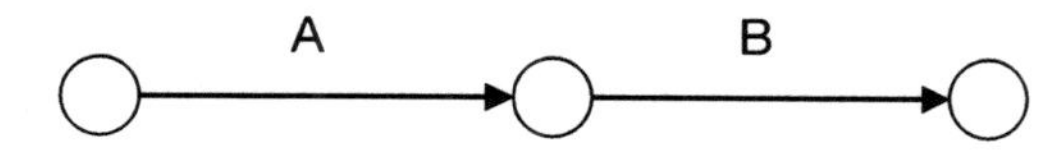

Figure 20.4 Connecting activities

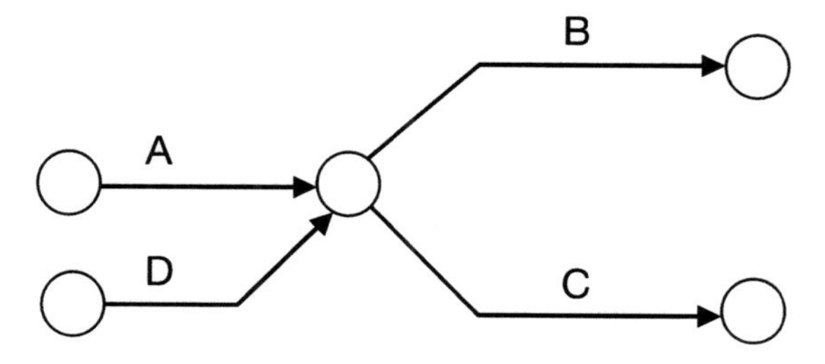

Figure 20.5 Several connecting activities

Note also that activities B, C and D are drawn with part of the arrow horizontal and the description, in this case B, C and D, written adjacent to it. While this is not mandatory, it is recommended as it eliminates confusion with 'dummy' activities (Figure 20.6), which would normally be draw at an angle.

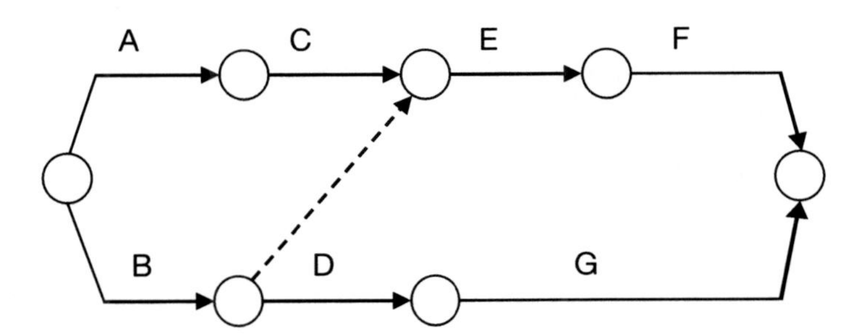

Figure 20.6 Network with dummy

Table 20.3 Activity precedence

Activity	*Preceding activity(s)*
A	Start
B	Start
C	A
D	B
E	B and C
F	E
G	D

With arrow diagrams a problem occurs when, as shown in Figure 20.6, activity E can only commence when activities B and C are completed. To overcome this problem the 'dummy' activity is introduced. In this case it is shown between the end of activities B and C drawn at an angle and with a broken line so as to distinguish it from an activity arrow. A dummy has no duration and is used purely as a mechanism to permit the logic to be shown. The information from which the network is drawn is shown in Table 20.3. It is normal at both the start and finish of a network to take all the start arrows from one node and all those that finish to one node.

In Figure 20.7, activity F can only commence when both activities B and D are completed, and activity G when B and E are completed. The information to draw this network is shown in Table 20.4.

An analogy that sometimes assists those learning how to draw arrow diagrams is to imagine the network represents water pipes, with the water flowing from the start to the finish, the nodes acting as valves. When a valve is opened water can flow, so in the example above, water enters pipe F from both D and B, and enters G from B and E. Therefore the network is drawn correctly.

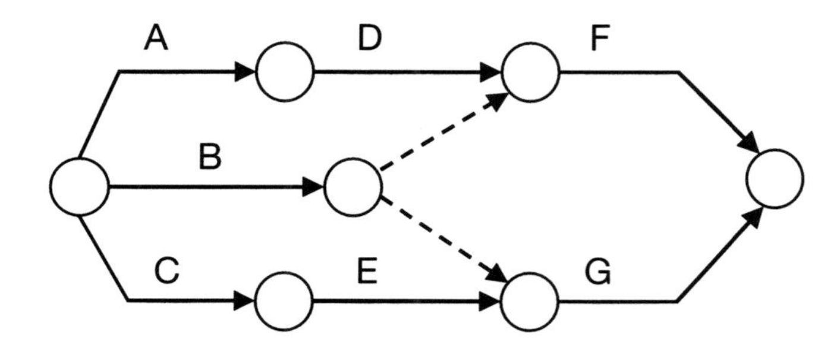

Figure 20.7 Network with two dummies

Table 20.4 Activities

Activity	*Preceding activity(s)*
A	Start
B	Start
C	Start
D	A
E	C
F	B and D
G	B and E

20.5.2 Definitions

Duration

This is the time that an activity is calculated to take to complete (20.3.1). It may be necessary to change the time of this duration when the network is being resourced (20.7), normally by reducing the time it takes by increasing the labour content, working extra overtime or changing the method of work.

Earliest starting time (EST)

This is the earliest time that an activity can start within the network. The first activity in a network is at the start so the EST is therefore zero unless the network is being produced for a later part of the contract when it would be from that point in time. For all following activities, the earliest the activity can start is the earliest the longest activity can start plus its duration. So, in Figure 20.8, if the EST for activity A is 0 and its duration is nine days then the EST for the following activities B, C and D is 9. However, if the activity is preceded by more than one activity the process is more complicated. Activity E can only commence when B, C and D, all of which start at the completion of A, have been completed. Clearly E can only commence when the activity of the longest duration is completed, i.e. D.

So what about B and C which are of shorter duration? Since D takes 12 days, B has 6 days spare and C has 7. This is referred to as float, which gives the planner various options. These are:

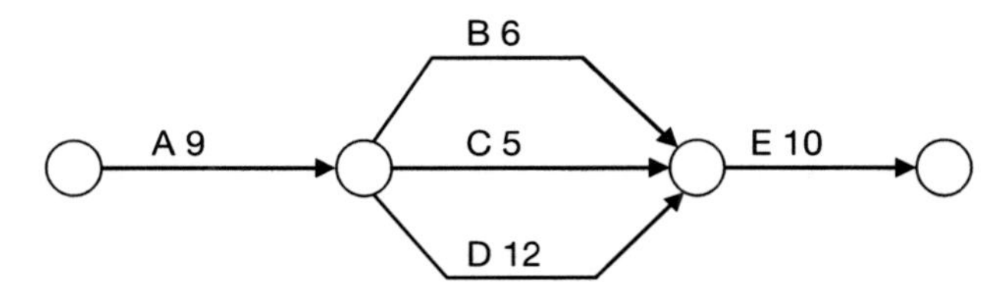

Figure 20.8 Network with durations

- Either or both B and C can start at the completion of A and if they begin to go behind programme it does not matter unless this delay is more than 6 and 7 days respectively.
- The start time can be delayed up to and until days 6 and 7 respectively and provided they are completed in 6 and 5 days the programme will be maintained.
- B can be started and then followed by C or vice versa. This might be a decision the planner might take if, for example, both activities required the same piece of plant, as this would be more resource efficient than having two identical pieces on site at the same time.

Earliest finishing time (EFT)

This is the earliest time that an activity can finish. It is calculated as its EST plus its duration. So using the example in Figure 20.8, the EFT for A is 9, the EFTs for B, C and D are 15, 14 and 21 respectively and because E can only start when D has been completed, its EFT is 31 (i.e. 21+10).

Latest finishing time (LFT)

This is the latest time an activity can finish without affecting the final completion time of the project. In Figure 20.9, it is assumed there are only two activities both starting at the same time, one with duration of 6 and the other 8. The latest finishing time both can end without delaying the contract or further activities is 8.

Latest starting time (LST)

This is the latest time an activity can commence without affecting the final completion time of the project. This is the latest finishing time minus the duration of the activity. In Figure 20.9, activity P could commence two days after the commencement of the contract and still not cause delays, but Q must commence at the start.

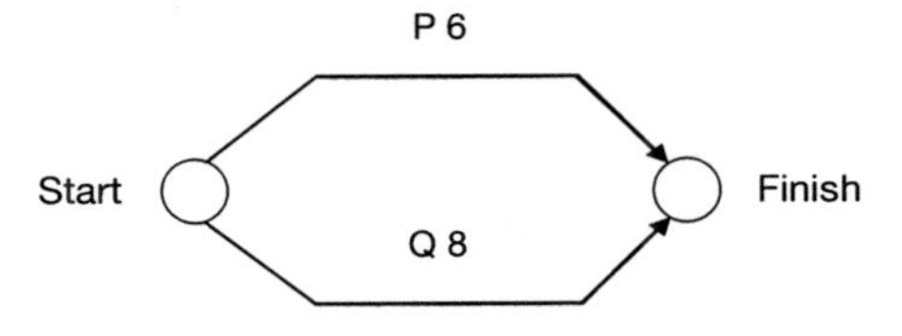

Figure 20.9 Latest finishing times

Some planners and other texts, rather than using the terms EST and LFT, refer to them as the Earliest Event Time and Latest Event Time; this is because the point in time represented by the circle in the drawn network is called either a node or an event.

The node or event

As indicated previously, the node is a point in time between the heads of activities that must finish there in the logic and the tails of the following activities that commence from there. This point is used in the calculation process when establishing the critical path through the network. The event number is given to this point in time so that each activity and the dummies can be given a unique reference. This becomes more relevant when using computer network programmes. There are various ways of dividing the circle. For the purposes of this text the format shown in Figure 20.10 will be used, but some writers divide the node into four sections using diagonal lines and insert the EST, EFT, LST and LFT in the quadrants.

The critical path

This is the pathway, or pathways, through the network that have no excess time associated with them. In other words, if any of these 'critical' activities are not completed within the duration they have been given, the overall programme will not be achieved and delays will occur unless critical activities further along the pathway are shortened by the same amount. Note, in this case other activities not previously critical may become so.

Float

On analysis of the network many of the activities will not be on the critical path. The time difference between the EFT and the LFT is the amount of float available. This means that the activity can commence after the EST by this amount providing subsequent dependent activities are delayed by the same amount. This permits the planner to resource the network. Float can be defined in different ways as shown in Figure 20.11. It demonstrates the relationships of float between two dependent activities A followed by B. The EST, LST, EFT and LFT are shown for activity A and the EST for activity B.

Total float is the total amount of float that the activity can be delayed without holding up the subsequent activity in the network logic. It is the amount of float that

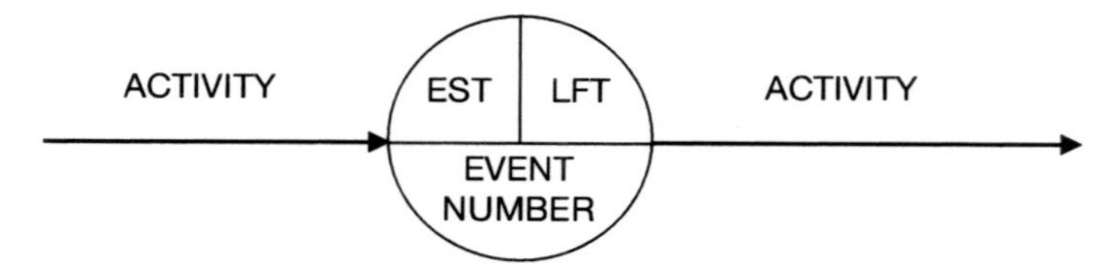

Figure 20.10 The node or event

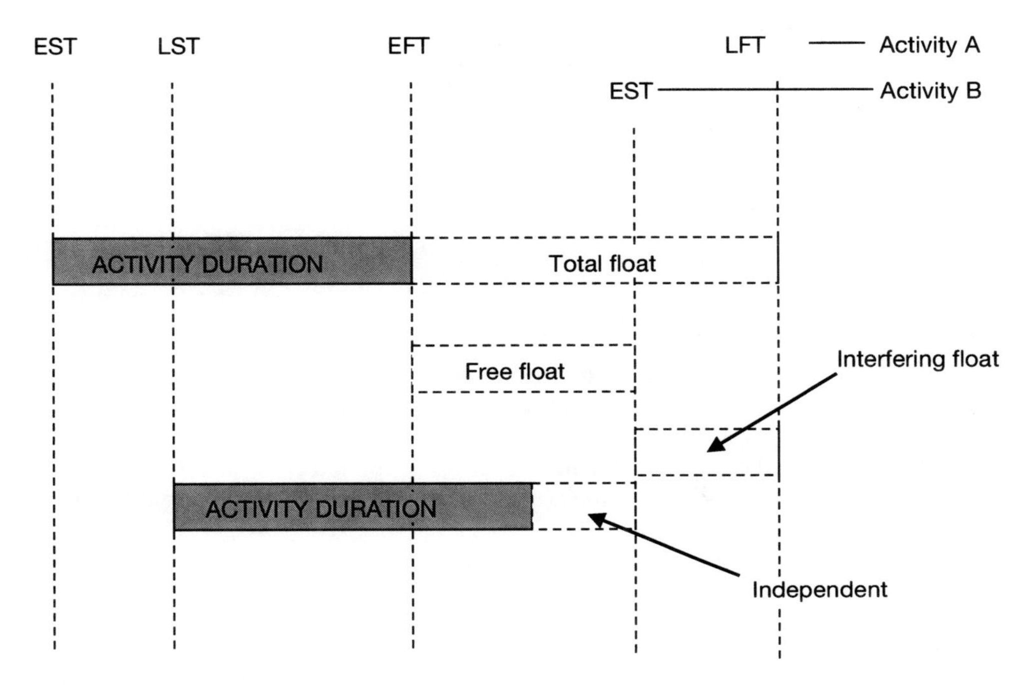

Figure 20.11 Types of float

can be used up before becoming a critical activity. It is therefore the difference between the EFT and the LFT.

Free float is the maximum amount of float that is available if activity A starts at the EFT and the next activity, B starts at its EST.

Interfering float is the difference between total and free float and is rarely used in any calculation when analysing the network.

Independent float is the float available if activity A commences at its LST and activity B starts at its EST. This is also seldom used in calculations.

20.5.3 Network analysis

The purpose of the network analysis is to determine the total duration of the project based upon the durations of the activities, to identify which are critical and establish the amount of float on the non-critical ones. If on completion the total time is in excess of that required, it can be further analysed and critical activities modified to comply. Those selected will be chosen based on practicality, reality and costs. For example, an activity may be too short in duration to be considered or too expensive to accelerate. There is a limit to the amount of time a critical activity can be reduced to as by doing so it may make another critical. For example, as shown in Figure 20.12, where activities A, B and C are 2, 3 and 4 days respectively, it can be seen that it takes 5 days to complete activities A and B, one day less than the critical activity C. If therefore C is reduced to 5, then activities A and B also become critical.

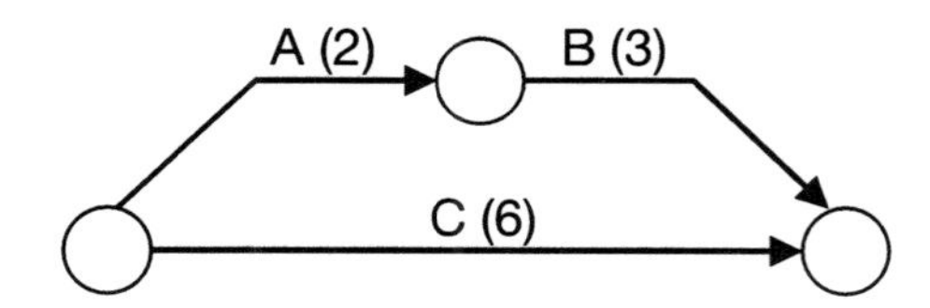

Figure 20.12 Critical path

20.5.4 Example of network analysis

Table 20.5 lists the activities, their duration and the preceding activity(ies) that have to be completed before the activity can commence. From this information the network can be drawn as shown in Figure 20.13.

After the logic has been drawn the durations can be entered alongside the description of the activities and the network analysis can start. This is done in Figure 20.14 with the durations in parenthesis.

To carry out the calculation, first the left hand upper quadrant of each node is filled in. These are the earliest start times. They are calculated commencing on the left hand side, which is the start of the project, and finish at the right, which is the end. The start is zero which means that the earliest that D can start is when B is completed, i.e. after 11 days, similarly F after 8 days. In the case of E, it can only start when both

Table 20.5 Activities

Activity	*Duration*	*Preceding activity*
A	10	Start
B	11	Start
C	8	Start
D	6	B
E	3	A, B
F	7	C
G	2	D
H	6	E
J	3	F, M
K	5	E, G
L	4	J
M	2	D

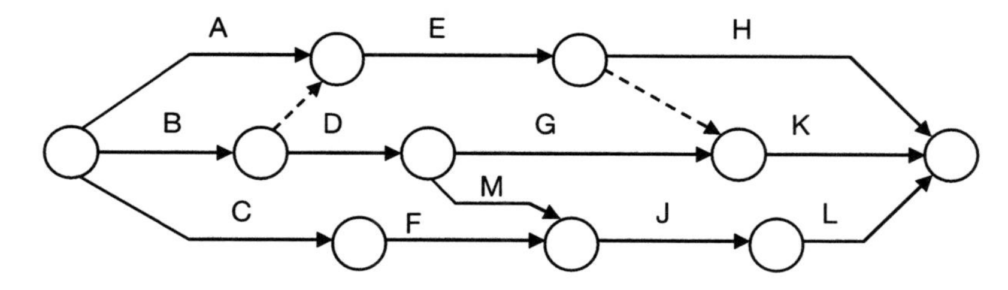

Figure 20.13 Network

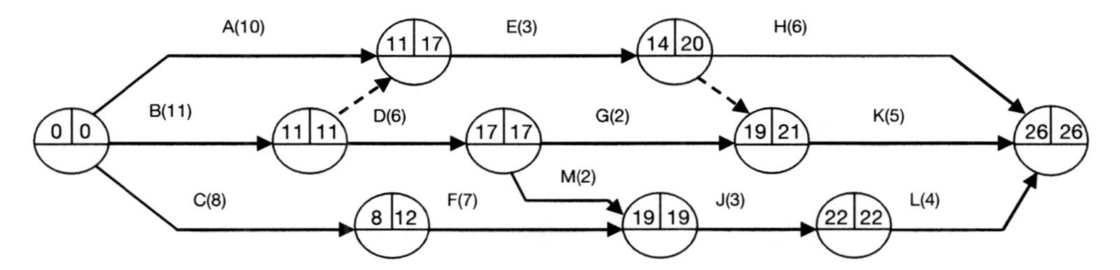

Figure 20.14 Network with EST and LFT

A and B are completed. Since B takes longer than A, E can only commence after 11 days. Since the earliest D can commence is at day 11 and it takes 6 days to complete, then the earliest activities both G and M can start is 17. This process is continued until the finish point where it can be seen that the earliest the completion of the project can be reached is after 26 days.

The second part of the calculation is to complete the upper right hand quadrant. In this case the pass is made from right to left commencing at the finish. If the earliest the finish can be attained is 26, then the latest the last activities (H, K and L) must finish is also 26, so this is entered in the finish node. Thus, the latest time J can finish and still permit L to finish by day 26 is four days earlier, i.e. day 22. In the case of E, it has to be finished before both H and K can commence; this means that the latest it can finish and permit the longest activity to start, i.e. H which takes 6 days, is 20 days. This process continues back to the start. It is important to note that the latest finishing time of the start is 0 and if this is not achieved in the calculation there is a mistake somewhere.

When this is complete, the data can be transferred to the network analysis sheet (Table 20.6). The EST and the LFT are abstracted from the network. The EFT and LST are calculated as follows:

EST + Duration = EFT
LFT – Duration = LST

Table 20.6 Network analysis sheet

Activity	*Duration*	*EST*	*EFT*	*LST*	*LFT*	*Total float*
A	10	0	10	7	17	7
B	11	0	11	0	11	0–critical
C	8	0	8	4	12	4
D	6	11	17	11	17	0–critical
E	3	11	14	17	20	6
F	7	8	15	12	19	4
G	2	17	19	19	21	2
H	6	14	20	20	26	6
J	3	19	22	19	22	0–critical
K	5	19	24	21	26	2
L	4	22	26	22	26	0–critical
M	2	17	19	17	19	0–critical

On inspection of the analysis sheet it can be seen that the total float can be calculated in two ways:

Total float = LST – EST and
Total float = LFT – EFT

If they don't match up then there is an error somewhere in the calculation of the EST and LFT on the logic network diagram. Where the total float is zero, these activities are identified as being critical.

20.5.5 Producing a bar line from a network diagram

A bar line is a relatively simple way of communicating the sequence of the construction activities to others, whereas a network diagram would be meaningless to most, so it is usual to convert the network accordingly. This is a relatively simple operation after the network analysis, but it is important to link activities together that are dependent upon one another. From the analysis a line can be plotted showing its EST, EFT, LST and LFT as shown in Figure 20.15. The shaded area represents the duration of the activity and the dashed line box, the total float available.

This means that the activity can start at any point between the EST and the LST without affecting the completion of the contract on time, unlike a critical activity where there is no float. The proviso to this is that another activity is not dependent upon it. For example, as shown in Figure 20.16, assuming that C is critical and both A and B each have float, A can only be started later if B is also delayed by the same amount, because it is dependent upon A being completed before it can start.

Using the data shown in the network analysis sheet (Table 20.6) a bar line can be constructed as shown in Figure 20.17.

Activities B, D, M, L and J are those lying on the critical path. They have no float as shown on the diagram and are all dependent upon one another in the sequence. All remaining activities have float, but some such as A and E, and E and H have dependent relationships. Activity A can be commenced one week after the start, but after that cannot be delayed unless E is also. Similarly, E cannot be delayed unless H is also.

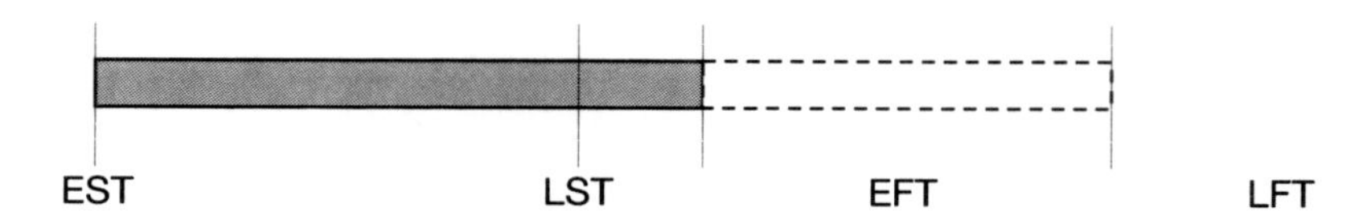

Figure 20.15 Bar line including float

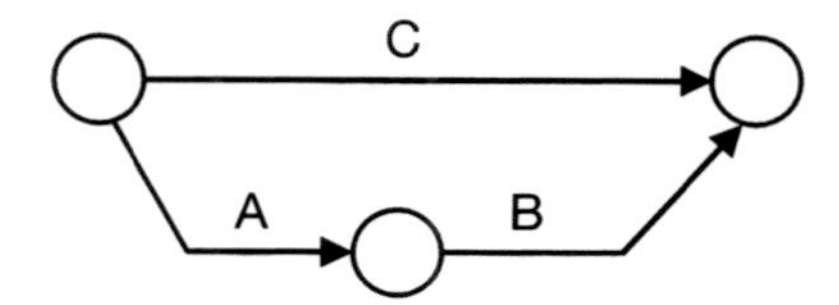

Figure 20.16 Network

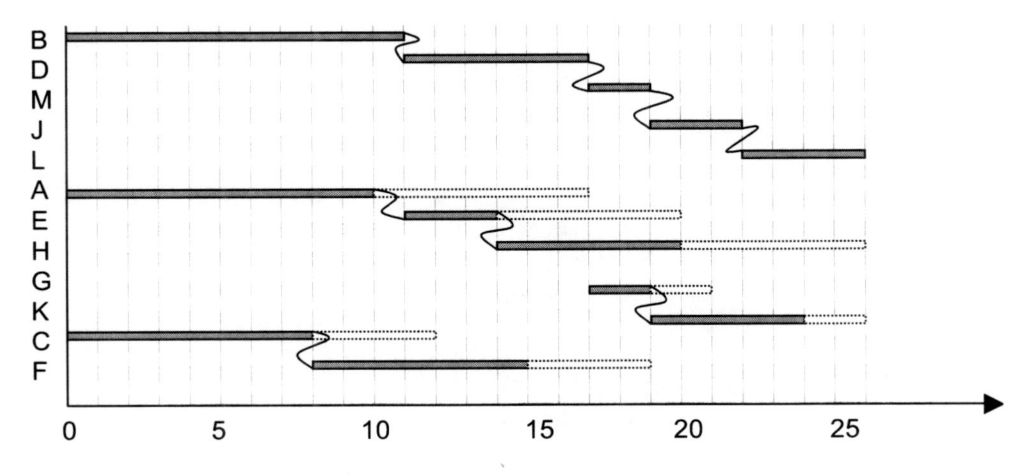

Figure 20.17 Bar line converted from network analysis data

20.6 Resourcing networks

Once the analysis has taken place on either arrow diagrams or precedence diagrams they can be resourced. The main considerations usually are to do with reducing the overall project timescale, carrying it out with less plant and labour, and trying to do these at the least cost. On the bar line shown in Figure 20.18, the amount of resource required for each week of each activity is entered above the bar line. These are then totalled at the base of the programme. The critical activities A to E have been drawn at the top of the programme, as their start time cannot be moved.

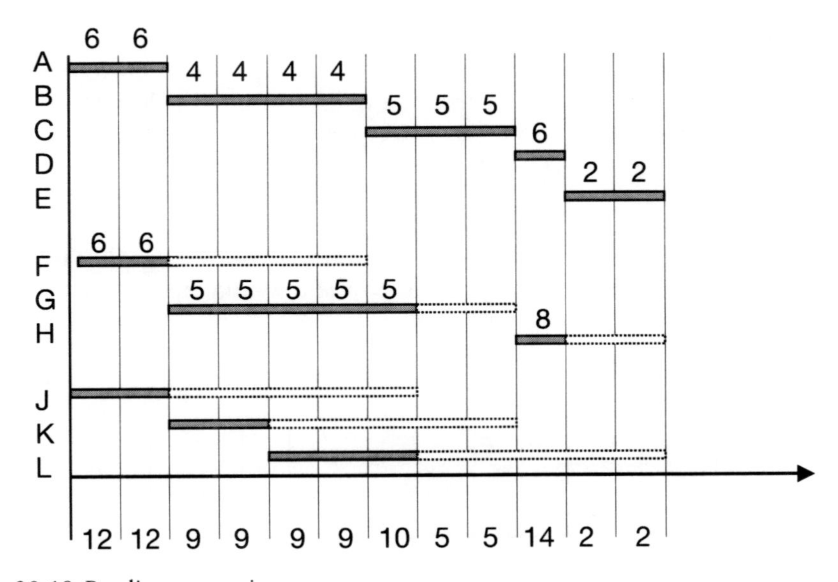

Figure 20.18 Bar line resourcing

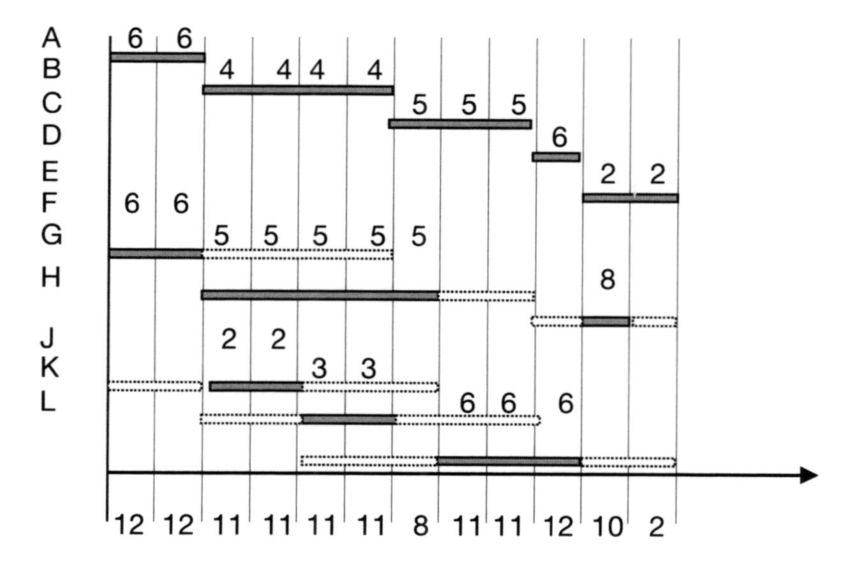

Figure 20.19 Resourced programme

Activities F, G and H are dependent upon each other as are J, K and L. It can be seen that the resources required for each week vary considerably and it would be an improvement if these could be made more constant. This is only a simplistic example assuming that the type of resource required for each activity is the same. Figure 20.19 shows a revised programme taking advantage of the float in delaying the commencement of some of the activities.

By moving some of the activities it is possible to improve the resourcing. It is unlikely that a perfect solution would be achieved in practice, but every little helps. Note that the dependent relationship between activities has been maintained. On a network with many activities it would be time consuming and difficult to achieve the optimum solution. The advantage of programme management computer software is that all the various solutions can be produced rapidly for consideration.

The same approach can be applied to resourcing plant. For example in Figure 20.20, A and B are excavating activities both requiring the same piece of excavation plant and have to be completed before C can start. It can be seen that by delaying activity B until A is completed, it is possible to use the same piece of plant for both operations.

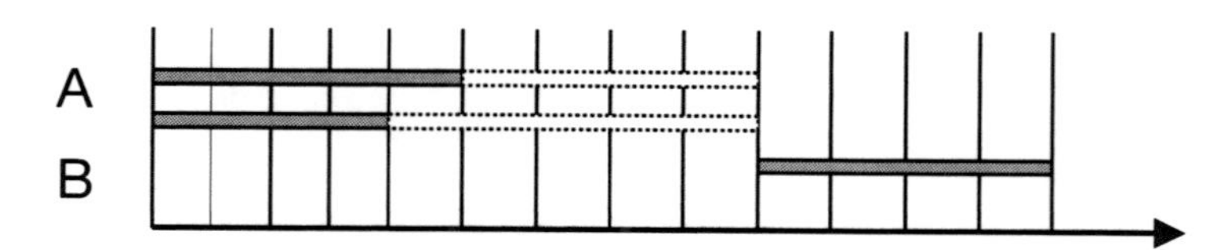

Figure 20.20 Plant resourcing

20.7 Cost resourcing networks

It is not unusual to want to reduce the duration of the overall programme. This can occur for a variety of reasons: the client wants the building earlier; the contract starts late, but the client still wants the original completion date: and the contractor wants to complete faster than asked for and believes that by offering an earlier completion date when tendering, this may give a competitive edge. Inevitably, an acceleration of the programme will increase the total cost of the project. Networks can be used to establish the lowest increase of costs. Figure 20.21 demonstrates a simple network. It can be seen that the critical path is A, C, E, G, H and J and that activities B, D and F have a total of one-day float between them. This means that if any of the activities C, E or G are reduced by one day then B, D and F also become critical. If another day's saving is required, then a day's reduction from both of these two routes will have to be made.

Having established the logic, now it can be calculated how much it costs to speed up an activity. Table 20.7 shows such costs for the activities on the network. Activities can be speeded up in a variety of ways. First, by increasing the amount of resources being used, second, by working longer hours, and finally by changing the method of work. However, not all activities can be reduced in practice. This is because they may be too short in duration, there is insufficient space available to permit an increase of labour or plant resource, or the length of the activity is already at its minimum. There will also be a limit to the extent that an activity can be reduced and the costs may well become progressively more expensive as further reductions in time are made.

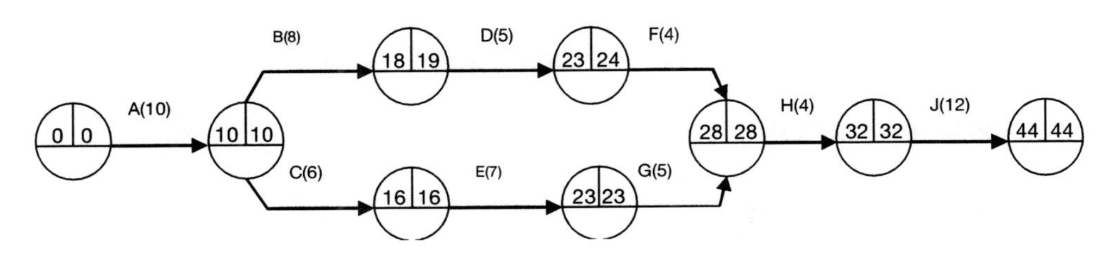

Figure 20.21 A simple network

Table 20.7 Costs of reducing durations

Activity	*Duration*	*Cost of 1 day reduction*	*Cost of 2 day reduction*
A	10	85	400
B	8	63	105
C	6	28	–
D	5	–	–
E	7	102	204
F	4	–	–
G	5	92	–
H	4	–	–
J	12	45	93

Table 20.7a

Activity	*Cost*
A	85
B	63
J	45

Table 20.7b

Activity	*Cost*
A	85
B	63
J	93

Table 20.7c

Activity	*Cost*
A	85
B + lowest of C, E and G which is C	105+28 = 133
J	93

To establish the cost of reducing the overall programme time the calculation is set out below. The cost of reducing by one day can be achieved by reducing one of the activities in Table 20.7a.

J is the cheapest. The cost of reducing the overall programme by a further day is from the activities in Table 20.7b.

J has already been reduced by one day hence the cost of the second day reduction is used. In this case the cheapest solution is B. This now means that route B, D and F has also become a critical path. Saving the third day is from the activities in Table 20.7c.

Therefore the cheapest is A. As with the resourcing in section 20.6, to carry out this manually on a large network would be laborious, and project management software packages have made this process much easier.

20.8 Ladder diagrams

The preceding discourse has assumed that as one activity finishes, another dependent one starts, but ignores the fact that in many cases dependent activities commence before the preceding activity has ended. For example, the cladding of a multi-storey reinforced concrete building normally commences long before the structure is completed, as do the various activities involved in laying a long length of drainage. In all these cases the activities overlap. This can be overcome by splitting each of the activities into smaller parts. For example, the frame of a 12-storey building could be divided into activities, ground to first, first to fourth and so on. This would work, but makes the network unnecessarily complex and large. An alternative method to show these relationships requires the use of ladder diagrams as shown in Figure 20.22. Here dummies are introduced to show the logic, but unlike before where they were of zero duration (20.5.1), in this case they are given a duration. The calculations for the EST and LFT are carried out as before.

The dummy activities 2–4 and 4–6 are referred to as lead time and dummy activities 3–5 and 5–7 as lag time.

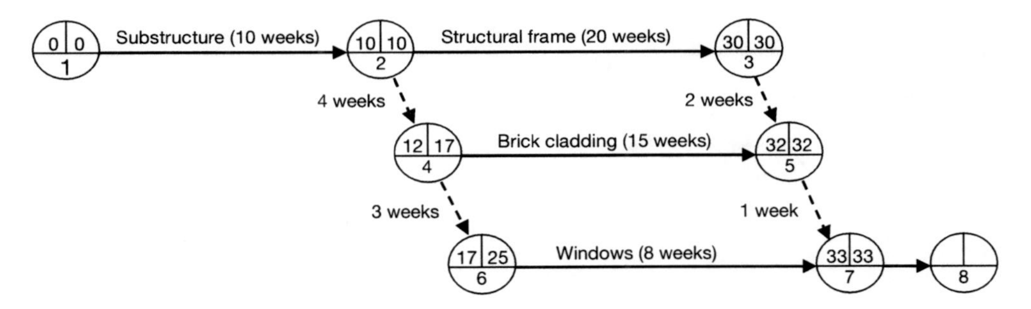

Figure 20.22 Ladder diagram

20.9 Precedence diagrams

Precedence diagrams generally match typical software programmes. The same definitions are used as in arrow diagrams and the same procedure is adopted in producing them. The difference is that the activity is shown using a box as shown in Figure 20.23 and the arrow between the activity (box) has zero duration like the node in the arrow diagrams. No dummy activity is required in precedence diagrams and more information is put in the boxes.

Although the logic is identical in activity and precedence diagrams, their appearance is quite different. Some examples of each are demonstrated in Table 20.8.

An example of a precedence diagram is shown in Figure 20.24. It is constructed from the information shown in Table 20.9. The EST, EFT, LST, LFT, duration and, in this case, total float have been calculated and entered into the activity boxes as demonstrated in Figure 20.23.

The network analysis is accomplished in exactly the same way as with arrow diagrams and can be manipulated to produce bar line charts and resource finance, plant and labour. However, they can also be used, unlike with arrow diagrams, to show relationships other than just that one activity must finish before the next one commences. Examples are shown in Table 20.10.

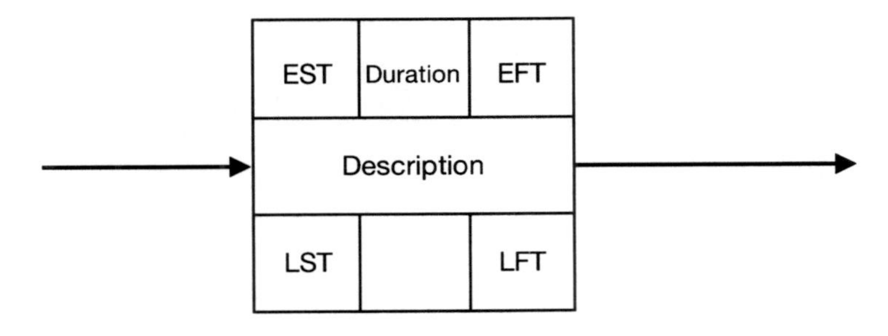

Figure 20.23 Precedence activity box

Table 20.8 Appearance of arrow and precedence diagrams

Example	*Arrow Diagram*	*Precedence Diagram*
1	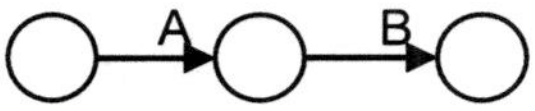	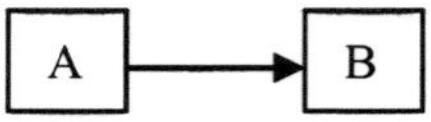
2	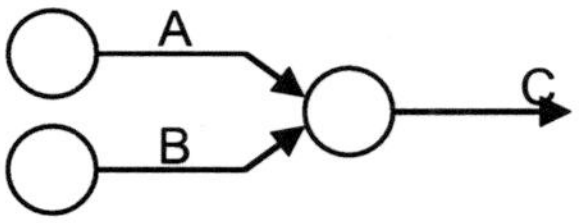	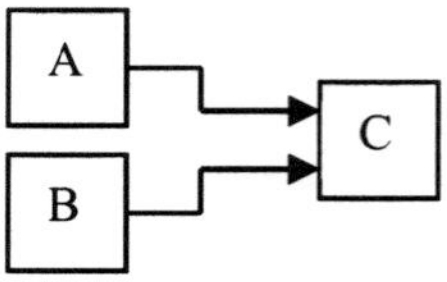
3	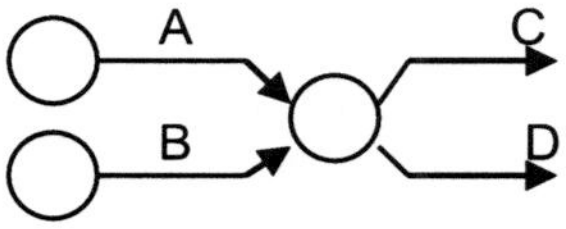	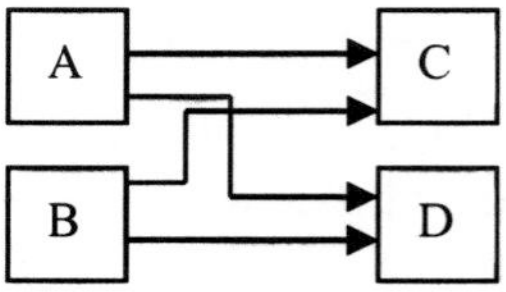
4	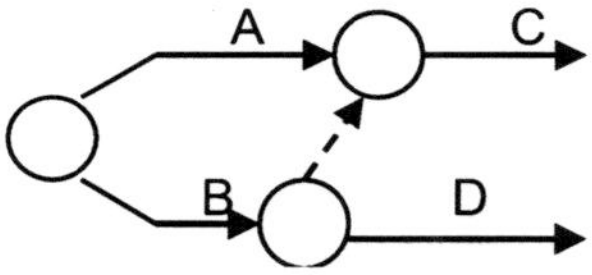	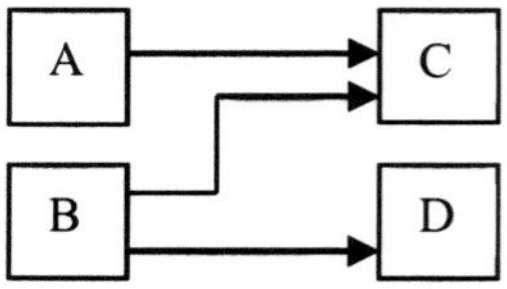
5	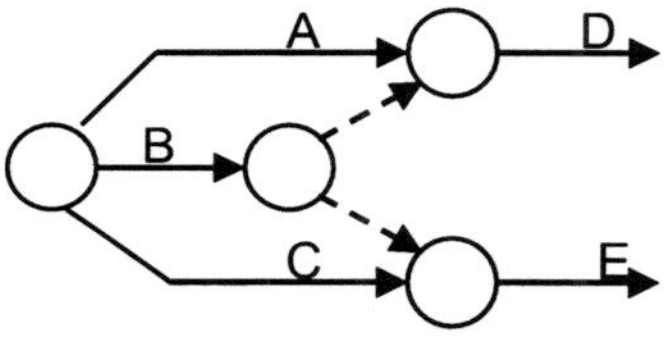	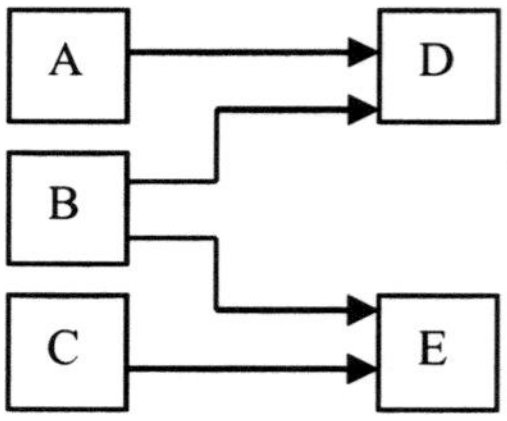

Table 20.9 Precedence data

Activity	*Duration*	*Precedent*
A	10	Start
B	4	Start
C	5	A
D	16	B
E	12	B
F	9	C
G	3	D
H	7	F
J	4	G, E
K	16	E

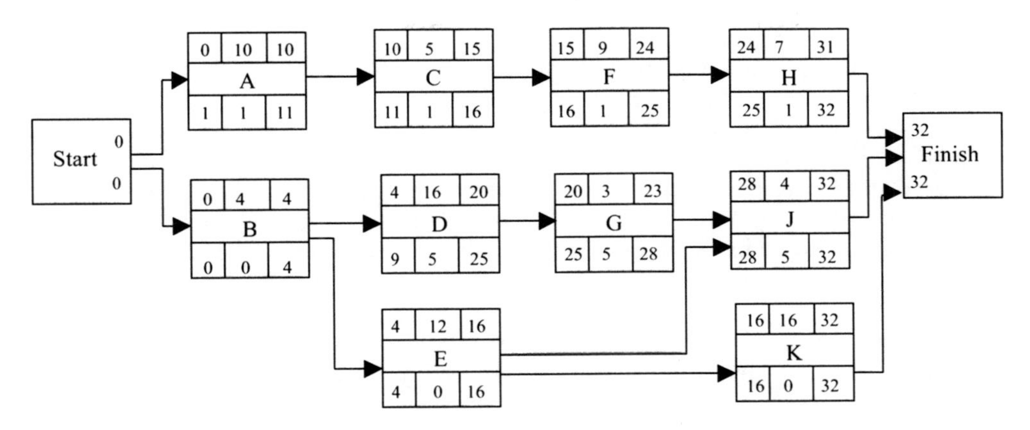

Figure 20.24 Precedence diagram

Table 20.10 Precedence relationships

Description	*Precedence Diagram*
Both activities can run concurrently, but the first activity must start before the second activity can start. For example, the design work commences before construction work, but they can still run partially concurrently as in fast tracking.	EST EFT / LST LFT → EST EFT / LST LFT
The second activity can start before the first has finished, but the first activity must complete before the second one can. An example of this is the laying of drainage that must be completed before the backfilling is completed.	EST EFT / LST LFT → EST EFT / LST LFT
This is theoretically possible but is usually avoided. The first activity must start before the second activity can finish. It would cover such examples of the inspection of a piece of plant has to start before the final installation is considered to be completed.	EST EFT / LST LFT → EST EFT / LST LFT
As with arrow diagrams it may be appropriate to utilise time lags. In the case of the arrow diagrams this produced a ladder network (Figure 20.22) But in the example shown opposite, there is a time lag of 5 days between the first activity and the upper of the two subsequent activities. This means, when carrying out the calculation, that the earliest it can start is 5 days after the first activity has been completed.	EST EFT / LST LFT → 5day lag → EST EFT / LST LFT; → EST EFT / LST LFT

20.10 Line of balance

This is a planning technique only relevant for repetitive work. Typical examples include multi-storey buildings, low-rise housing units where the majority of the units are the same or similar, and occasionally for drain laying and motorway construction. The finished programme is easy to interpret by all and most important of all it clearly demonstrates trends and what is likely to happen if action is not taken and thus gives management the opportunity to take corrective action depending upon circumstances. In this case, as shown in Figure 20.25, the lines shown represent one of the production activities, such as a floor of a reinforced concrete building followed by the brick cladding.

Since the vertical axis reflects the number of storeys of the building, the graph is seen to be user friendly and it is clear to all the rate at which the building is to be built. The brickwork starts at the first floor. This is because it is usual to use the ground floor for storage of materials and sometimes as a place to carry out certain pre-construction tasks such as cutting timber to length. When this is no longer required the brickwork from ground to first floor is constructed.

The dashed line on Figure 20.26 shows the actual progress of the structural frame and it can be seen that the first two floors were constructed slower than was planned for, but subsequently there has been steady progress in clawing back this deficit.

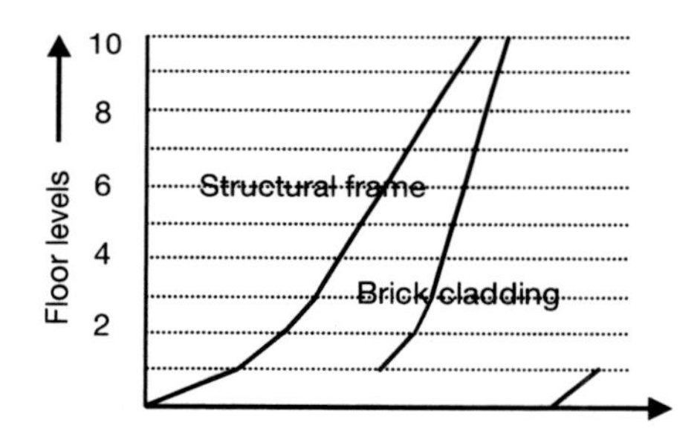

Figure 20.25 Line of balance

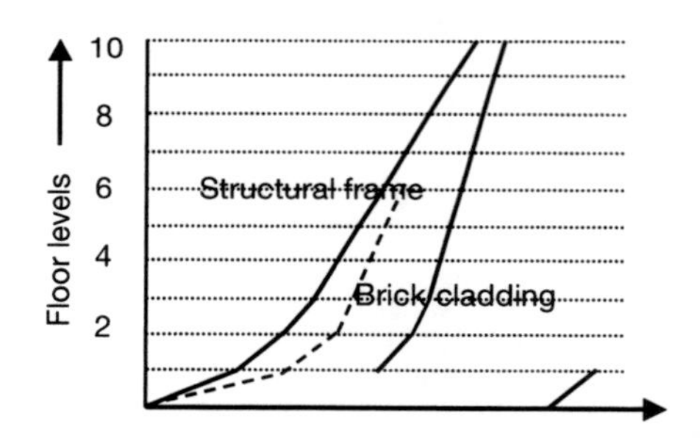

Figure 20.26 Line of balance

Bibliography

Fisk, E.R. (2013) *Construction Project Administration*, 10th edn. Englewood Cliffs, NJ: Prentice-Hall.

Forster, G. (1994) *Construction Site Studies*. London: Longman.

Griffith, A., Stephenson, P. and Watson, P. (2000) *Management Systems for Construction*. London: Longman.

Harris, F. and McCaffer, R. (2013) *Modern Construction Management*, 7th edn. Oxford: Wiley-Blackwell.

Oxley, R. and Poskitt, J. (1996) *Management Techniques Applied to the Construction Industry*, 5th edn. Oxford: Blackwell Science.

21 Health and safety

21.1 Introduction

There are four good reasons why safety should be taken seriously. First, by doing so you increase the likelihood of not having an accident yourself; second, the legal obligations, which, if flouted, could mean the offender would finish up with a jail sentence or significant fine; third, whenever anybody is injured, irrespective of severity, it costs the company money; and fourth, all of us, employees and employers, have a moral obligation to look after each other.

The industry has one of the highest accident rates of all industries. In 2014/15 there were 35 worker fatalities, and about 3 per cent sustained a work-related injury. It is unacceptable to talk about accidents as being an occupational hazard; the industry is not just striving to reduce the accident figures, but is also aiming to achieve an incident- and injury-free place of work.

21.2 Definitions

The terms *risk* and *hazard* are commonly used terms when dealing with safety and are regularly misused, so at the outset the reader should comprehend the difference. A hazard is something with potential to cause harm. The effect can either be to cause a physical injury or impact on one's health. A risk is the likelihood of that potential being realised.

21.3 Legal obligations – background

There are numerous pieces of legislation concerned with health and safety, but the key one is the Health and Safety at Work etc. Act 1974 (HSW), and it is within this Act that there is the potential for managers to be sent to jail for negligence. Currently this seldom happens, but there has been increasing pressure for this and/or corporate manslaughter to be utilised. The Health and Safety Executive (HSE) have pressed for one senior person in the company to be responsible for all matters concerning safety, but employers are resisting this, no doubt preferring to hide behind the shield of corporate responsibility, which makes it more difficult to apportion blame. A more detailed look at the implications of the HSW and other key regulations is given in section 21.9.

The United Kingdom (UK) has to comply with European Union legislation though acts of parliament and enabling acts. This is because under the Treaty of Rome (1957), Article 118A, member countries have to 'pay particular attention to encouraging improvements, especially in the working environment, as regards the health and safety of workers'. Directive 89/391/EEC was produced as a result of this and in turn was enacted in the UK through the Management of Health and Safety at Work Regulations 1992, revised 1999.

Other Directives followed:

- Directive 89/654/EEC: the workplace Directive
- Directive 89/665/EEC: the use of work equipment Directive
- Directive 89/656/EEC: the personal protective equipment Directive
- Directive 90/269/EEC: the manual handling of loads Directive
- Directive 90/270/EEC: the display screen Directive.

These have been enacted in UK legislation as:

- The Workplace (Health, Safety and Welfare) Regulations 1992
- The Provision and Use of Work Equipment Regulations 1992 revised 1998
- The Personal Protective Equipment at Work Regulations 1992
- The Manual Handling Operations Regulations 1992
- The Health and Safety (Display Screen Equipment) Regulations 1992 (amended 2002).

A further Directive of special interest to the construction industry was produced:

- Directive 92/57/EEC: the construction sites Directive.

This was implemented in the UK as:

- The Construction (Design and Management Regulations) 2015.

There are important further regulations that have a direct bearing on the construction industry. These are:

- The Reporting of Injuries, Diseases and Dangerous Occurrences Regulations 2013
- The Construction (Health, Safety and Welfare) Regulations 1996
- The Control of Substances Hazardous to Health (Amendment) Regulations 2004.

21.4 Financial costs of an accident

Every accident costs money. Even a small injury resulting in the need for minor treatment, such as the application of a plaster, results in the operative stopping work with loss of production as well as the first aid worker's time. It has been suggested that all these minor incidents annually equate to the cost of building one mile of new motorway. If it is necessary to go to the hospital's accident and emergency department, the costs rise and the operative could be off work. A fatality can work out at many tens of thousands of pounds.

In the UK in 2014/2015 there were 35 fatalities; 3 per cent of the workforce reported injuries, of which 23 per cent were slips, trips or falls, 22 per cent by lifting and handling,

19 per cent from falling from heights and 11 per cent from being struck by objects. A further 3 per cent of the workforce said they suffer from work-related illnesses, of which 20 per cent was stress, depression or anxiety and 64 per cent musculo-skeletal disorder.

In UK industry as a whole, 23.3 million days (2014/2015) were lost from work-related injuries and ill health costing £4.9 billion (2013/2014). In 2014/15 construction represented some 6.3 per cent of Gross National Product (GNP) and, being more dangerous than the majority of other industries, it can be seen that the cost to the construction industry is considerable.

Fatal accidents can be very costly to the company depending upon the amount of negligence apportioned to the management of the organisation. Other costs include:

- the loss of production resulting from the trauma and hence the overall morale of employees;
- a cessation of production due to an enforcement order being placed by the Health and Safety Executive (HSE) if they are not satisfied that the work situation has been made safe or they need to conduct further investigation;
- costs of staff involved in sorting out the accident and further investigation costs;
- costs involved in attending inquests and other meetings associated with police, coroner, trade unions, insurance companies, and one's own safety department;
- short-term payouts, without prejudice, to relatives to assist them through the immediate aftermath of the accident;
- compensation awarded by the courts – while this might be covered by insurance the premiums may well rise the following year as a result;
- any damage caused to property as a result of the accident;
- the costs involved in training a new member of the team;
- reduction in the company's reputation.

In the case of incidents where the operative is off for three days or more, they will have to be replaced, or the gang will continue to work inefficiently. The replacement member of the gang has a learning curve and there is the administrative load in managing this change.

Another cost, which does not affect the company directly, is the financial burden placed upon the deceased's relatives through the loss of income, even if compensation is eventually awarded, as this may take several years to happen. Due to this the Lighthouse Benevolent Club Benevolent Fund was founded in the 1960s. This charity gives aid and assistance to construction workers who suffer an accident or ill health and to their families during this waiting period.

Finally, there is the intangible cost: the loss of a family member and colleague can have untold emotional impact.

21.5 Moral obligations

There is clearly a moral responsibility on managers and companies to look after the welfare of those employed. To place someone into an unsafe environment is clearly unacceptable and yet this is done and not always unwittingly. In the post-war years it was still commonplace to talk about industrial diseases and accidents as occupation hazards. This changed especially as legislation was brought in to protect the workforce. It would now be unacceptable to consciously put workers into a known hazardous work situation unless it was unavoidable and sensible protective measures had been adopted; for example, the removal of asbestos.

There is, however, still one notable exception, that of stress, although even this is now beginning to be taken more seriously and steps taken to reduce it (see 29.11.3). We continue to allow employees to be subjected to intolerable stress conditions as a result of long hours, pressure of work, harassment and the employee's own choice. The result is lost time through illness, domestic unrest, and reduced quality of output or in extreme cases, burnout. In certain work environments it is thought to be macho, so peer pressure plays its part.

21.6 Self preservation

If safety is taken seriously, actions will be put in place to make the workplace safer and to improve attitudes to safety and the risk to everyone will eventually decrease. However, the site remains a potentially dangerous place and there is concern that those who are not 'street wise' can place themselves in greater danger. Two examples follow:

Incident 1:

While working on a sheet roof with two experienced operatives, a young trainee fell through the roof to his death. The client had approved the subcontracting firm's method of safe work. This involved the use of cat ladders on the roof with clear instructions that no work was to be carried out except from these ladders. It is believed that the two experienced workers walked along the roof from one cat ladder to the next avoiding the need to climb down to ground level and re-ascend at the next ladder. They walked along the bolts securing the roof sheeting to the purlin, which in turn supported their weight. The trainee on the other hand walked on the unsupported roof.

Incident 2:

In Hong Kong it is common practice to use hand-bored caisson piles. The operative carrying out the excavation excavates a circular hole some half a metre deep and then, using a metal former, concretes a ring around the perimeter of the excavation. This process continues to the required depth, which can be up to 30 metres below ground level. A hoist is positioned over the top of the excavation to lift out excavated materials and to lower the concrete for the sides of the caisson. A pump and compressed air line are also lowered into the excavation, the latter to reduce the risk of methane gases getting into the working area. These are often husband and wife teams, the man being in the excavation and the wife operating the hoist.

During one of these excavations a scream was heard and a young UK engineer rushed to the scene and, on seeing the operative unconscious at the base of the excavation, instructed the hoist operator to lower him down to assist. He was overcome by fumes and died. A second young engineer repeated his colleague's ill-advised actions and also perished. An experienced person would have realised the probable cause of the unconscious operative's plight, stood back, rationalised the situation and if breathing apparatus was not available along with a qualified user, would have called the fire brigade.

The final goal is to get to a point where safety awareness is part of everyone's daily life, not just at work, but also in their own personal lives. So at home, for example, using the correct sized fuse in appliances, checking the polarity of sockets, closing doors at night to stop the spread of smoke and fire, installing smoke detectors and generally living in a safe manner should be the norm. If an employee visits a site and observes an unsafe practice, action can be taken by reporting it to the appropriate supervisor or by taking direct action if appropriate. Visitors, however, can only bring it to the attention of the site management, but should be encouraged to do so and not made to feel that they are interfering.

Direct action can be taken by all to ensure that any, even minor, indiscretions are dealt with, such as nails protruding from wood or loose reinforcement tie wire, and the perpetrators educated not to do it in the first place.

21.7 The impact of an accident on others

It should be remembered when a serious accident occurs there is a knock-on effect on colleagues at work and the victim's family, and to demonstrate this the scenario discussed poses the question 'What would you feel like if put into that position?'

Scenario

In the incident in question an operative was out of sight underneath a machine making a modification while the rest of the gang were having their tea break. This was a daily event each morning. He had been instructed to take the ignition key to the machine with him so that the machine could not be started inadvertently while he was still underneath. On the day in question he had not done this. After the tea break the foreman instructed the charge hand to start up the machine, the result being that the operative underneath was impaled though the chest with two 100mm diameter steel tubes driven in with a 50 horse power electric motor. He died within a few minutes after having been extracted from the situation.

Questions:

- Even though it was not his fault, what would you feel if you were the charge hand and had started the machine up?
- How would you feel if you were the foreman having given the instruction?
- How would you feel as one of the other members of the gang, since they had all worked together for in excess of three years?
- What goes through the mind of the first-aid worker down underneath the machine trying to keep the injured man alive?
- How does the manager feel and react, while having to take charge of the situation until the emergency services come, and after they leave continue their command?
- What goes through the mind of the person going to the injured man's home to tell his wife that he had been injured and then escort her to the hospital?

The point of this example is it shows how far reaching the impact of an accident is. It is not just the loss of a loved one and income for the family. The incident does not just stop after the day of the accident. As the questions imply, the trauma for those involved continues, not helped by the fact that they will be interviewed several times more over the next couple of years while a case is being prepared either for court or an insurance claim for compensation. Further, it is not hard to imagine as a result of the trauma that, it takes time before production reaches its pre-accident rate.

21.8 What is the problem? – A statistical analysis

Frank Bird produced the information shown in Figure 21.1 indicating the relationships between the different types of event. For every fatal or serious accident there are ten minor injuries, 30 incidents of property damage (in the case of construction, the building, materials, components or plant) and 600 'no-injury accidents'. So, for example, it takes on average 600 bricks to be dropped off the top of the scaffolding before a person is eventually hit and seriously injured or killed.

Delving deeper, by looking at the statistics of accidents, it is possible to discover the number of accidents, the types and causes of accidents and trends, as well as comparisons with other industries. The prognosis is not good.

The main source of statistical data is from the HSE who publish annually Health and Safety Statistics. These documents provide information going back over the previous ten years showing the accidents occurring in all workplaces. This means that comparisons can be made between different industrial sectors as well as whether any trends are occurring within the construction industry. The comparisons are based on the number of accidents per 100,000 employees working in an industry or sector. All the statistical data in this section is sourced from various HSE publications.

The data as presented should not be considered in isolation, as there are other issues that need to be brought into the analysis. For example, just because accident trends are going in a particular direction, does not in itself mean that the safety of the industry is getting better or worse. It also depends upon the annual volume of turnover and the type of work being carried out. The greatest causes of death in the construction industry are from falling persons or falling objects landing on people. It might be argued that the greater the percentage of the turnover comprising tall buildings the more likely there will be an increased number of fatalities. Alternatively, tall buildings are more likely to be built by the more safety-conscious contractors and therefore the numbers injured working on these buildings is fewer.

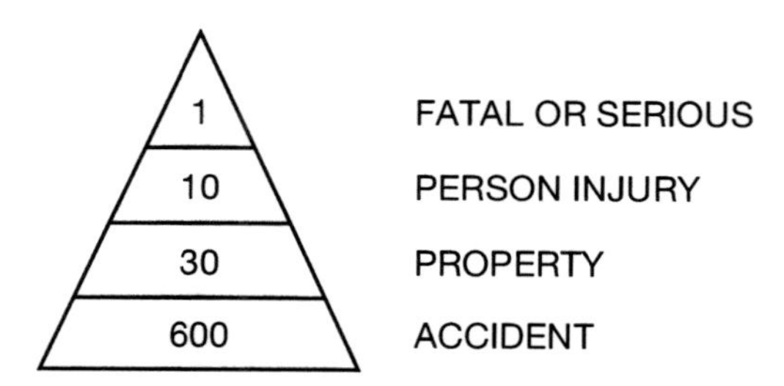

Figure 21.1 Relationships of accidents

Other variables include: whether there is any difference between the larger contractors' safety record and that of smaller operators; whether some trades are more susceptible to accidents than others; whether the age and experience of the operators affects the likelihood of an accident; and whether there are any regional differences in accident rates. If there are found to be any differences, then an analysis as to why needs to be conducted.

Statistics are based upon information provided as a result of the Reporting of Injuries, Diseases and Dangerous Occurrences Regulations 2013 (RIDDOR) in which the classifications used are: fatal and serious accidents, those that cause at least seven consecutive days off work, and diseases and dangerous occurrences (21.9.2).

From the statistical information trends in the industry can be established. These statistics vary from year to year and it is suggested that, if the reader needs the most up-to-date data, they consult the HSE website. The rate of fatalities is calculated on the number of fatalities per 100,000 workers and is significant as it more clearly demonstrates the trend in the safety performance of the industry rather than the numbers killed. Similar data are collected for reported major injuries and over-seven-day injuries. By transposing the rates of fatalities, major and over-three day accidents onto the same graph, the ideas proposed by Frank Bird are confirmed that there is a relationship between the numbers and types of accident.

The HSE breaks down this data further into the causes of fatal, major and over-three-day accidents and identifies that over the years, the main causes of fatal accidents on construction sites are falls from heights, being struck by an object or moving vehicle, contact with electricity, and being trapped due to collapse of excavations. More alarming is the even higher percentage of 70 per cent among those who are self-employed.

In the case of major accident, there are similar causes, but accidents caused by handling and lifting, slips, trips and falls, become more significant. This is reflected in over-seven day accidents.

Each of the categories of types of accident can be broken down into the causes of accidents and this gives management the knowledge as to where to concentrate improvement efforts. This is necessary as, while the aim should be to eliminate all accidents, this does not happen overnight, so it is sensible to concentrate on the areas of greatest risk (see Pareto's law, 16.2).

Tables 21.1 and 21.2 are concerned with fatal accidents associated with maintenance. Table 21.1 shows the main types of accidents that can cause fatalities, and the approximate percentage of each.

Table 21.1 Causes of fatal accidents in maintenance work

Type of accident	*Percentage*
Falls	50
Crushed or entangled	24
Asphyxiation	9
Electrocution	7
Burns	6
Impact	3
Falling objects	1

Table 21.2 Types of falls from places of work during maintenance

Type of accident	*Percentage*	*Type of accident*	*Percentage*
Flat/sloping roofs	22	Plant	6
Ladders	22	Scaffolding collapse	4
Fragile roofs	22	Window cills and ledges	3
Work platforms	11	Cranes and gantries	1
Scaffolding	8	Others	1

The two most serious types of causes of fatalities are falls and being crushed or entangled in machinery and plant, and it would be sensible to further investigate the causes of these first. Falls is a broad description and as can be seen in Table 21.2 there are many different causes, some of which are more frequent than others such as falling off ladders, roofs and through fragile roofs. Each of these can be broken down further. For example, the causes of falls from ladders could be as a result of not tying the ladder properly at the top, insufficient gradient, or slipping while ascending or descending. From this further detailed analysis, steps can then be put in place to reduce the risk of these occurrences. The process can then be applied to the remaining types of accidents.

A further investigation can reveal who is responsible for the cause of the accident. Blame can be apportioned to that of the management, workers, a combination of both, third parties or it is unknown. What is found to be significant is that it is not unusual to find management is responsible in full or in part for nearly 70 per cent of all accidents associated with maintenance.

21.9 Key legislation and regulations for the construction industry

While not a complete summary of all the legislation, the Acts summarised in this section are the ones most commonly cited and referred to.

21.9.1 Health and Safety at Work etc. Act 1974

This is an extensive piece of legislation so only the main elements of the general duties and enforcement are briefly mentioned.

Section 2 – General duties of employers to their employees

It is the duty of every employer to ensure, 'so far as is reasonably practicable', the health, safety and welfare of all their employees. In particular this includes:

- the provision of plant and equipment, its maintenance, and systems of work, that are safe and without risks to health. On site this includes all the static and mobile plant as well as portable power tools and hand tools. Examples in the office environment include all electric goods such as kettles, photocopiers, shredders, printers and computers;
- the safe use, handling, storage and transport of all articles and substances;

- provision of such information, training and supervision as is necessary to ensure employees can carry out their tasks safely;
- ensuring that access and egress to and from the work are safe and without risks to health which means keeping these routes, well maintained, clean and clear from obstructions;
- provision of a safe working environment and adequate welfare facilities.

Employers are also required to prepare and update a written safety policy (see 21.14.2), consult with any recognised safety representatives and to establish a safety committee (see 21.15).

Section 3 – General duties of employers and self-employed to persons other than their employees

Employers must ensure that any persons not in their employment who may be affected by the work are not exposed to health and safety risks. This includes official visitors and those delivering goods to the site as well as third parties passing by. Note that there is a duty of care to trespassers. Self-employed persons also have a duty to ensure that other persons are not exposed to risk as a result of their actions. In both cases they have an obligation to provide information about their work where it might affect the health and safety of others.

Section 4 – General duties of persons concerned with premises to persons other than their employees

This is similar to section 3 except that it is concerned with third parties other than employees who use the premises for work. Particular attention needs to be given to plant or substances provided and access and egress.

Section 5 – General duties of persons in control of certain premises in relation to harmful emissions into atmosphere

It is the duty of the owner of any premises, and this includes the contractor during construction, to use the best practicable means for preventing the emission into the atmosphere from the premises of noxious or offensive substances or rendering them harmless and inoffensive before emitted. At first reading, it appears that this is not a problem for construction sites as it might be thought to be about chemicals, odorous gases and so on, but the interpretation also includes smoke, grit, dust and noise, and construction sites are major generators of dust and noise.

Section 6 – General duties of manufacturers etc. as regards articles and substances for use at work

It is the duty of any person who designs, manufactures, imports or supplies any article at work to ensure that it is safe to use when used properly. This means testing and research has to be carried out to ensure that risks to health and safety have been eliminated or reduced to a minimum. Those who install the article must ensure it is safe to use when handed over. They must provide adequate information to the user to ensure it can be used safely. There are implications for the main or managing contractor in construction in ensuring this is carried out. Most buildings have equipment installed as part of the building services element, almost invariably installed by subcontractors who will have sourced the equipment from some other supplier. The systems of control therefore have to be robust.

Section 7 – General duties of employees at work

All employees have an obligation to take reasonable care for the health and safety, not just of themselves, but also for others who may be affected by anything they do or don't do at work. For example, leaving a piece of wood with a protruding nail for others to walk on, spilling coffee on the floor and not wiping it up, ignoring the risk of someone slipping or just generally fooling about. Employees are also obliged to cooperate with the employer to enable them to comply with their legal obligations.

Section 8 – Duty not to interfere with or misuse things provided pursuant to certain provisions

It is an offence for anybody to deliberately or recklessly interfere with or misuse anything that has been provided in the interests of health, safety or welfare.

Section 9 – Duty not to charge employees for things done or provided pursuant to certain specific requirements

Employers are not permitted to charge an employee for doing anything or for the provision of safety equipment and clothing that is required to carry out work safely.

Enforcement

There are three stages of enforcement at the disposal of the Factories Inspector on a standard visit or as a result of a visit resulting from an incident occurring at the workplace. The first is called an improvement notice that details a specific period of time in which the contravention must be corrected. The length of time will be a function of the future risk associated with the hazard and the cost and time it would take to rectify the situation. The second, called a prohibition notice, is where it is adjudged that there is a serious risk of injury if the activity is continued. In this case the activity will be stopped until the specified remedial action has been completed. Finally, the HSE will prosecute if it is considered the infringement is as a result of negligence. This can result for offences after March 2015 in unlimited fines. The courts have the authority to send the guilty party to jail for up to two years.

21.9.2 Reporting of Injuries, Diseases and Dangerous Occurrences Regulations 2013(RIDDOR) (adapted from The RIDDOR explained, *HSE)*

The regulations require that any incident covered by the regulations that occurs on a construction site has to be reported to the HSE Incident Contact Centre, by phone, fax, the Internet, email or by post depending upon the nature of the incident. It is the responsibility of the principal contractor to do this. If it occurs in the head or regional office in the UK, then it *can* be reported to the environmental health department of the local authority. This reporting procedure gives the HSE the opportunity to identify where risks arise and to investigate serious accidents as well as providing statistic evidence on the safety performance of all types of commerce and industry and this in turn informs the legislature in making decisions about new or revised regulations.

A report is required in the following circumstances:

Death or serious injury. If an employee, directly or indirectly, is killed or suffers a major injury (including any caused by physical violence) or a subcontractor, or a member of the general public is killed or taken to hospital, then the enforcing authority must be notified without delay by telephone or online. If not done online it must be followed up within ten days by a completed accident report form. Examples of a major injury are specified in the regulations and include:

- fractures other than to fingers, thumbs or toes;
- amputation;
- dislocation to shoulder, hip, knee or spine;
- loss of sight, temporary or permanent;
- any serious injury to the eye;
- injury from electric shock or burn leading to unconsciousness or requiring resuscitation;
- injury leading to hypothermia;
- injury requiring resuscitation;
- injury requiring hospital admittance for more than 24 hours;
- unconsciousness caused by asphyxia or exposure to harmful substances;
- acute illness requiring medical treatment resulting from inhalation, ingestion, absorption through the skin;
- exposure to a biological agent or its toxins.

Over seven day injury. An over seven-day injury is classed as one that is not major, but results in the injured person being away from work or not capable of carrying out the full range of normal work activities for more than seven consecutive days. This does not include the day of the accident, but does include weekends, rest days and holidays. A report must be submitted within ten days.

Disease. If a doctor notifies the employer that a worker has a reportable work-related disease, a completed disease report form must be sent to the enforcing agency. The full list can be obtained from the HSE, but examples include:

- Carpal Tunnel Syndrome (CTS), where the person's work involves regular use of percussive or vibrating tools;
- cramp in the hand or forearm, where the person's work involves prolonged periods of repetitive movement of the fingers, hand or arm;
- occupational dermatitis, where the person's work involves significant or regular exposure to a known skin sensitiser or irritant;
- Hand Arm Vibration Syndrome (HAVS), where the person's work involves regular use of percussive or vibrating tools, or the holding of materials that are subject to percussive processes, or processes causing vibration;
- occupational asthma, where the person's work involves significant or regular exposure to a known respiratory sensitizer;

- tendonitis or tenosynovitis in the hand or forearm, where the person's work is physically demanding and involves frequent, repetitive movements;
- any cancer attributed to an occupational exposure to a known human carcinogen or mutagen (including ionizing radiation); or
- any disease attributed to an occupational exposure to a biological agent.

Dangerous occurrences. When an incident occurs not resulting in a reportable injury, but which could have done, this could be classed as a dangerous occurrence and has to be reported immediately followed up within ten days if not done on line. Examples of dangerous occurrences are:

- a collapse of part of the building during its construction, alteration or demolition – this would include the temporary works such as scaffolding and formwork;
- major collapse of the soil into an excavation;
- plant coming into contact with overhead power lines;
- malfunction of breathing apparatus;
- a sudden uncontrolled release of flammable liquids and gases above a level set by the HSE;
- accidental release of any substance that may damage health.

Records of any of the above reportable incidents have to be kept by the employer for three years after the date on which they occurred. This includes the date and method of reporting; the date, time and place of the incident; the personal details of those involved; and a brief description of the nature of the incident or disease.

21.9.3 The Control of Substances Hazardous to Health Regulations 2004 (COSHH)

At first glance it may be thought that the construction industry does not use many materials and chemicals that are hazardous to health, but on closer inspection not only are many chemicals such as adhesives, timber preservatives and solvents commonly in use, but seemingly innocuous materials such as some species of timber and concrete can, while being processed, cause dusts that are hazardous. Further processes such as welding give off hazardous fumes, and waste at the end of the process can create a hazardous situation on disposal such as the residue in containers. There are exemptions to the regulations such as asbestos, lead and radioactive materials as these have their own regulations.

There are steps required under the regulations, but often in practice they are not all carried out as diligently as they should be. The first step is to assess the risk, which means identifying all the hazardous substances present in the workplace, and in the case of construction, all those that are to be purchased to complete the project. The hazards associated with many of the materials are identified with a warning label on the containers and all suppliers must provide a safety data sheet for them. Assessing the risks is about using experience and common sense in establishing how likely it is that the material will affect the health of the operatives and third parties when used for a particular application in a specific location, and over what period and how frequently. For example where fumes are the potential hazard, is the place of application well ventilated or enclosed?

Having established the type and nature of the hazards and ranked their risk, the second step is to decide what precautions need to be taken. If significant risks are identified, a plan on how to remove or reduce the risk needs to be implemented. A risk is seen to be significant if the published workplace exposure limits (WELs) are lower than that which the worker is exposed to or the method of work does not comply with usual construction industry good working practices and standards. A WEL is the maximum concentration of an airborne substance to which workers may be exposed by inhalation. It is necessary to take note of the supplier's advice on correct use, handling, storage and disposal recommendations. Besides inhalation, substances can also enter the body orally and through the skin, and this risk should also be considered. Finally, the company's control and monitoring systems must be working and shown to be effective and this includes recording and maintaining a record of the decisions made.

If there is little or no risk, then no further action need be taken, but if there is, then step three and onwards comes into effect: to prevent or adequately control exposure to hazardous substances if it is reasonably practical to do so. A simple rule is that if you have two substances that perform equally well and are of similar in cost, then use the less toxic. Far too often, the safety data sheets and warnings on the containers are accepted and only the precautions necessary to satisfy these needs are taken. Considering alternative methods of executing the process can also result in not using the substance, or the materials could be used in an alternative form or applied differently such as with a paintbrush rather than a spray gun. If prevention is not possible or reasonably practicable, then the exposure levels have to be controlled. This can be achieved by totally enclosing the process, which is difficult in construction activities as they almost invariably require hands-on labour; partially enclosing the process and using mechanical ventilation extraction equipment; ensuring there is good natural ventilation; reducing the numbers of workers exposed to the substance, especially those who are not needed in the process such as those employed on another trade; and developing safer methods of work so that the likelihood of the materials leaking or spilling is minimised. The last resort is to use personnel protective clothing if the aforementioned methods are insufficient. These include respirators, facemasks and other protective clothing such as gloves, aprons and full body coverings. The reasons for all these different approaches is the same in all cases, that of reducing the exposure to a level that a normal healthy worker can safely work at every day without their health being adversely effected. The levels of WELs that are acceptable vary depending upon the material and are found on the safety data sheets.

It is a requirement under the regulations that employees use the control measures installed and report if they are not working properly, and management has a responsibility to see that they do and are encouraged to do so. This is step four, and requires that all equipment, personal protective clothing and procedures should be checked at appropriate intervals. While not common in the construction work environment, step five requires that if the risk assessment concludes that there could be a serious risk to the health of the employees if the control measures put in place fail or that the WELs could be exceeded, then it is necessary to monitor the exposure levels and records kept for a minimum of five years. In step six, if there is a risk that a substance the employee has had contact with is linked to a particular disease or adverse health effect, it is necessary to carry out health surveillance. These records have to be kept for 40 years.

In step seven, if there is a risk of an accident that is higher than the risks associated with normal day-to-day work, the employer must have a plan in place to respond to such an event. This is a sensible approach to all potential emergencies, even when not covered by the COSHH regulations. There should be safety drills carried out from time to time to ensure that, in the event, those involved are properly prepared. This is especially important in construction as the workforce is continually changing.

Finally, in step eight employees involved in the use of hazardous materials and processes must be properly supervised, trained and informed about the materials they are using. Employees cannot be expected to carry out their role of using the control measures (step 4) if they are not properly trained and informed. They need to know what the hazard is, how risky the employer believes it to be, what work procedures have to be complied with, including the use of protective clothing and equipment, and, in the event of an accident, what the emergency procedures are.

21.9.4 Construction (Design and Management) Regulations 2015

These regulations, commonly referred to as the CDM regulations, require that health and safety issues be considered at every stage of the process including design, construction, alterations, repair and maintenance. This means that everybody involved in the process, including the client, designers and the managers of the construction processes, is covered by the regulations.

In the original 1994 Act, two new roles were created, appointed by the client, called the planning supervisor, responsible for the coordination of the health and safety aspects of the design and planning phase, and the principal contractor responsible for the planning, control and management of the health and safety of the construction process. However, all the main participators are involved in the process. The 2007 Act resulted from concern for the amount of bureaucracy in the 1994 Act and the role of the planning supervisor. The role of the planning supervisor was replaced by the CDM co-coordinator in the 2007 Act. The latest Act (2014) has revised this further and refers to all involved in the process collectively as CDM duty-holders. These are:

- *clients* – organisations or individuals for whom the project is being constructed. Their main responsibilities are to make suitable arrangements for managing the project which includes making sure the other duty-holders are appointed; sufficient time and resources are allocated; relevant information is produced and provided to the other duty-holders; the principle designer and principle contractor carry out their duties; and that welfare facilities are provided. They must ensure that before the construction phase begins, a construction phase plan (see Table 21.3) is drawn up by the contractor if there is only one contractor, or by the principal contractor; the principal designer prepares a health and safety file for the project, which is revised if appropriate as things change; and the principal designer and principle contractor comply with their duties. They should also monitor performance in all of these, by setting key milestones and standards. When the building is being handed over they must ensure the new user and third parties are protected from risks caused by ongoing contractors' work, such as snagging. There are also *domestic clients* who have construction work carried on their or their family members' home;

- *designers* – those who prepare or modify designs for a building, product or system related to the construction work. In their designs they are to eliminate, reduce or control foreseeable risks that might arise during the construction process, and for maintenance during the life of the building. This includes the interface between the general public and the construction process. They must provide information to other members of the project team to assist them in fulfilling their duties in managing risks, such as stability problems during the construction process usually shown on drawings or within models such as BIM, and for the health and safety file;
- *principal designers* – appointed by the client in projects involving more than one contractor. They are responsible for planning, managing, monitoring and coordinating health and safety in the pre-construction phase of a project by identifying, eliminating or controlling foreseeable risk. They are to ensure that designers carry out their duties and prepare and provide relevant information to other duty-holders and provide relevant information to the principal contractor to help them plan, manage, monitor and coordinate health and safety during the construction phase. They must establish good communications between parties and liaise throughout the contract with the principal contractor assisting in the production of the construction phase plan. They help the client bring together all currently available information that affects health and safety such as asbestos surveys or previous health and safety files and fill the gaps where necessary. They prepare the health and safety file for the project and update it depending upon changing circumstances;
- *principal contractors* – appointed by the client to coordinate the construction phase when more than one contractor is involved. They are responsible for the health and safety on the site and the interface with the general public; preventing unauthorised access during the construction phase of a project and preparing a health and safety file and construction phase plan for the project. This includes: liaising with the client and principal designer; preparing the construction phase plan; organising cooperation between contractors and coordinating their work; providing site inductions courses; consulting and involving workers in securing their health and safety; providing welfare facilities and ensuring contractors do the same for their employees (see 19.13 Considerate Constructors Scheme). They demonstrate a safety management structure on site identifying those responsible for safety, ensure sufficient supervision is in place and monitor safety perfor-mance. Under CDM 2015 the principal contractor has a duty to involve the workforce in matters of health, safety and welfare. They are to liaise closely with the principal designer;
- *contractors* – defined as those who do the actual construction work and can be either an individual or a company – often referred to as subcontractors. They are responsible for health and safety under their control, which they should monitor, and ensure welfare facilities are available. They also have a responsibility to coordinate their activities with others in the project team and in particular to abide by the instructions given to them by the principal designer or principal contractor. They must only employ qualified people and provide suitable induction courses if not provided by the principal contactor. If it is a single-contractor project they prepare the construction phase plan;
- *workers* – the people who work for, or are under the control of, contractors on a construction site. They must be consulted about matters that affect their health,

safety and welfare; be responsible for their own health and safety and that of others who may be affected by their actions; report anything they see that is likely to endanger either their own or others' health and safety; and cooperate with their employer, fellow workers, contractors and other duty-holders.

21.10 Construction phase plan

The Approved Code of Practice in Managing Health and Safety in Construction (HSE 2007) states 'the health and safety plan should include or address all the following topics where they are relevant to the work proposed'.

It is easy to forget that health and safety issues should be thought about during the design process and that it is not just a construction process problem as this is where the accidents occur. However, thought and careful planning during the pre-tender phase can assist in the overall reduction of accidents and is in line with the philosophy underpinning the CDM Regulations. Its purpose is to provide information for those bidding for the work, which can include both the main contractor and subcontractors. The construction phase plan is to show how safety and health will be managed during this process. Table 21.3 is adapted from *Managing Health and Safety in Construction: Approved Code of Practice and Guidance* (2007) HSE.

When drawing up a plan it saves time later if the principal designer and principal contractor consider the issues that will have to be used in the health and safety file that has to be handed over to the client or the building's end-user.

21.11 Health and safety file

It is the responsibility of the principal designer to produce this file which is a record of information for the client or the end-user, alerting them to any health and safety matters they might have to manage in the future during subsequent maintenance, alterations and extensions. The file includes:

- a brief description of the work carried out including drawing and plans of the building and works;
- any information about residual hazards and how they have been dealt with, such as any surveys in refurbished buildings showing asbestos, which has still not been removed or has been repaired;
- data on structural design such as safe working loads for floors and roofs and identifying elements that have been pre- or post-tensioned;
- hazards associated with any materials that have been used that could cause a problem if not maintained or removed in a certain manner;
- information advising on the methods needed to remove or dismantle installed plant and equipment;
- manuals outlining the safe operating and cleaning or maintenance procedures for installed plant and equipment;
- details of the location and types of services, including emergency and fire-fighting systems;
- information and as-built drawings of the building, its plant and equipment (for example the means of safe access to and from service voids and fire doors).

Table 21.3 Construction phase plan

Construction phase plan
Description of the project
General particulars, including a description of the project; details of the client, design team and consultants, principal designer; the extent and location of existing records and plans
Communication and management of the work
Management structure and responsibilities; health and safety goals and arrangements for monitoring; arrangements for regular liaison between parties on site, consultation with the workforce and the exchange of design information between the client, designers, principal designer, principal contractor and contractors; handling design changes; selection and control of contractors; exchange of health and safety information between contractors; security, site induction and on site training; welfare facilities and first aid; reporting and investigation of accidents and incidents including near misses; production and approval of risk assessments and method statements; site rules; fire and emergency procedures
Arrangements for controlling significant site risks
a) Safety risks Services, including temporary installations; preventing falls; work with fragile materials; control of lifting operations; dealing with services e.g. water, gas; the maintenance of plant and equipment; poor ground conditions; traffic routes and segregation of vehicles and pedestrians; storage of hazardous materials; dealing with existing unstable structures; accommodating adjacent land use; other significant safety risks b) Health Hazards Removal of asbestos; dealing with contaminated land; manual handling; use of hazardous substances; reducing noise and vibration; other significant health risks
Health and safety file
Layout and format; arrangements for the collection and gathering of information; storage of information

There should be enough detail to allow the likely risks to be identified and addressed by those carrying out the work. However, the level of detail should be proportionate to the risks.

21.12 Risk assessments and method statements (RAMS)

A risk assessment (section 21.17) is undertaken to identify (and if possible eliminate) risks and then a method statement is produced to control the risks identified. These are referred to as RAMS and are, for example, provided by the subcontractor before arriving on site to carry out their work. These RAMS and the main contractor's may be requested by the client. The RAMS should be monitored at regular intervals otherwise it becomes little more than a paper exercise.

21.13 Welfare requirements

The CDM 2015 regulation lays out the minimum requirements for the physical welfare provision on a construction site as summarised below. Note that this is the minimum only and further discourse on this can be seen in section 29.10.

Sanitary conveniences should be provided in sufficient numbers in rooms, adequately ventilated and lit, and must be kept clean. Separate rooms must be provided for women and men with doors that can be locked from inside.

Washing facilities should be provided in sufficient numbers, including showers if required by the very nature of the work or health reasons, in ventilated and lit rooms, positioned in accessible places in the immediate vicinity of sanitary conveniences or changing rooms. They must include a supply of clean hot, cold and warm water, soap or other cleaning materials, towels or other means of drying. Separate facilities must be provided for men and women unless intended for the use of one person at a time; the rooms must be kept clean.

Drinking water must be supplied or made available at readily accessible in suitable places. Every supply of drinking water must be a conspicuously marked by an appropriate sign and a supply of suitable drinking vessels unless supplied in a jet from which people can drink easily.

Changing rooms and lockers must be provided if the worker has to wear special clothing for the purposes of work and cannot be expected to change elsewhere because of health or propriety. Separate facilities are to be provided for women and men. The changing rooms must have seating and include drying facilities for clothes and effects, and facilities provided to lock away any special clothing not taken home, clothes not worn during working hours and personal effects.

Facilities for rest rooms must be provided with an adequate number of tables and chairs for the numbers likely to use them at any one time and include, if necessary, suitable facilities for any woman at work who is pregnant or who is a nursing mother to rest or lie down.

There are further calculations to be made about the size and quantity of the facilities to be provided which is a function of the number of personnel employed on the construction site.

21.14 Managing safety within the construction industry

21.14.1 Introduction

While everybody in an organisation has a responsibility for their safety and that of their colleagues, it is essential management take ownership of safety to the highest level, otherwise it is unlikely that the best safe working environment will be achieved. The more senior the management, the more likely they are to be involved in ensuring standards are set, maintained and improved. Some companies have been considering making the annual bonus of their board members' determined in part, if not in whole, by their section's or the business's safety record.

The key elements of successful health and safety management are shown in Figure 21.2, taken from *Successful Health and Safety Management* (HSE 2007).

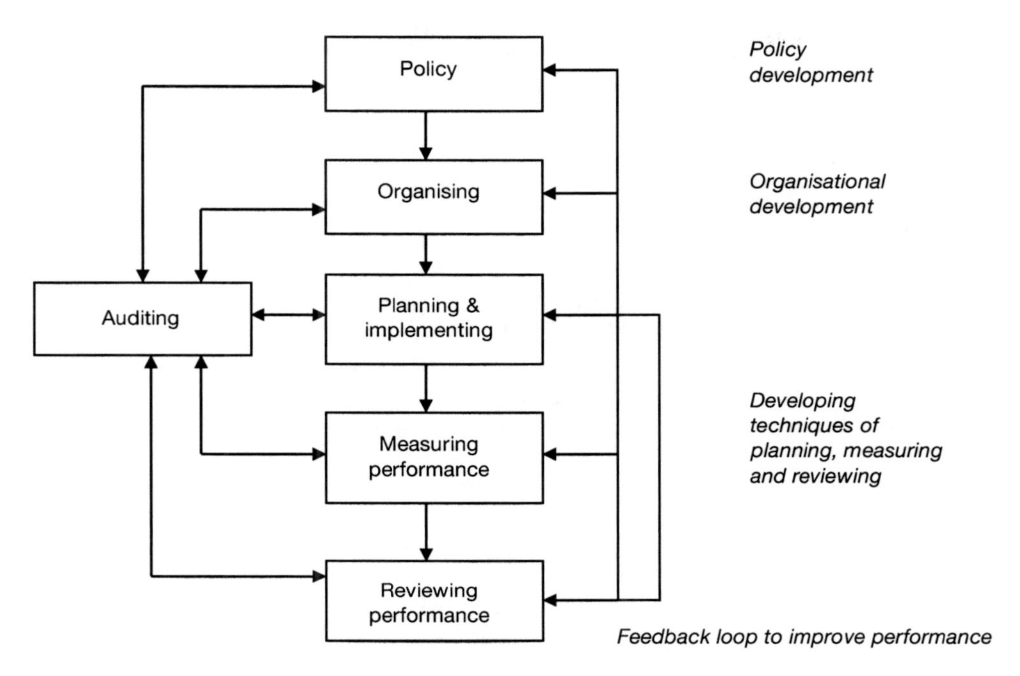

Figure 21.2 Key elements of successful health and safety management

21.14.2. Safety policy

The first stage in any safety strategy is to produce a safety policy. Its prime purpose is to improve the safety of the work environment by reducing the number of injuries and accidents and engaging the workforce in becoming safety conscious. It is a published statement reflecting the organisation's intentions in relationship to the management of health and safety matters. It should also define the organisation's corporate philosophy towards health and safety matters in the context of the business activities and be clearly presented in the form of a policy statement, originating from the board of directors. It is against this that the performance of the organisation can be compared. A typical policy would include similar statements to the following:

1 The company is committed to providing a safe working place for its entire staff. The company will pursue a policy of continual improvement in its safety management and will review legislation and introduce procedures as required to meet its safety requirements.
2 It is the company's intention to take all steps necessary as are reasonably practicable to meet its responsibilities in providing a safe and healthy environment for its entire staff, visitors, suppliers, contractors and the general public.
3 The company will seek to provide: (i) Systems of work that are safe and without risk to health; (ii) safe arrangements for the use, handling, storage, transport and disposal of articles and substances; (iii) adequate supervision, instruction and training, to enable all employees, contractors and visitors to avoid hazards and

to contribute to their own safety and that of others; (iv) a safe place to work, including safe access and egress from the company's offices and sites; (v) the provision and maintenance of a working environment for *all* visiting or working on sites and offices.

4 The company expects *all* working on sites or in the offices: (i) to be responsible for their own safety and that of others who may be affected by their acts or omissions at work; (ii) to cooperate with the company in fulfilling its statutory duties; (iii) not to interfere with, misuse, or wilfully damage anything provided in the interests of safety; (iv) to report accidents and near misses to the person responsible for safety in the workplace.
5 To ensure that the company's Safety Policy is effective, the company will: (i) communicate any changes to the safety policy to *all* working on site or in the offices; (ii) review its Safety Policy regularly; (iii) maintain procedures for communication and consultation with staff on matters pertaining to safety.

Often accompanying the published statement will be further information outlining the company safety organisation structures and the responsibilities of those involved in safety matters, the arrangements for identifying hazards, assessing risks and controlling them.

This document needs to be distributed within the organisation to all current and new employees, but if this is all that is done, it will probably just be filed away and forgotten by the majority of the recipients. It is important that all employees attend some form of training session so not only is the document introduced to them, but their part in ensuring a safe and healthy working environment is made clear. The document must always be kept up to date and staff notified of any changes.

21.14.3 Organisation of safety and lines of communication

The important issue is to demonstrate to the employees of the company how seriously safety is taken. This involves the active involvement of the board and having lines of communication so that senior management can react rapidly. Figure 21.3 demonstrates a typical mechanism for this to happen. In this case, the Managing Director is made responsible for co-ordinating safety issues, but it could equally be another member of the board.

If an accident occurs on site then the following procedure is followed depending upon the severity of the accident. Minor reportable accidents are reported to the regional safety officer, usually on a monthly basis, who then collates all the information from all the sites in the region and submits it to the chief safety officer who presents the whole company record to the Managing Director or Board.

Over seven-day injuries (21.9.2 RIDDOR) are immediately reported to the regional safety officer at the end of the seven days or at the end of month and then to the board via the chief safety officer. Death, serious injuries and dangerous occurrences, as defined for RIDDOR, are notified immediately by phone to the regional safety officer who quickly notifies the regional director and the chief safety officer who, in turn, immediately advises the Managing Director. All reports of accidents will also be sent to the Regional Director on the same timescale.

Under most circumstances, in a large organisation, the site management rarely has direct contact with the senior members of the company other than site visits, but in

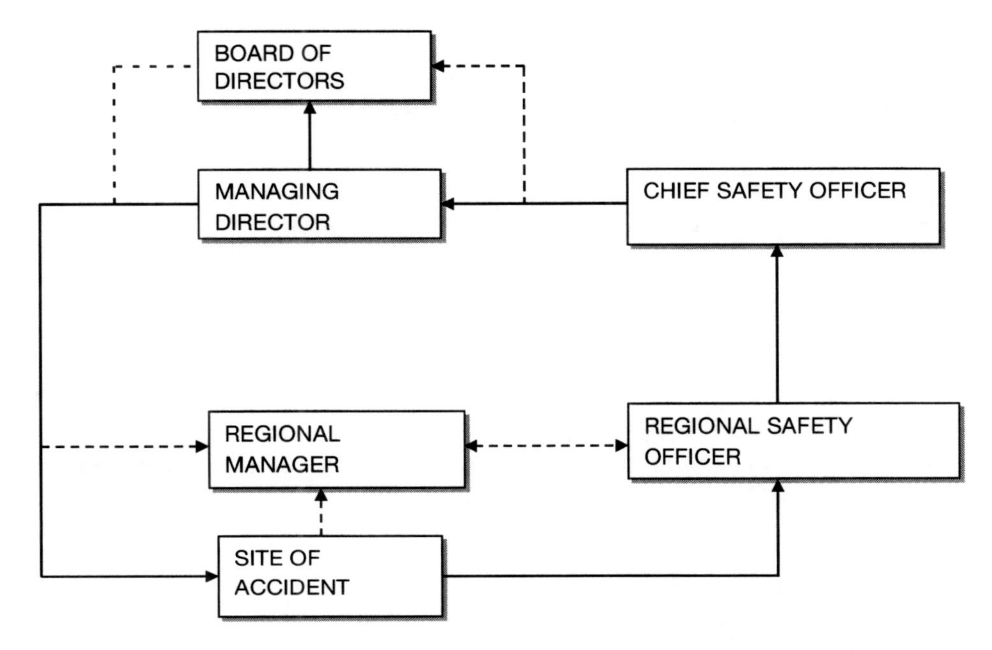

Figure 21.3 Safety lines of communication

this case senior management will contact the site directly if the safety performance is unsatisfactory or if it is excellent. Site management naturally wishes to avoid the former and therefore becomes focused on trying to reduce the number of accidents. Since this pressure exists, the company must ensure that the records produced on site are a true reflection of reality and are not watered down or misleading.

All employees, including temporary staff, must be competent to carry out a task safely, which means that all tasks need to be assessed to establish the skills required to accomplish them without injury or impact on the employees' health. If an employee is deficient in any of the skills required they should be trained accordingly. In all cases the means should be provided to instruct staff on safe working practices and procedures and to give advice and help when requested. If a task is known to be especially dangerous and there is no alternative safer working method, the employee must be selected as being the most qualified and experienced to carry out the work and given the appropriate training.

Management must always lead by example, in their own action and performance, but also in the instructions it gives so employees have clear direction and are not asked to work in an unsafe manner. All supervisors must know and understand their responsibilities, and those they are in charge of must know what is expected of them and the implications of failing to comply, both in terms of the potential danger and disciplinary action. Management should communicate all relevant information about the hazards and risks associated with the task and identify the precautionary measures to be taken.

The safety of any work situation can be improved if the staff are involved and consulted as often they have a more detailed understanding of the task in hand. They should be involved in planning the way the work is to be carried out and in resolving problems at the planning stage and when the task is being executed. If the task has to be repeated and requires written procedures, they could contribute to this process as well. Health and safety issues should be discussed regularly either in a formal setting, if necessary, or as an integral part of the day-to-day management of the works.

21.14.4 Planning, implementing and setting safety standards

The first stage in the plan is to identify the hazards and assess the risks. On a construction project this should be carried out at the tendering stage and when the contract is awarded. The production of the method statement is an ideal place to start as the production process is being looked at in detail for the first time and the level of risks involved could have an impact on the price. If the risk is high for any part of the process, it may mean a change of method so as to eliminate or reduce the amount of risk. To this is added information given by the principal designer generated from the design process. Once the contract is awarded, further detailed hazard identification can be made from the latest information available. Risk assessments are covered in more detail in 21.17 and Chapter 30.

Having established the risk likely from the hazard, and confirmed that there is no alternative method, a detailed analysis of the process and the safety precautions that have to be invoked, takes place. Any method developed must comply with the current health and safety laws and regulations that are appropriate to the business on site and in the offices. This involves protective clothing and equipment needed, limitations on how the operative can work, such as the positioning of guards on cutting equipment, and the protection of third parties such as when welding. From this can be produced standards of performance against which monitoring can take place (see 21.14.5). These health and safety targets should be discussed with the managers, supervisors and all the subcontractors in order to agree that they are practical and achievable. On occasions the work on the construction site can impact on the safety and health of the neighbours and, in the case of alterations and extensions to existing premises, to those working in close proximity to the construction process. Where this is identified, third parties should be consulted so appropriate solutions can be found and implemented. The site requirements should be communicated to the purchasing department so when ordering materials the methods of delivery, packaging and unloading comply with the site's safety requirements.

While the reason for putting all these systems in place is to eliminate accidents, they may still occur, so procedures on how to deal with anything that may go wrong have to be considered. All the procedures must be written down so nothing is left to chance.

Much of this is purely administrative and when implemented will have some effect for the good, but to derive full benefit the overall culture of the business has to be directed to making safety a normal management instinctive practice and not simply seen as an added task that has to be considered from time to time.

Setting standards is one of the ways that the culture of the organisation can be influenced because it becomes engrained in the minds of the employees. However, this will not happen unless the standards set are measurable, achievable and realistic. Wherever there are standards commonly in use it would be usual to adopt them (e.g. HSE regulations), but where not, these have to be developed, taking advice where appropriate.

Standards should be set and applied across the full range of the business's activities. The workplace includes the site, offices, workshops and the regional and/or head offices. Standards are not just about the methodology of the work, but involve environmental controls such as ranges of temperature, humidity, lighting and dust levels. It includes ergonomic issues such as seating and proximity and ease of materials to hand. There are special issues resulting from using plant and equipment often requiring designated training such as using percussion hand tools or walking along the jib of a tower crane. Mobile plant can potentially collide with people or other objects and must be stored so nobody can operate it other than the designated experienced operative. Methods and the frequency for checking guards on machinery or scaffolding guardrails and toe boards, etc. have to be considered. Raw and unfinished materials have to be off-loaded, stored, transported to the place of work, processed, and unused or scrap material recycled or tipped. If the process produces hazardous waste or toxic emissions, the maximum acceptable levels have to be set, as have levels of emissions of dust and noise. The hazard and risk has to be assessed for each of these operations. Finished components have to be unloaded, either stored before being placed in the building or lifted directly into position requiring specialist lifting gear and access. The systems for assessing design solutions, especially amendments during construction, need to be in place. Standards and targets need to be set for the training of all those employed in the workplace and, where relevant, visitors. Agreements should be in place to consult staff, or their representatives, at specified intervals.

It is easy to fall into the trap of having generalities in the descriptions of the standards and targets, which are difficult to either measure or achieve. For example:

- 'Staff must be trained' is a meaningless statement as it does not define what training means and involves, nor who carries it out.
- 'All machines will be guarded' on the surface appears to be specific, but how is this to be achieved and to what standard is the protection, as the ultimate protection will probably make the machine unusable?
- 'All employees will wear safety helmets' again seems reasonable, but does this mean they have to wear them when sat at their office desk and is it acceptable if the helmets are ten years old?

21.14.5 Measuring performance

As in any other part of the business it is necessary to measure performance to discover whether or not the company is successful. The method of monitoring has to be determined, as has the frequency at which it takes place. The frequency is a function of the level of risk and the temporary nature of its use. For example, scaffolding can often be changed to suit production requirements and, because the risk of people falling from heights or the dropping of objects is high, the frequency of monitoring would be much higher than say checking to see whether or not the kettle in the office is still electrically sound.

The aim of monitoring is not just to ensure that targets are being met, but also to continually improve performance. There are three basic sets of questions that need to be addressed when monitoring performance. They are:

1 *Where* are we now? In other words are we reaching the target set?
2 *Where* do we want to be? The answer to this is either on or approaching target, or if having attained targets, to raise the threshold further, but to achievable and realistic levels.
3 *What* is the difference and *why?* If below target it is essential to establish the reasons for this.

There are two types of monitoring of performance: that of being active and that of being reactive. Active monitoring is monitoring before things go wrong. This is seeing whether the standards set are being implemented and their effectiveness monitored. Reactive monitoring occurs after things have gone wrong. Examples of this include investigating the causes of injuries and dangerous occurrences, cases of illness, instances when property has been damaged, and near misses; in each case, identifying why the performance was sub-standard and instigating change, if necessary, so as to improve the situation for the future.

21.14.6 Learning from experience: auditing and reviewing

Whereas monitoring provides the information to be able to review activities and decide on how to improve performance, auditing complements monitoring by looking to see whether the policy, organisations and systems are achieving the correct results. A combination of the results from measuring performance and auditing can improve health and safety management.

The key issues to look for are whether the results demonstrate a compliance with the standards set; whether there are standards to measure against; whether the standards are set too low; and whether the improvements set have been achieved in the timescale set. Finally, an analysis of the injury, illness and incident data can identify any underlying trends, their causes and whether there are any common features such as a high occurrence of back injuries.

It is important that the company learns from the mistakes made rather than giving excuses for the causes. It should take action as a result of the audit findings and this information be fed back to advise future policy and improve performance by increasing the standards.

21.15 Safety committees

The Safety Representatives and Safety Committee Regulations 1977 (SRSCR) came into effect in October 1978 (revised 1997), and were modified by the Management of Health and Safety at Work Regulations 1992 and subsequent amendments. These regulations and Section 2(7) of the Health and Safety at Work etc. Act 1974, provide for the appointment of safety representatives by recognised Trade Unions, and the setting up of safety committees at the request of these representatives. There is no reason why the company cannot set up safety committees where there is no trade union representation, nor do they have to wait for a request from the safety representatives. Under the SRSCR 1997, the trade union safety representatives are to investigate possible dangers at work, the causes of accidents and general complaints by employees on health and safety and welfare issues, and to take these matters up with the employer. They should carry out inspections of the workplace particularly following accidents, disease or other events. Besides attending safety committees, they represent employees in discussions with health and safety inspectors. In carrying out this role they are protected against dismissal or other disciplinary action when taking part in health and safety consultation.

The composition of a safety committee depends upon the way it is set up. To work effectively there needs to be representation from the majority of, if not all, the facets of the workplace. Clearly in a very large industrial works, the committee may become too large and unwieldy if every part of the business is represented, although there is no reason why there cannot be sub-safety committees reporting to the main committee, to deal with this problem. If the employees are trade union members, then the membership of the committee will be selected through the trade union appointment process, but if not, then non-trade union members must be consulted by their employers. On construction sites with a high percentage of subcontractors, representatives from the key trades should be enlisted.

The Guidance to the SRSCR states that the size, shape and terms of reference of the committee must depend upon discussion and agreement between employers and unions. The recommendations are that the committee should be compact; there should be a 50/50 split between management and employee representatives; safety officers and any other advisors should be ex-officio members; and safety committees could also provide a link with the enforcing authorities such as the HSE and local environmental health officer. There is nothing to stop the split between management and employee representatives being different to the 50/50, but it is suggested this should only be in favour of the employees.

The success of a safety committee depends much upon the tone set by and the attitude of management. If they go in with a positive attitude and demonstrate their willingness to listen, accept reasonable arguments for improving safety, and then enact the changes, the committee can make a successful contribution to safety performance improvement. It is not unusual for a member of the management team to chair the meetings because of their chairmanship skills and experience, but this is not mandatory.

A typical agenda for a safety committee could include the following items:

- reviewing the progress of recommendations made at the previous meeting(s). This acts as a demonstration of management's willingness to listen and act;
- discussing the causes of accidents or incidents that have arisen since the last meeting;
- inspecting and reviewing accident and ill health trends with a view to improving working methods and reducing the number of accidents. This highlights the overall trends, but also can be used as a way of establishing whether there are similar accidents that regularly occur that can focus the members on resolving this issue. Illness and absenteeism, if confined to a particular work activity, can indicate higher than normal levels of stress, but may also be about the quality of the management. Sometimes an individual's name comes up with regularity and if not perceived as a malingerer, needs special attention as the operative may require special training;
- examining the health and safety implications of the next phase of construction, or of a new piece of plant or equipment that is going to be introduced;
- examining the safety audits and making recommendations;
- discussing the reports made by the safety representatives;

- reviewing the content of the health and safety training for employees, subcontractors and visitors;
- reviewing risk assessments, especially if a new phase of the construction work is about to commence;
- considering any reports and information of relevance sent by the HSE;
- reviewing any safety publicity and campaigns and their effectiveness.

The frequency of the meetings depends upon the nature, size and rate of changing activities of the work, but once a month would not be unreasonable.

21.16 Instruction and training

Instruction means telling people what they should and should not do, whereas training means helping them learn how to do it, with the emphasis on helping and learning.

The first stage is to establish who needs to be trained and this starts with the person initiating the training programme. Have they the full understanding of the implications and needs of the organisation in terms of the legal requirements, its current culture and image, what the company wishes to achieve and why, and how this can be brought about? The managers and supervisors have responsibility for safety at the workplace, designers need to understand their responsibilities because of the CDM regulations, new recruits have special needs as do part-time employees. In the office, estimators and planners have to comprehend the safety implications of any methods and sequence of operations they adopt, and purchasing officers have to understand how materials are handled and stored on site safely. Should the company also be involved in training subcontract labour, especially if their organisation is small and does not have the knowledge or resources to carry it out themselves?

The next stage is to determine what training is needed and the outcomes expected in terms of the knowledge and experience needed to work safely when carrying out their role, i.e. levels of competence. To give people the wrong training, or too much, is a waste of time and money. This has to be designed to take account of the current state of the company's safety performance and should consider the need for refresher training and updating of existing skills and knowledge, especially if the regulations change or recommendations are made by the enforcing authorities.

Having decided upon the need, the method of how to carry out the training has to be determined. There are two approaches: first, to use in-house trainers, providing they are competent in health and safety matters; and second, to use the expertise of external training providers such as the Construction Industry Training Board (CITB).

It is necessary to prioritise training needs and produce a plan of training. This has to be costed to ensure that adequate resources are made available. The CITB may contribute towards training costs. There are key times and occasions when employees should be trained. These are when commencing work for the organisation, if they have a work or responsibility change, if they have not used their skills for a while, and if new or changed risks occur at the workplace.

To complete the cycle, the effectiveness of the training must be monitored to establish whether the standards of competence set have been achieved and whether there has been any improvement in the company's health and safety performance. A system of permitting feedback from managers and supervisors and those who have been trained, contributes to improving future performance.

21.17 Risk assessment

Risk assessment is discussed in fuller detail in Chapter 30, but as it is an integral part of the improving safety in construction, it would be unfortunate not to include it here. There are key elements in assessing risk in safety: hazard identification, determining who might be harmed and why, evaluating and recording the risk, installing preventative and protective measures, and reviewing and updating the assessment if necessary.

21.17.1 Hazard identification

As indicated earlier, hazard is something that presents a potential to cause harm and, by its very nature, construction is a process where there are a high number of hazards. Table 21.4 gives an indication of just a few of the many typical hazards to be found on a construction site.

It is necessary to systematically assess all the production stages of the process and assess where potential hazards are. This also means calling on the safety data sheets required for COSHH (section 21.9.3).

21.17.2 Evaluation of risk

Once the hazards have been identified, the risk is assessed, using one of the various methods available. There are two factors that affect the degree of risk, the level of harm that might occur as a result of the hazard and the likelihood of the occurrence or the frequency at which it might occur. For example, crossing a busy road is clearly a hazard and the amount of harm received as a result of being hit by a car is high, but the likelihood of this occurring if using a pedestrian crossing, reduces the risk considerably. If the severity of the hazard is multiplied by the likelihood of it happening, a degree of risk is produced. To do this, criteria have to be chosen for them both. Table 21.5 shows a typical example.

Using the assigned values in Table 21.5, the degree of risk associated with an event that is *likely to occur* resulting in an accident *needing first aid attention* would be: $4 \times 2 = 8$, whereas an *unlikely* event that, if occurring, would result in *death* would be: $2 \times 5 = 10$.

Using these criteria, it can be seen that the maximum degree of risk is 30 (5×6), so the values calculated above, 8 and 10, could be expressed as a percentage of the maximum risk, i.e. 26.67 per cent and 33.33 per cent respectively. This calculation

Table 21.4 Typical hazards on a construction site

Working at heights	Electrocution
Falling from heights	Mobile plant and machinery
Objects falling from heights	Hand tools
Excavation collapses	VOCs
Asbestos	Nails poking through wood
Temporary works	Chemicals, adhesives and wood preservatives
Operatives not used to working together	Dust, fumes and fibres
Fragile roofs	Ladders
Handling heavy materials	Loose reinforcement tie wire

Table 21.5 Criteria for assessing risk

Criteria for measuring hazard severity		*Criteria for likelihood of occurrence*	
Description	*Assigned value*	*Description*	*Assigned value*
Minor injury requiring no first aid attention	1	Remote- highly improbable	1
Accident needing first aid attention	2	Unlikely – in exceptional circumstances only	2
Over three-day reportable injury	3	Possible – where certain circumstances might influence occurrence	3
Serious injury	4	Likely	4
Death	5	Probable	5
		Highly probable	6

is carried out because, in risk assessment, a priority rating often is preferred rather than degree of risk (see Table 21.6a).

Note that the percentages chosen for each category are arbitrary and could be any figures deemed appropriate. A variance on this is to use a colour code of red, amber or green to identify the severity of the risk and its likelihood for each hazard identified.

An alternative way is to compare the likelihood of occurrence with the severity of harm of the hazard if the event took place, as shown in Table 21.6b. In this case a scale of 1 to 10 is used to categorise the likelihood and severity, 1 being the lowest

Table 21.6a Priority of risk

Priority of risk	
Low priority (L)	<10%
Medium priority (M)	10–50%
High priority (H)	>50%

Table *21.6b* Priority of risk

Priority of risk *Likelihood of occurrence*	*Severity of harm*	*Percentage risk*
Low	High	10
Medium	High	50
High	High	100
Low	Medium	5
Medium	Medium	25
High	Medium	50
Low	Low	1
Medium	Low	5
High	Low	10

and 10 the highest. The two numbers are then multiplied together to give a percentage risk. The table could be broken down into smaller bands than low, high and medium.

These are but a few ways of assessing risk, but the important issue is it is essential to have some mechanism to do this so that one can focus on potential danger areas in the design and construction processes, during the building's lifetime and final demolition.

21.18 Reflection

It is not unusual to hear one or more of the following:

- 'It will never happen to me' – hospital wards are full of these people.
- 'What I am doing won't take a moment' – but they live with the consequences for a lifetime.
- 'I can't afford the time to build a scaffold so I'll just use this ladder' – you will have plenty of time to reflect on that in your wheelchair after you fall.
- 'I can't afford the cost of all that protective equipment' – so when you are off work you won't be able to afford to put food on the table for your family.

Bibliography

Bird, F.E. (1974) *Management Guide to Loss Control*. Atlanta, GA: Institute Press.

Coble, R.J., Haupt, T.C. and Hinze, J. (2000) *The Management of Construction Safety & Health*. Rotterdam: A.A. Balkema.

Griffith, A. and Watson, P. (2004) *Construction Management Principles and Practice*. Basingstoke: Palgrave Macmillan.

Health and Safety Commission (2007) *Managing Health and Safety in Construction: Approved Code of Practice and Guidance*. HSE.

HSE (1996) *The Construction (Health, Safety and Welfare) Regulations*. HSE.

HSE (2004) *Comprehensive Statistics in the Support of Revitalising Health and Safety Programmes: Construction*. HSE.

HSE (2004) *The Control of Substances Hazardous to Health (Amendment) Regulations*. HSE.

HSE (2005) EH40 *Workplace Exposure Limits*. HSE.

HSG 65 (2007) *Successful Health and Safety Management*. HSE.

HSE (2013) *RIDDOR Explained: Reporting of Injuries, diseases and Dangerous Occurrences Regulations*. HSE.

HSE (2015) *Construction (Design and Management) Regulations*. HSE.

Ridley, R. and Channing, J.(2008) *Safety at Work*, 7th edn. London: Butterworth Heinemann.

22 Waste management

22.1 Introduction

There was a time when the standard joke of an occupier of a new house was that 'there are more bricks left in my garden than were needed to build my house'. While this was an exaggeration, the reality was that a considerable amount of building materials that could have been used were wasted. Indeed, the Cambridge University Centre for Sustainable Development has suggested that there is the equivalent of enough waste left from 13 houses to construct a fourteenth. This is no longer acceptable and the industry increasingly has woken up to the fact that it is both expensive to create waste and environmentally unacceptable.

The UK construction industry consumes the equivalent of 6 tonnes per head of population each year. In the United States of America it is nearer 10 tonnes. It is estimated that the total annual waste created in the construction and demolition stages within the United Kingdom is 72 million tonnes. It is accepted by those working in the environmental field that consumption of materials, as well as energy, has to be dramatically reduced. A significant contribution to this target can be achieved in the construction industry by significantly reducing waste, both at the manufacturing and construction stages, as well as developing ways of recycling the materials used in buildings and infrastructure when coming to the end of their useful life.

There are also legal implications for anyone involved in the production or handling of construction and demolition waste, as there is a duty of care under Section 34 of the Environmental Protection Act 1990. This also includes those that are contracted to dispose of or recover waste. The contractor has a duty of care to ensure that no unauthorised handling or disposal of waste occurs. Further, in transferring the waste to an authorised person, this must be accompanied by a written description of the waste. An authorised person is usually either a waste collection authority, the holder of a waste management licence or a registered waste carrier, all of whom, in England and Wales, will be registered with the Environment Agency who issues them with a certificate of registration. The procedure recommended by the Environment Agency (adapted from *Construction and Demolition Waste: Your Legal Duty of Care* (2003)) is to:

- check the registration certificate of the waste carrier;
- ask where they are taking the waste to and check that the destination is authorised to accept it;

- ensure that both the contractor and the waste carrier have signed the waste transfer note;
- keep a copy of the signed waste transfer note: they need to be retained for two years and in the case of certain hazardous materials, for three;
- produce a description of the waste and state an accurate six-figure waste classification code obtainable from the European Waste Catalogue. Chapter 17 is concerned with construction and demolition wastes (including excavated soils from contaminated ground). Examples of the classification are concrete and bricks, which are 17 01 01, 17 01 02 respectively. There are also classifications for mixtures such as 17 01 07 which is for concrete, bricks, tiles and ceramics not containing dangerous substances;
- if the wastes are hazardous, carry out extra legal responsibilities and possibly complete detailed waste consignment notes;
- be alert for any evidence or suspicion of illegal disposal of waste, or using an unauthorised disposal site in which case the waste carrier should not be given the waste material to handle and the Environment Agency should be advised.

22.2 The cost of waste

The true cost of waste is not just the replacement cost or the purchase price of the excess material. It includes the administrative cost of placing the order, progress chasing, checking on arrival and arranging payment. There are further costs in managing the waste disposal and monitoring that it is done correctly and there may be an insurance liability implication affecting premium costs. The cost of landfill is based upon the skip rather than its composition. Since 60/70 per cent of waste disposed of is actually air voids, the less waste sent, the better. Figure 22.1 demonstrates the flow of materials from their material source until returned back as landfill, but does not indicate the return flow of recycling or reuse. It can be seen there are many stages where costs can be incurred such as that of transporting the material to site from the manufacturer or extraction supplier, storage, processing on site, removing from site, landfill charges as well as the potential for reuse or recycling. The latter two can occur at any of the stages of extraction, manufacture, construction and demolition. The activities contained in the highlighted box are concerned purely with the extraction and manufacturing processes.

There are also environmental impact implications in waste disposal directly related to the extraction, manufacture or processing operations that can be lessened if the correct amount of material has been ordered and used. These include the raw material, ancillary materials necessary for the process, wear and tear on plant, consumables used in the office, wear and tear on protective clothing, any packaging for protection during transportation, energy and water. There is an environmental impact from emissions to the atmosphere and effluent treatment resulting from the extraction and manufacturing processes. There can also be a flow of materials from demolition back to the construction and manufacturing processes.

Finally, there is the impact on the company's reputation. This can be as a result of not carrying out disposal correctly or the visual impact of seeing so much waste on the site making the site look untidy and potentially less safe. It can also affect the morale of the employees on the site, especially if the waste is excessive and clearly unnecessary.

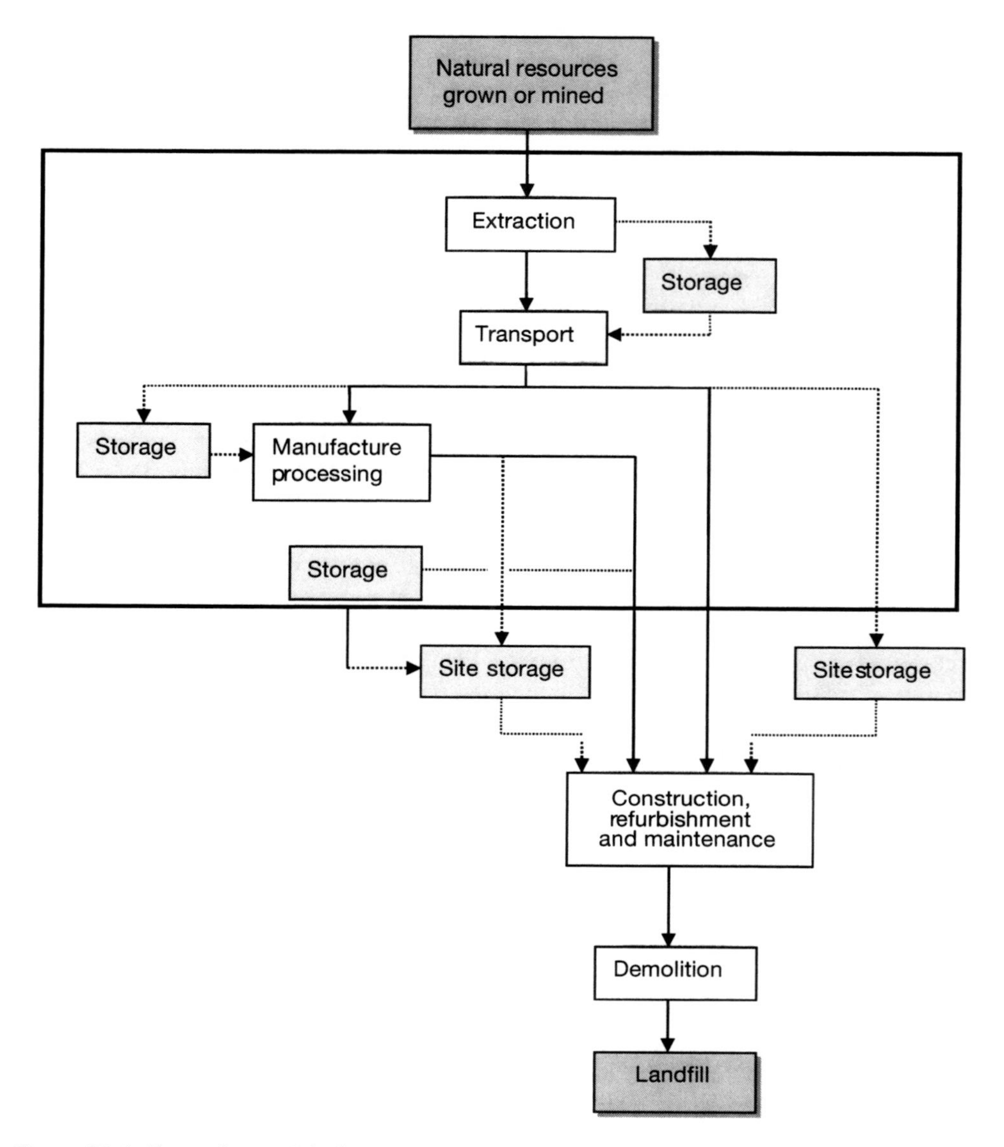

Figure 22.1 Flow of materials from source to landfill

22.3 Defining waste

Waste is defined in the Waste Framework Directive (2008/98/EC) as 'any substance or object which the holder discards or intends to, or is required to discard'. Waste is anything that is discarded, even if it has value, until such time as it has been fully recovered after which it ceases to come under the rules of the Directive. This deals with safe disposal of waste that exists and continues to occur. However, the fundamental question to be addressed is why not try to eliminate waste at every stage of the process from extraction to demolition. It is outside the remit of this text to consider the stages prior to construction, although as will be seen, designers and contractors occasionally can have an influence, especially on the manufacturing process. Figure 22.1 shows materials can be sourced either directly from the extraction stage, notably sand and aggregates, or from manufacturers and processors. These products can either be delivered directly as and when required in the construction process known as 'just-in-time', or they can be brought and stored on site prior to use. For the purposes of this discourse, the latter category can also include the intermediate stage of purchasing from builders' merchants where materials, often bought in bulk, are broken down into smaller units or packages. In these instances, the material has to be unloaded, stored and then redistributed to the place of work on the site. This increases the risk of damage.

22.4 Causes of waste

Having established the reasons and need to control waste, it is necessary to categorise the main causes of waste for analysing its material content, and how this should be dealt with. Figure 22.2 divides the waste into three distinct categories. It is not drawn to scale, as the amounts would vary from site to site depending upon the materials being used and the quality of the waste management. In describing these, there are occasions where some waste could be assigned to more than one category.

22.4.1 Natural waste

There will always be some waste that is unavoidable. Examples of this include waste resulting from cutting bricks to bond, residue left in a can, small quantities left over from a batch size, cutting out a shape from plywood or making a hole in an element of the building for a service pipe. Precisely when natural waste becomes salvageable is a moot point, and difficult to give a definitive answer. Some waste is impractical to reuse or recycle, but other waste could be, except that the costs of doing so could be greater than the value of the material saved and the total costs of disposal. Where the line is drawn depends upon the economic climate of the country at the time. However, this decision should be made based upon calculations rather than just assumption. It would normally be taken account of in the tender price.

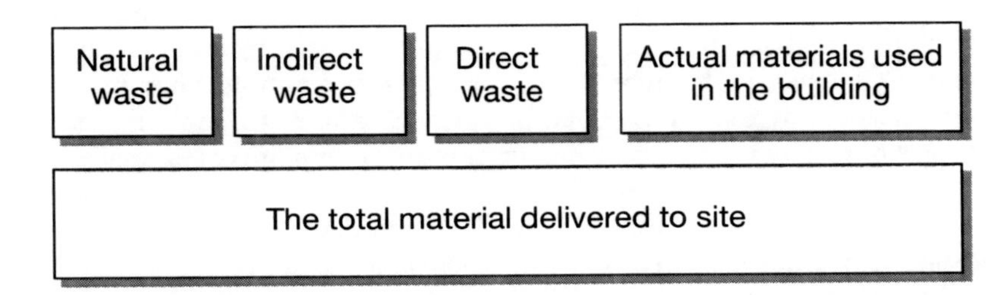

Figure 22.2 Waste on building sites

Source: Adapted from Skoyles and Skoyles (1987).

22.4.2 Direct waste

This is waste that can be prevented with good management, planning and foresight. It occurs as a result of materials being lost on the site, such as bricks submerged in the mud, materials having to be removed from site because they have been over-ordered or being stacked out of sight and forgotten about, in which case they have to be reordered.

Typical examples of direct waste include:

- *transportation waste*: every time materials are transported, loaded, or unloaded there is a possibility of breakage. There is greater risk when transportation occurs on site because loads are not necessarily as well packaged and the terrain is more uneven;
- *site storage*: damage occurs as a result of poor preparation of the storage areas, bad stacking and storage. Inclement weather can degrade materials if not properly protected, and certain materials with limited shelf lives, such as bagged cement, can become unusable if not used within a stated period;
- *conversion waste and cutting waste*: the dimensions of many construction materials do not match the actual size needed in the building. These include timber and plasterboard sheets, roofing sheets, timber, plastic and metal tubes, reinforcement steel bars and meshes, rolls such as plastic sheet and felts, and drainage products. Waste occurs as a result of cutting material to size, bond and into irregular shapes;
- *fixing waste*: during the construction process, losses will occur as a result of errors made by the operative, such as damage caused by dropping material or a component, and incorrect fixing requiring remedial work entailing partial or complete replacement;
- *application and residue waste*: this includes materials left in unsealed containers, materials dropped or spilt, and others, such as plaster and mortar, left over at the end of the day surplus to requirements and allowed to set;
- *waste due to uneconomic use of plant*: ideally plant and equipment should be used continuously throughout the working day. Any time it is left idle or with the engine left running when not in use is a waste. Plant should be selected to be 'fit for purpose' and not be over- or undersized for the job in hand;
- *management waste*: these are losses arising from poor organisation or lack of supervision resulting from bad or incorrect decisions;
- *waste caused by other trade*: the construction industry requires many different trades to follow each other. Sometimes following trades cause damage to the previous trade's work through lack of care or understanding. This can require remedial work to be carried out. It can also include damage caused by others such as driving dumper trucks over kerbstones or grazing concrete columns when manoeuvring;
- *criminal waste*: any materials stolen or vandalised have to be replaced;
- *waste due to wrong use*: it is not uncommon to observe incorrect materials being used. Sometimes this is acceptable, such as a steel reinforcement fixer using an off-cut left at the end of a cutting schedule of larger diameter than required, because otherwise it would be wasted. On the other hand, using bricks in lieu of blocks to avoid cutting is a waste of the bricks although it may have advantages from a productivity point of view. Using materials of too high a quality is also wasteful. Materials of lower quality used will have to be replaced;

- *waste stemming from materials wrongly specified*: standard specifications may have been used in the B of Q without analysing the required quality resulting in over-specification. This can mean either using too high quality materials or expecting too high a level of quality. For example, the quality of finish does not have to be as high, if it is to be covered up as against seen, providing all other appropriate considerations are met, such as structural needs;
- *learning waste*: this occurs when trainees are learning, not properly qualified tradesmen are used and qualified tradesmen are carrying out operations for the first time.

22.4.3 Indirect waste

Indirect waste is the difference in costs between that estimated for and the actual costs incurred that cannot be passed on to the client just because it costs more. Examples of this include:

- substitution of another material – this can occur in several ways, such as where facing bricks that have been over-ordered are used in lieu of commons; using bricks to save cutting blocks to make up a bond at a perpendicular end such as at a door entrance or butting up to another wall;
- use of materials in excess of quantities allowed for in the specification, or on the drawings such as the size of the bucket determining the foundation width, or uneven surfaces having to be plastered;
- builder's errors, such as over-digging foundations depths and widths, and incorrect setting out;
- returning to complete unfinished work.

22.5 Waste arising outside the contractor's organisation

22.5.1 The design stage

The design team can contribute to waste in various ways, particularly if not fully understanding the production and manufacturing processes. This is not surprising as this is not part of their normal education and training and is not meant as a criticism. Examples of how waste is created are:

- using dimensions that are not compatible with machine, material or component sizes. In the former case an example would be selecting a strip foundation width different to the excavator bucket sizes available and in the latter selecting the centres of ceiling joist not compatible with standard plasterboard sheets;
- using specifications higher than are required such as the quality of the finish of concrete surfaces that are to be covered, over-designing reinforcement in a structural element and requiring complex details that in practice will never be noticed by the users or general public;
- designing solutions difficult to construct and more costly than the design advantages gained. This is of course a subjective judgement and difficult to quantify;

- making variations to the design as the building is being constructed, causing both interruptions to the works and flow of production, and often extra or different materials being required. Sometimes work done has to be dismantled to accommodate the changes;
- a lack of understanding of how the different trades are coordinated to carry out their work – this can mean them returning to the work several times to complete it. A more complete understanding could mean the design is carried out in such a way as to permit completions of the works with fewer visits and less likelihood of damage caused by other trades;
- the potential vandalism – materials and layouts should be selected to reduce the chances of this occurring. Examples of this are avoiding the use of copper pipes and lightning conductors on the exterior of a building at ground level unless protected, not providing easy access onto roofs by conveniently positioned downpipes, and not using light-gauge cladding materials at ground level, deterring ram raiders from entering.

The clients also cause waste by changing their minds as the project progresses, placing emphasis on unnecessary high standards for the purpose of the building, or specifying lower standards, which serve the function of the building in the first instance but result in heavy maintenance costs and replacement of materials and components at more frequent intervals.

22.5.2 Building materials and components

Materials manufacturers often do not seem to take account of the reality of construction in deciding the dimensions they make their components. The common example of this is plasterboard sheeting, which comes in 2.4 m × 1.2 m boards. Yet the floor to ceiling height in the majority of low-rise residential buildings is no more than 2.3 metres, which means a minimum wastage, taking this dimension alone into account, of 12.5 per cent. Bovis Lend Lease in 2001 estimated that 20 to 30 per cent of all London construction site waste was plasterboard. On large contracts, contractors have now been able to insist that the sheets come in the lengths they require, because they have the purchasing power, but smaller builders still rely on the local builders' merchants for their supply. To change the width of boards is more problematic for the manufacturer, but variations in the length are relatively easy.

Rigid and semi-rigid pipes come in standard sizes. In many cases this is not a major problem, because the pipes can be jointed and much of the waste material used, although a wider range of lengths would reduce the number of connectors and hence reduce waste. Pipes such as rainwater goods that are seen, need to be jointed as infrequently as possible for aesthetic reasons, but the current range of pipes' sizes is limited and still results in significant offcuts even though the BRE recommended as long ago as the 1980s that manufacturers should produce half and quarter sizes. There is also a similar problem with sawn timber.

The design of packaging used for protection of goods during transit and to permit bulk handling at the destination is another source of potential waste. Bulk packaging such as used for bricks, blocks and pipes inevitably means that waste will occur as the builder may not require the total content of the package. The percentage waste where large quantities of these materials are being used is small, but on smaller orders this increases. On the other hand, the advantages of this form of packaging from a production point of view are self-evident, because of the speed of unloading, the lack of labour resource needed and the ability to deliver on time and with more flexibility, but it has been suggested that smaller packages of materials could also be accommodated.

The trade literature provided by some manufacturers does not always include the full range of sizes available. This means that purchasers will tend to select from those shown rather than investigate further and reduce waste as a result.

The materials used in packaging can be excessive, as is well documented in the retail trade, often to make the purchase more attractive to consumers. There should be no need for this with building products, except perhaps in the DIY stores for certain items, although this is debatable. Some of the packaging such as strapping and cartons can be recycled but not reused, but dunnage and pallets are usually reusable. Some 10 to 15 per cent of all construction waste is packaging. The problem is how best to return them to the supplier.

Wastage also occurs as a result of poor packaging, loading, stacking or unloading by the supplier. Not only is there waste material, but it can also affect production if the materials are 'just-in-time'. Thus, it is important to note, as the industry moves in this direction, that it is essential that all goods arrive intact.

The contractor must be advised by the supplier of any special requirements for the unloading of the materials or components. This may be where the lifting points are, the lifting equipment needed, the way it should be stacked or stored, and the way up the product or container should be kept. There should be a clear understanding between the site and the supplier of the sequence in which products are required so they are packaged correctly. For example, precast concrete or steel structural components should be loaded so they can be taken straight from the vehicle into the building without causing any undue stresses on the vehicle due to the load becoming unevenly distributed.

22.5.3 Plant sizing

There is a conflict of interests between the needs of an individual construction site and those of the plant manufacturer. The former has specific issues to be addressed whereas the latter has to produce plant that will satisfy a wide range of customers whose construction operations are not identical. Examples of this include excavator bucket sizes not compatible with the width of trenches resulting in producing wider-than-needed trenches, cranes having over-capacity or loads having to be split for lifting purposes, and a limitation in the range of tasks the plant can be used for.

Other plant may be limited in its manoeuvrability, for example not being able to easily steer around obstacles, or have inadequate suspension to cope with the uneven nature of construction site terrain resulting in spillage, damaged material or slower operation because of the caution needed.

22.5.4 Communications

In all the issues discussed above, it becomes clear that good communication between all the stakeholders is needed. Further, if each understands the others' needs and limitations, this will improve the situation.

22.6 Construction site waste

There has been a significant change of attitude within the construction industry in recent years to the disposal of waste due to increasing awareness of environmental issues and the introduction of landfill taxes that have increased the costs of disposal several times over since introduced and will almost certainly continue to rise.

Waste can be divided into three categories that can then be subdivided further. The three types are:

- *inactive waste* – these are materials that do not undergo any significant physical, chemical or biological reactions or cause environmental pollution when deposited in a landfill site under normal conditions. The kinds of construction materials that fall into this category include soils, rocks, concrete, ceramics, masonry and brick and minerals;
- *active waste* – these undergo change when deposited. These include acids, alkaline solutions, pesticides, wood preservatives, oils, asbestos, timber, plastics, bitumen and batteries. Active waste attracts a higher landfill tax than inactive;
- *special wastes* – these are classified as being deemed to be dangerous to life. Some of these overlap with substances classified as active waste. It depends on the composition and relative risk. They are materials that are toxic, very toxic, corrosive, reactive, explosive, carcinogenic or flammable. Typical examples include acid and alkaline solutions, waste oils and sludges, and wood preservatives.

22.7 Waste recycling and reuse

Materials should be segregated at the place of work. To attempt to do it at the collection points is difficult once the materials have been mixed because of the nature of the materials especially those used in wet trades. The categories of material from new-build operations normally segregated are the inert materials such as concrete, brick, block, timber, paper and cardboard, metals and some plastics. Soils, ceramics, plaster, insulation materials, bitumen, cans and containers used for toxic materials will usually finish up in some form of landfill. In maintenance and refurbishment other materials will be involved, notably asbestos which will have to be dealt with carefully, the method determined by the category of the fibres.

To segregate on site at the place of work requires space for the bins collecting the materials, methods for getting the material from upper storeys to the bins, usually chutes, and methods of transporting the materials on the horizontal plane. Finally systems have to be in place to ensure that the segregation process at the workface takes place, bearing in mind that the majority of operatives will be subcontractors. This can be accomplished either by having a gang directly employed by the main contractor, or as part of the contractual arrangements with the subcontractor. The advantages of segregation on site are that it is easier to see what types of waste are being generated and monitor the qualities, leading to focusing on where waste reductions should be targeted.

However, all of this fails if there is nobody locally who wishes to collect and recycle the materials. Most contractors will have the names and addresses of companies who will do this, and the local Environmental Health Office can usually help. Some businesses advertise on the web.

There is a move towards leasing the material for the duration of its life, then collecting and recycling it when the end of its life is reached. Two examples of this are road surfacing materials and carpets. The main materials where there is a proven market for recycling are concrete, blacktop, topsoil, excavation spoil, timber, metals, bricks, stones and pipes. There are also specialist markets for reusing architectural features such as fireplaces, rainwater goods, plaster covings and ceiling roses. Plastics is a developing market and some packaging can be reused providing the supplier wishes to participate. A word of caution with timber: increasingly timber used in buildings has had a preservative treatment, using chemicals that are toxic and not always obvious, so reuse of these materials, especially if previously structural, needs to be carefully thought out. Burning for fuel could result in toxic emissions.

Reductions in waste can be achieved by analysing the amount of waste being produced, determining the cause and putting in place an action strategy to reduce or eliminate it. For example, setting waste targets, changing the method of transportation and packaging, arranging for materials to be pre-cut before delivery to site and modifying a design. Table 22.1 gives an indication of the breakdown of construction waste.

22.8 Implementing a waste minimisation policy

A waste minimisation programme comprises various stages and starts by concentrating on attitude and the culture of the organisation. As with any company policy, to be successful, all members of the organisation have to eventually engage in the process. This can only occur if the senior management are committed to making it happen. While individual site managers can have an impact, without the support from above it is unlikely they will be as successful as they might be. This is because to minimise waste other key players, such as the purchasing department, have to play their part.

To engage and convince the senior staff, they will need to see the data that demonstrate the implications and cost savings that will be made, and contrary to popular belief, it would be rare if savings were not possible. Once convinced and committed, senior management will have to commit resources, usually people's time, to the various stages of the process.

Table 22.1 UK Construction waste

UK construction waste *Materials*	*(%)*
Packaging	25.0
Timber	13.8
Plaster and cement	11.5
Concrete	10.2
Miscellaneous	9.6
Ceramic	8.6
Insulation	7.5
Inert	7.1
Metal	4.0
Plastic	3.2

Source: BRE.

The next stage is to engage the staff in the process. There may be some reluctance to change, as it can appear as yet another senior management initiative requiring more time and effort for an already time-pressurised employee. Further, if asked to look for areas of wastage, if found, an employee can be afraid that they will be criticised for not dealing with it earlier. Using several methods including posters, newsletters, team briefing, workshops, information enclosed with pay slips, and suggestion schemes can raise awareness.

It is not unusual for waste to be considered simply as the material left after being fashioned, but by stepping back and investigating the whole process it can be seen to have wider implications as shown in Figure 22.3, showing a flow diagram of the production process, and Figure 22.4 showing the flow diagram of the construction of a one-brick-thick brick wall, assuming no scaffolding or plasticisers are required.

If the process is then changed to that of an external cavity wall, the basic materials would additionally include wall ties, blocks and insulation material. The waste stream would also include strapping from block packs, block offcuts from bonding, damaged blocks and blocks surplus to requirements, insulation offcuts and insulation surplus to requirements. If the wall included openings, then the use of lintels and formers to form the opening would be added.

An analysis of the causes of waste is necessary to establish whether the waste is as a result of the supplier (standard size components and materials etc.), the type of packaging, or the production process including how the materials are handled and stored on site, so suggestions on how this can be improved is thought about and action taken prior to production commencing.

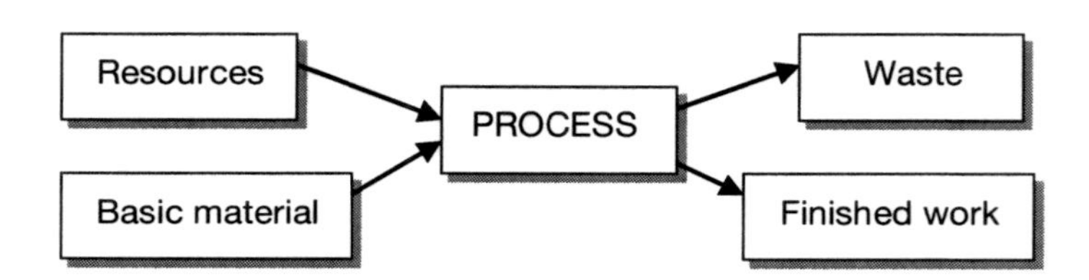

Figure 22.3 Flow diagram of production process

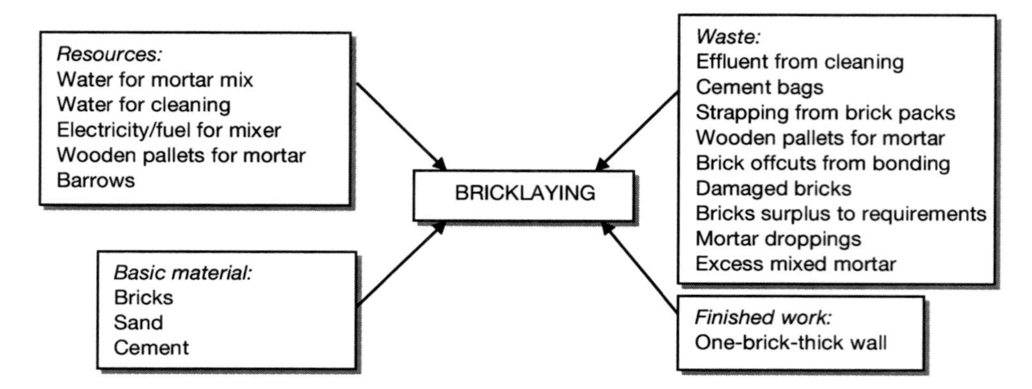

Figure 22.4 Flow diagram of the production of a one brick thick wall

The collection of data can be divided into two parts, the theoretical waste and the actual waste. The former can be established by taking the data from the estimate, the amount of wastage allowed for and the quantities requested on the orders. From this can be calculated the theoretical waste for all the basic materials. The latter can be established by observing what happens to the waste stream and its content. If these are not the same, analysis of why should be undertaken. However, if similar, complacency must not be allowed to set in. The next stage is to set realistic targets for improvement and continue the monitoring process with the aim of eliminating waste as far as is practical and possible. Note should be taken of any extra ordering that is required in excess of the initial order.

Part of this process is also to prioritise which materials should be concentrated on initially in reducing waste and then recycling at disposal. Questions that should be asked include:

- Which waste is the most expensive to replace?
- Which waste has the greatest recycle value?
- Which waste is the easiest to collect?
- Which on disposal has the greatest environmental pollutant effect?
- Is it a process that causes a nuisance to others?
- Is the best method of work being utilised?
- Are there obvious ways of reducing waste?
- How easy is it to segregate the waste?
- Are there recycling businesses available in the area to take away the waste?
- Is there another waste stream created in the process such as the production of effluent?

22.9 Disposal of hazardous waste

The fact that they are hazardous wastes means there is a risk of contamination and environmental pollution if not disposed of correctly. All such wastes should be clearly identified and all those who might be exposed to them should be made aware of the implications and the methods the site has put in place to deal with them.

Storage

All hazardous wastes should be stored safely in correctly labelled containers. It is uneconomical to include non-special wastes in these containers because of the cost of disposal of special wastes. Different special wastes should not be mixed, unless it is known that the mixing is harmless, otherwise there can be reaction between two or more materials and/or the subsequent management of the waste at disposal made more difficult.

Transport and disposal

The Environmental Agencies have to be informed in advance if hazardous waste is to be disposed of using a special waste consignment note. This note must also accompany the transport of the special waste in lieu of the usual waste transfer note (22.1). An appropriate disposal route has to be identified so as to reduce risk in the event of an accident en route and the documentation has to be stored safely.

Bibliography

Biffa. (2003) *Future Perfect*, ch. 11.

Coventry, S. and Woolveridge, C. (1999) *Environmental Good Practice on Site*. London: Construction Industry Research and Information Association (CIRIA).

Curwell, S.R., Fox, R., Greenberg, M. and March, C.G. (2002) *Hazardous Building Materials*. London: E & FN Spon.

Environment Agency (2001) *Waste Minimisation: An Environmental Good Guide for Industry*. Environment Agency.

Environment Agency (2003) *Construction and Demolition Waste: Your Legal Duty of Care*. Environment Agency.

Environment Agency and The Natural Step (2002) *Towards the Sustainable Use of Materials Resources*, 2020 Vision Series No. 4. Environment Agency.

Hazardous Waste Regulations (2005).

Skoyles, E.R. and Skoyles, J.R.(1987) *Waste Prevention on Site*. London: Mitchell.

Waste Framework Directive 2008 (Directive 2008/98/EC).

23 Stock control and materials management

23.1 Introduction

This chapter can be read in conjunction with Chapter 24 Supply chain management and Chapter 15 Purchasing. When a high percentage of the site construction work was carried out by the main contractor using directly employed labour, the control of stock, or lack of it, was a significant issue. Much of this obligation now has been transferred to the subcontractors. The contractors still have to appreciate the implications of stock control as they have to procure materials and components for labour-only subcontractors and they can also be the subcontractor on a large contract. They have the responsibility to ensure the materials brought onto site for use in the building are of the correct quality, as rejected material can affect progress. As the industry moves towards increased supply chain management, there is a further role in supporting subcontractors in their control processes.

The purposes of stock control can be summed up in two simple statements:

- to control the number of items necessary to run the operation without carrying unnecessary stock; and
- to act as a buffer to insure against delay and uncertainty.

23.2 Types of materials

It is necessary to categorise materials into different types, as they require different treatments in the way they are managed and controlled. In Chapter 1, the concern was about protecting them from damage, but the emphasis is different for stock control as one of the key issues is the duration the materials are needed on site before being incorporated into the building as this has a direct impact on the date the material is ordered. Categories include:

- raw materials such as bricks, blocks, aggregates, cement, uncut reinforcement steel and timber that have to be processed to produce the final component of the building. In some cases they will have to be stored and then transferred to the place of work, in others, distributed to the place of work on arrival;
- partially processed materials such as cut and bent reinforcement steel, ready mixed concrete that in the case of the former has to be stored, assembled either in situ or prefabricated and lifted into position, or in the case of the latter poured directly into the required location. Other examples are softwood timber window- and doorframes that have to be painted;

- completed components such as uPVC windows, precast concrete cladding units, kitchen fittings, hardwood door frames, boilers, and air conditioning units;
- small items used in the construction process such as nails, screws, nuts and bolts, ironmongery, paints, adhesives and preservatives;
- office supplies such as paper, pens, pencils, paper clips, envelopes and printer cartridges;
- maintenance materials such as cleaning fluids, light bulbs and refuse bags.

Inevitably some materials can fit into more than one category; for example, is an off-site manufactured component completed or partially completed on arrival on site because fixing materials are not included?

23.3 Stages of stock control

To fully appreciate the subject of stock control it is necessary to start at the manufacturing stage and understand that manufacturers of different materials produce their output in a variety of ways that in turn can affect the purchaser's requirements. There are five different examples:

- Rolled steel is a continuous process that only stops when the rollers controlling the size of the section wear out and reach the outer limit of tolerance sizes permitted. At this point the rollers are changed to new ones either to continue making that section or start another. If steel is in short supply and little is in stock, it may be some time before the size required is made again.
- Precast concrete components are not made in the sequence they are to be erected on site, but in runs of the same type of component because to change the mould to another type may mean a day or more lost production. On completion the components will have to be stored in the yard until they are properly cured. One of the effects of this is that if a component is damaged on site it cannot always be replaced immediately unless the supplier has one of the same types in stock.
- Kitchen units on the other hand are largely made from the same-size component parts, the only difference being the facings such as drawer fronts and cupboard doors, so the supplier can react more rapidly.
- Bricks and block manufacturers are continually making the same products and therefore the only problem likely to be met in replacement or increasing an order is if stocks are very low because of demand elsewhere. However, special bricks will take longer to replace as it is unlikely they are in stock.
- Plant and equipment are often purpose made and sometimes designed by the supplier. Depending upon the complexity of the piece, this process can take several months to complete.

Manufacturers require different lengths of time from the order being placed until they can deliver to site, known as the lead-time, and many not be able to react rapidly in the case of design changes or replacements. It should be remembered, except in certain circumstances, the manufacturer is servicing many other clients, some of whom may be more important to them long term than others.

Once on site a further period of time may have to be added depending upon what processing has to be carried out before fixing into the building or to build up a stock of materials or components before this occurs. For example, a section of the building may comprise several large elements, each individually off-loaded by a mobile crane brought in for this purpose. When all components have been delivered a larger and very expensive crane is brought in to erect the components that may only take a short time to accomplish.

Bearing these points in mind, the quantity of materials required for the contract, when they are required, and the quality and specification of the goods can be established. The selection of the suppliers is covered in Chapter 24.4.

23.4 Problems of excessive stock

While the construction industry is shielded to a certain extent by the provision in most contractual documents for the client paying for goods on site even though not incorporated into the building, there are still many reasons why excessive stock is a problem and should be controlled; for example:

- Excessive stocks take up capital that could be better used, and in spite of the shield mentioned above, most small items in the stores, such as nails and screws are rarely costed in the monthly valuations.
- They take up space on the site and absorb further capital for all the supports and protection against the weather that has to be provided.
- They are more likely to be damaged, resulting in unnecessary expenditure and possible delays incurred in replacement.
- It may be necessary to double handle stock to elsewhere on the site and miss the opportunity to have just-in-time delivery when the components go directly into place in the building.

23.5 The storage function

This discourse includes all materials arriving on site that are stored or, in some of the issues, for goods delivered just-in-time. The value of the materials incorporated into a building represents anything from 40 to 60 per cent of the final cost of the building, so tight control is essential. This involves receiving the goods, inspection, issuing, stocktaking, dealing with replacements and disposal of damaged or excess goods. There is a danger on construction sites that this responsibility gets shared between various persons; for example, the secure store accommodation being in the control of a store-keeper, who will also be responsible for the receiving of goods, and the storage areas around the site being the responsibility of another. In the latter case, that person will have many other duties besides looking after stored goods with the result they are not looked after and controlled as they should be.

23.5.1 Receiving

There is certain information the receiver of goods must have in their position to carry out their function properly. They need a copy of the order indicating the quantity and types of goods, the date of delivery and in the case of scheduled deliveries, all the dates when goods are planned to arrive. They should be furnished with information as to where the materials are to be stored or in the case of just-in-time deliveries, where the materials are to be sent and any special advice and instructions on checking procedures for a particular load. This information gives the storekeeper the knowledge to be able to plan for space and the methods of storage such as shelving and bins in the internal sheltered secure stores.

When goods arrive they need to be checked against the order to ensure all that is expected has arrived and is undamaged and to the correct quality. If for any reason it does not comply, action has to be taken to rectify the situation as production could be held up as a result. There need to be clear procedures as to what must be done and at all times records kept so the story can be tracked at a later stage if necessary. Decisions such as if and when damaged goods are returned to the supplier, who notifies the supplier of this or other discrepancies, should have been decided when setting up the site. It should be noted that the delivery ticket may not tally with the order requirements and this should be checked. The date and quantity of accepted goods received should be recorded in such a way as to allow the quantity surveyor to calculate the value of the goods received on site for the monthly valuation. With the development of computer software this process has been simplified.

23.5.2 Inspection

It is not possible to check every item on every lorry as it arrives as in many cases it would take too long or mean unpacking items to inspect the contents, so procedures have to be in place to account for this. There are four ways checking can be used to cover most eventualities.

- Experienced staff carrying out visual inspections. The condition of materials delivered in bulk, such as bricks, aggregates and sand can readily be seen to be of the right standard. Bricks and blocks will show signs of damage and if sand has excessive moisture content, water will be seen coming out of the tailgate.
- The complete contents of the load are inspected if practical. These are usually components, such as precast concrete, structural steel, window and door frames, manufactured cladding components and kitchen units. These usually are readily visible.
- A random selection of smaller items can be taken and checked and if these are correct, the assumption made that the rest are also in order and the load accepted as per the delivery note.
- Deliveries can be tested on arrival. Samples of ready-mixed concrete can be taken and tested for slump before pouring and cubes made for analysis later. Bulk loads of sands and aggregates can be sent to a weighbridge to check the load is as heavy as stated.

23.5.3 Storage and location

This is discussed in materials storage and handling (see 19.6), which looks at types of materials and the criteria for storage. The key issues are that material should be stored correctly to stop damage by the elements, third parties, contamination, distortion or breakages; it should be made difficult for it to be stolen and the flow of materials out of storage should to be on the principal that the oldest in stock should be the next to be issued.

23.5.4 Issuing

It is essential there is strict control of the issuing of materials. Clearly, how much control is put in place depends upon the type of goods. If a cladding component is being lifted from a storage area to be fixed to the façade of the building, then it is not necessary to have an issuing procedure as the panel has only one purpose and can be identified easily. At the other extreme, the issuing of items such as door furniture needs to be carefully monitored otherwise many private homes could finish up with new door handles. This is relatively easy to control, providing goods are properly issued. The danger is that employees go into the store and take whatever they need without it being booked out properly.

The procedure starts with the request for goods to be authorised by the supervisor. The employee can then go to the stores, the goods issued against this requisition, the date and amount recorded by the storekeeper. This means that the storekeeper has a 'paper' stock count at any point in time and, if stocks are dropping below a pre-defined level, can call up further materials.

On a construction site, the most difficult items to monitor are materials such as timber that can be used in many different locations and in different sizes. Take any finished room and inspect the different lengths of skirting that have been fixed. Remembering the material arrived in standard lengths, imagine how to control the issuing of the correct amount required to complete the task so that minimal waste occurs. It is very easy for wastage to occur, unless consideration is given to a cutting schedule to obtain the optimum usage of the timber. This is standard practice for reinforcement steel. The usage of these materials should be monitored regularly and wastage measured and compared against set standards.

23.5.5 Stock control

There are three aspects of stock control. First, to ensure there is always enough in stock to support the production process; second, to test the systems and procedures used to control stock are working; and third, to ascertain the value of goods being held. Part of this process is to check to see whether goods are deteriorating in stock by inspecting shelf life dates, the state of goods, such as rusty nails in the store and outside storage areas, the general condition of bulk materials and factory-made components. Where deterioration is found, steps need to be taken to eliminate this in the future and to re-order replacement goods and materials.

There are a variety of reasons for having stock on site where it is not practical to have just-in-time delivery. By buying larger quantities at a time it may be possible to negotiate a lower price by having full loads delivered. It can also reduce the administrative costs of purchasing, resulting from fewer orders being placed, receiving the goods, checking and handling.

23.6 Just-in-time deliveries (JIT)

Much has been written about JIT deliveries. Much of the original work was carried out by Toyota and because of their success, the rest of the world has followed suit. However, it is not a new concept, nor is it new to the construction industry. In the 1960s many of the industrialised prefabrication systems relied upon components being delivered to site and incorporated into the construction on arrival or within a few hours. The majority of concrete is also delivered on the JIT principle. It is debatable whether or not the construction industry could go totally to JIT, because of uncontrollable factors when working outside, such as the weather and finding the unexpected when working in the ground, but once the building is watertight, then this is another matter.

The logic behind the principle is as follows: the reason for holding stock is to cover for short-term variations in both demand and supply. These stocks serve no useful purpose and are only in existence because of poor coordination of materials by management. If stocks are allowed to remain, there is no incentive for management to improve, resulting in the problems being hidden by the stocks. Therefore the best option is to improve management to find the reasons why there are differences between supply and demand and take the necessary action to deal with it. The result of this is to demonstrate that JIT is not just about delivering materials just-in-time, but also to improve management as a whole.

On further investigation, besides reducing or eliminating stock held, other benefits can accrue. Normally, if a piece of plant breaks down, production is reallocated elsewhere, and incoming materials are stored until the plant is repaired. The question to be answered is why the plant should break down in the first place, because if it didn't there would be no need to make these changes and production would remain at a higher level. There is often some suspicion between the contractor and suppliers, based often on the experiences of both. JIT means the contractor has to totally rely on the supplier and this means a build-up of trust and cooperation between all parties working together for a common aim. The same cooperation is required between the employees and management to eliminate friction between the two parties. Finally, extra stock is sometimes kept for the eventuality of defective materials. JIT encourages there to be no defective supplied materials by working closely with the supplier to ensure this cannot occur.

The principles of JIT appear very simple, but of course it is not so easy to implement in practice, as it requires a change in attitude. There need to be in place some key elements if it is to be used successfully. These include:

- maintaining continuity of production without switching from one operation to another, otherwise the planned flow of materials will be interrupted;
- encouraging the standardisation of components as this enables JIT to work more effectively. The more variations the more difficult is the control required, increasing the likelihood of errors;
- to control delivery costs, packaging the work in such a way as to utilise full loads on delivery if possible. If a partial load, the cost per item increases and there is a loss of control in delivery times as the vehicle has to go to other locations;
- careful consideration of the distances over which goods have to be supplied. Usually the longer the distance, the greater the risk of delay. This needs to be built into the programming of deliveries;
- suppliers ensuring that goods arriving are as meant and have no defects in any way. This requires stringent controls at the point of loading, appropriate packaging, correct equipment available for unloading on site and properly trained and supervised employees.

The relationship with the suppliers is paramount to the success of JIT. This can require change in attitude by both parties as contractors have traditionally looked for the lowest tender rather than building up any long-term relationships. Suppliers have had little customer loyalty, seeking out any customer available and trying to make the highest profit. JIT requires that both sides have the same objectives that are beneficial to both and the need to develop long-term partnerships. A word of caution though, this should not be a sole partnership whereby the supplier only has one customer. The construction industry is too volatile to give a guaranteed demand.

23.7 Communications

There needs to be a communication link between the person placing the order, the person controlling the flow of materials through the site, the projected programme requirements and the actual site production, otherwise control will be lost. There are two 'circles' of communication in the process as shown in Figure 23.1. The planner, after producing the programme for the contract, advises both the purchasing department and the site so that orders can be placed and the work planned on site. The purchaser agrees with the site a schedule of deliveries after which an order can be placed with the supplier. The supplier delivers to the site, relying on more up-to-date delivery requirements from the site. If there are any problems about the reliability of service that are not amicably resolved, the site advises the purchaser so that consideration can be made as to whether to place further orders in the future. At all times the site progress is monitored.

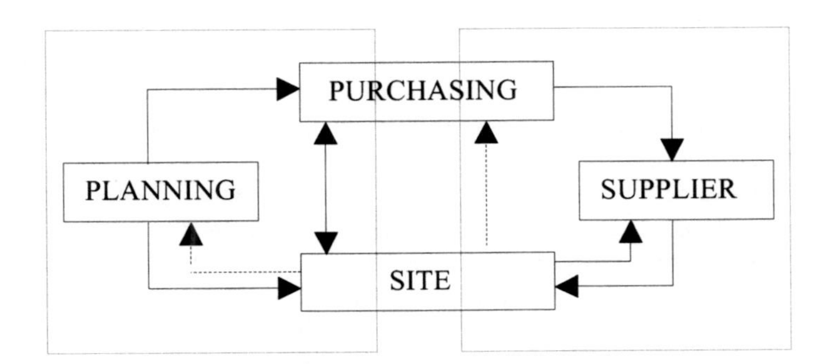

Figure 23.1 Communication – control of materials

Bibliography

Waters, D. (2002) *Operations Management*, 2nd edn. London: FT Prentice Hall.
Womack, J.P. and Jones, D.T. (1996) *Lean Thinking: Banish Waste and Create Wealth in Your Corporation*. New York: Touchstone Books.

24 Supply chain management

24.1 Introduction

Supply chain management as a subject has come to prominence in construction post the Egan (1998) report *Rethinking Construction*, but was in evidence long before, especially as a result of the work done in the Japanese motor car industry. Many academics have taken the position that the work done in manufacturing can be transposed into construction, but this should be treated with a degree of caution as the working environments are different in several significant ways. In manufacturing or retail, the factory or store is in a fixed location with a projected long life, whereas in construction the site activity usually spans between six months and two years and could be located anywhere within the area of the company's operations which may encompass large regions or all of the country. Subcontractors who have to be physically on the site carry out a significant percentage of the construction activity. This adds a further dimension to supply chain management, as there may not be a match between where the contractor and the supplier are working. It may be uneconomical to use subcontractors located at a distance from the site activity. Further, the contract conditions may require that a minimum number of employers on the site be resourced from the local community as a sustainability community issue.

Suppliers of materials on the other hand are similar to those servicing the retail and manufacturing business, in that materials and components generally are stored or made in fixed locations, the materials and components being brought together for assembly. The major differences are that for some of the suppliers, the product required for each contract will be purpose made, requiring considerable changes to plant and equipment at the supplier's base. Other materials and components such as timber, aggregates, reinforcement steel, structural steel, window frames and door frames will require no modifications to the supplier's processes and only minor adjustments to their plant. Delivery problems are similar other than that the delivery point is continually changing as one contract closes and another commences. The activities where some of these variances can be eliminated or reduced is where there is a contractual relationship between the client and a contractor for multiple contracts, as for example with some clients in the retail sector, or where the client and the contractor are one and the same as in speculative housing.

Construction plant for earthmoving, lifting, transportation, power provision and mixing has to be procured for the construction of the project, unlike manufacturing where the equipment will be in place. Similarly, temporary works such as scaffolding are required.

There has been some reluctance by construction clients to enter into supply chain relationships with main contractors, which makes it more difficult for supply chains to be developed with subcontractors and suppliers because of lack of guaranteed orders. Further, the client, perhaps because of the types of contracts in use, tends to communicate to subcontractors via the main contractor thus keeping out of the supply chain management at the lower levels.

This reasoning in demonstrating differences is not meant to dissuade against the argument for supply chain management but simply to show there are differences and one cannot always just transfer the practice of one industry to another and expect it to work as effectively.

24.2 What is supply chain management?

The supply chain, or supply side, comprises those organisations that supply materials, components, labour, plant and equipment, to enable the site operations to be accomplished.

Supply chain management is the active engagement of management in the activities of those involved in the chain to ensure best value for the customer and to achieve a sustainable competitive advantage. This may mean supporting an organisation in the chain that does not have the resources or systems in place necessary for the needs of construction processes such as conformity to ISO 9000 (Chapter 25) and could involve some financial investment by the contractor. Above all it is about the relationships and trust between those in the supply chain and the contractor. Traditionally this has often been poor because of the emphasis placed on the conditions of contract made between the parties and the protection of personal interests rather than the common good.

It should be remembered that the contractor is part of the customer's supply chain and that when the building is handed over the customer may in turn be part of somebody else's supply chain, often referred to as the demand side. This total picture is referred to as the supply chain network.

24.3 Supply chain network

If the business looks solely at those suppliers and subcontractors that are servicing the organisation there is the possibility these could be let down by a failure of one of their suppliers. Therefore all the links in the supply chain need to be considered. Figure 24.1 illustrates a typical supply chain network for a building project on the supply side. For simplicity it has ignored the supply of plant and equipment, and temporary works. The term source is used as the place of extraction and processing. From this it can be seen there are several levels before those that directly impact on the site activity. There could be more levels as well. This figure shows the relationship to a specific contract, but can be developed to look at the supply chain to the company as a whole. Figure 24.2 illustrates the demand after the building has been handed over. The third level only shows the network for the shopping tenants.

The advantage of producing the network is it can assist in making the company more competitive as it demonstrates the needs of the customers and highlights potential weaknesses on the supply side; it assists in identifying significant links in the network, thereby focusing on where breakdowns in the link will have the most disadvantageous

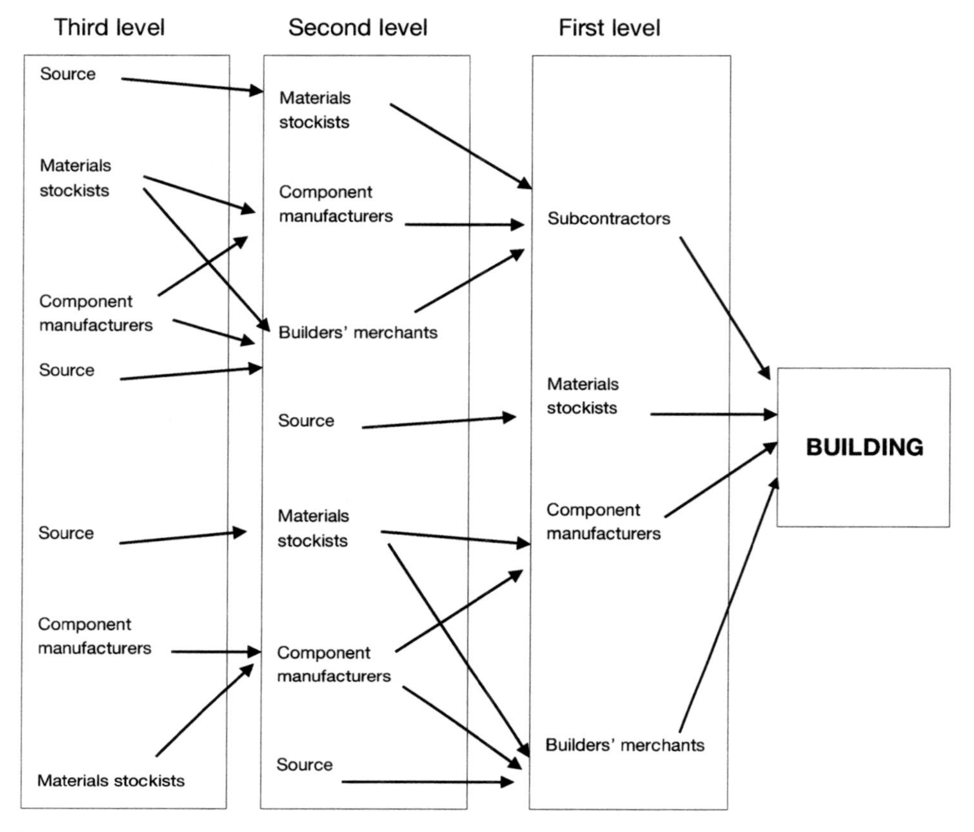

Figure 24.1 Supply side

effect; and the long-term effect of weak leaks can be analysed with a view to strengthening or replacing them.

When looking at the supply network for the whole of the company's business, it will be seen there are a wide variety of links between the suppliers and subcontractors on level 1 and their suppliers and subcontractors on levels 2 and 3, as shown in Figure 24.3. For example, similar businesses on one level may be sourcing the same product from different companies on the lower level. If it can be arranged that fewer businesses on the lower level providing this product can supply all the needs of those on the higher level, as illustrated in Figure 24.4, there will be economies of scale and the product should become cheaper as a result. Further, because the numbers of companies in the supply chain are fewer, but have a greater contribution, they become more important and relationships can be more readily developed for the common good.

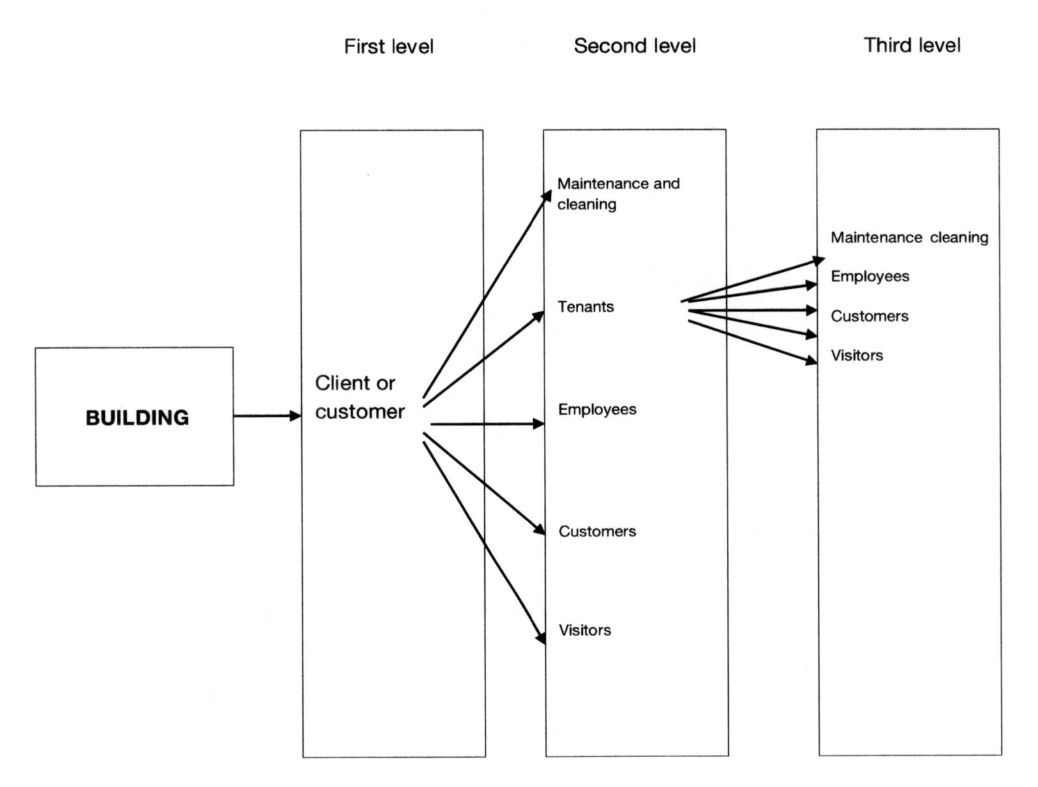

Figure 24.2 Demand side for a retail development

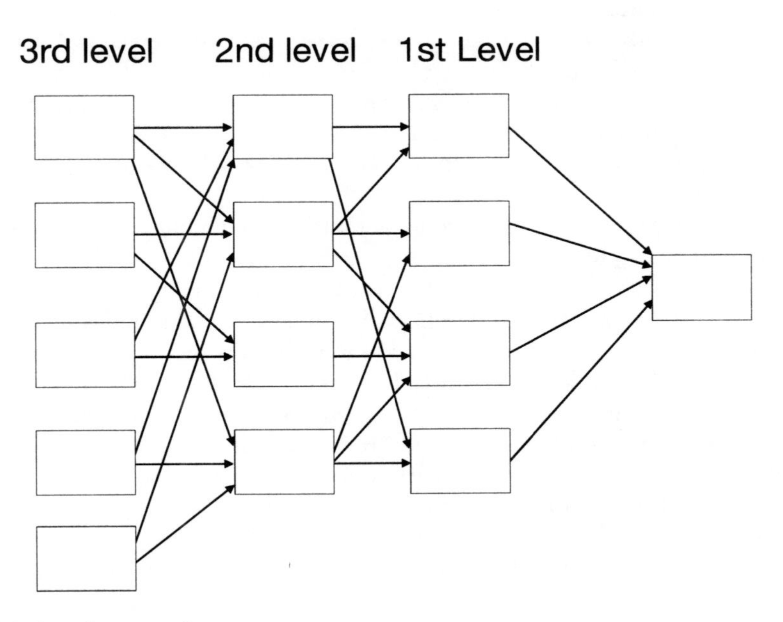

Figure 24.3 Supply network

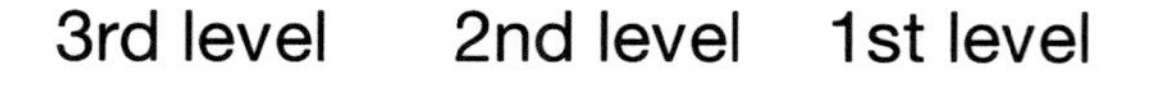

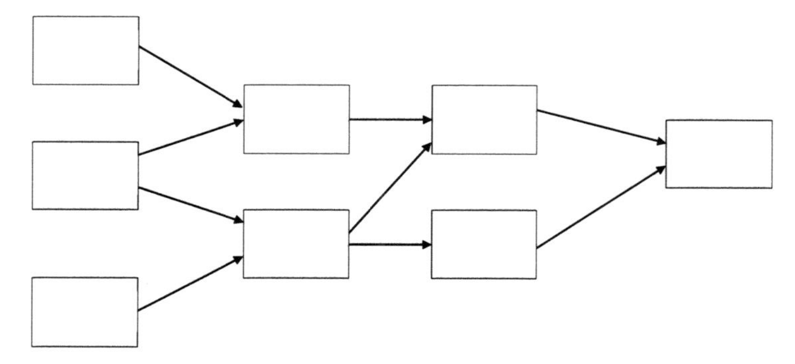

Figure 24.4 Modified supply network

24.4 Subcontract or in-house

In any supply chain strategy the question of what should be subcontracted and what should be done in-house, needs to be addressed. Post-war most contractors would directly employ joiners, bricklayers, steel fixers, scaffolders, plasterers and general operatives. The only subcontracted work would be for the mechanical and electrical services. There has been a steady trend from the 1970s to subcontract nearly everything, keeping just a small gang of general operatives to unload, tidy up and other such tasks. There have been some pockets of resistance to this in certain parts of the United Kingdom, but these are few and far between. In recent years some of the site staff functions have also been outsourced. In the immediate future this trend is unlikely to change, but it is argued that periodically the organisation should revisit this issue, not least because of the concern that smaller subcontractors are less likely to offer apprenticeships, so who trains for the future? On the supply side, some contractors had as part of their history a manufacturing arm, often in timber components, but this has generally died out. Some contractors have a mineral extraction operation, but these are generally run as separate companies and normally would be treated as any other supplier.

Table 24.1 looks at some of the advantages and disadvantages of in-house provision and outsourcing. Companies engaged in design and build and other forms of procurement where they are responsible for the design element, may well contract out parts or all of this process. This is especially the case for the building services, and may include architectural and structural design if the contract is of a specialist nature and the company cannot source within this level of competence and expertise.

The decision-making process on whether to outsource or not is a series of progressive questions. Is the activity of strategic importance to the company? If not, then does the company have the specialist knowledge to carry out the work? If not, does the company carry out the work better than its competitors? And if not, is it anticipated that in the future it will improve its performance? If the answer to the final question is no, then outsourcing should be considered. On the other hand, if the answer to any of the previous questions is yes, then serious consideration should be given to keeping the activity in-house as it would probably be unwise to outsource.

Table 24.1 Advantages and disadvantages of in-house and outsourcing of work

In-house	*Outsourcing*
Planning and resourcing work	
Requires detailed planning so as to assure all employed are working to their capacity. It can affect the flexibility of sequencing work as a result. More difficult to respond to the need for speeding up output.	Providing that the subcontractor is reliable, labour of sufficient quantity can be brought in as required to suit the needs of the project. This is one of the main reasons for outsourcing.
Quality	
Easier to trace the causes of quality problems and to take action quickly as a result.	While the subcontractor may have the expertise that is required, they are also under pressure to make a profit. They may not have as good a total quality management system in place as the main contractor.
Control	
Management can determine any action needed. Requires extra supervision and the company takes over the total responsibility for morale, motivation and discipline.	Some control is lost as the operatives are managed and supervised by the subcontractors and all instructions have to go through a previously defined chain of command.
Reliability	
A core of employees may have long service records, know the way the company works, be compliant with its needs and more likely to look out for its interests.	The contractor is at the mercy of the subcontractor in terms of punctuality, keeping to programme and quality. It is in the subcontractor's interest to perform well if new work is to be obtained. There may be some flexibility in being able to move employees from one contract to another proving there is enough critical mass.
Costs	
Any contribution the operatives make to the company profit is kept. However, when the contract is completed, there can be costs involved if the 'key' operatives are kept on while waiting for the next contract to commence.	The other main reason for outsourcing is that the overall costs to the project are likely to be lower as subcontractors are in competition with each other to obtain the work. The main administrative costs in terms of wages, transportation to site, accommodation, etc. is borne by the subcontractor.

24.5 Location of outsourcing suppliers and subcontractors

One of the basic questions to be asked here is whether or not the work is to be carried out on site or is manufactured and then fitted on site. Traditionally, for most components such as window and door frames, doors were made elsewhere and then fitted on site. However, more recently some of the fitting has been carried out by the supplier; for example, windows arriving on site already glazed, doors already hung in the frames, complete with locks and reinforcement steel cut and bent. Historically, there have been periods when considerable amounts of the building have been made elsewhere, including the structure, as occurred during the 1960s when 'industrialised building' was used almost exclusively in the provision of housing. Timber-frame construction for housing has recently started to make a comeback after reaching a peak of approximately 25 per cent in the 1980s. Currently there are significant moves to off-site production.

While the location of the construction site is predetermined, although speculative builders may have some choice, not all companies in the supply chain may be appropriately located for servicing that project. Suppliers may be less of a problem than subcontractors, as the movement of labour is not involved, the main consideration being the costs of transport to site and environmental impact implications. There are several factors that should be considered when selecting a subcontractor in the supply chain or one still not in the chain, besides the criteria that have been used to select them to be part of the chain. These are:

- *location*: is the subcontractor based within the area and if not do they have experience in working there?
- *labour*: can the subcontractor provide enough properly qualified people living in the area or will they have to be brought in for the duration of the work?
- *accommodation*: is there ample accommodation in close proximity to the project if personnel need to stay over during the week?

24.6 Forecasting

Without knowing one has a salary cheque coming in each month it would be imprudent to take out a hire purchase agreement. Similarly, without knowing the likely workload of the company in the future, it is difficult to negotiate with prospective or current members of the supply chain, unless they can demonstrate there is some guarantee of continuity of work for them.

There are many factors that can influence the potential for obtaining work in the future as illustrated in Figure 24.5. The ability to forecast future trends has a direct bearing on the capability of the company to prepare for the future and to manage the business to obtain work. This is often referred to as having vision.

The nature of construction means there can be a significant time gap from the client's conception of the idea to the time the contractor becomes involved, especially in traditional forms of procurement. Unlike manufacturing, contracting is about tendering for work against competition, rather than competing against other products in the market place whose success is often being determined by the ability to market products, on the assumption the product is well made and the belief it is what the customer requires. Construction output is largely a function of the economic climate determined

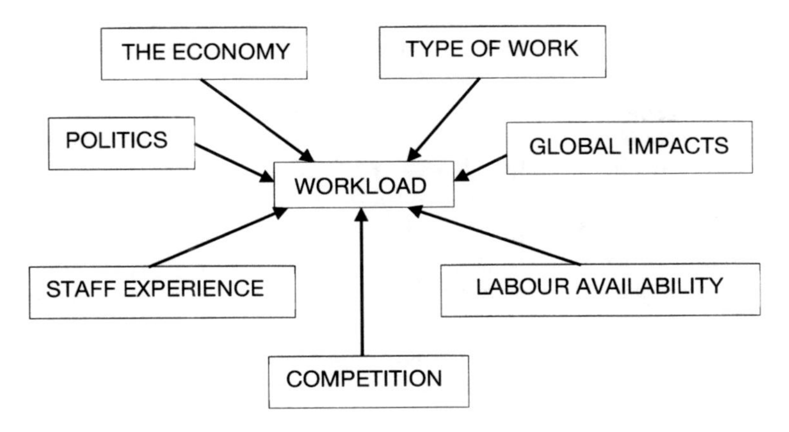

Figure 24.5 Factors affecting future workload

by government policy and global influences. The type of work in the public sector is primarily linked to the government of the day's political philosophy. Since there is only a finite amount of trained labour, if the predictions are there is to be a boom in construction work, difficulties arise in obtaining labour that can result in imaginative design and construction solutions to overcome these shortages. Accurate forecasting is therefore not easy.

24.7 Managing the supply chain

Supply chain management is not just about managing the supply of materials and labour between the various operations in the chain, but also about managing the flow of information. The information flow should not just be about the current contract but should include information about, for example, the future of the contractor's workload so that those in the chain below are kept fully in the picture. This is a reasonable expectation since their future is to some extent linked to that of the contractor. The supply chain can be considered as a series of activities with storage periods in between. For example, as shown in Figure 24.6, material is extracted, then stored before being made into a product after which it is stored until built into the project where it remains until the building is handed over to the customer.

This is very simplistic as many different materials will often have to be extracted to enable the manufacture of a product, and once in the building, they may immediately

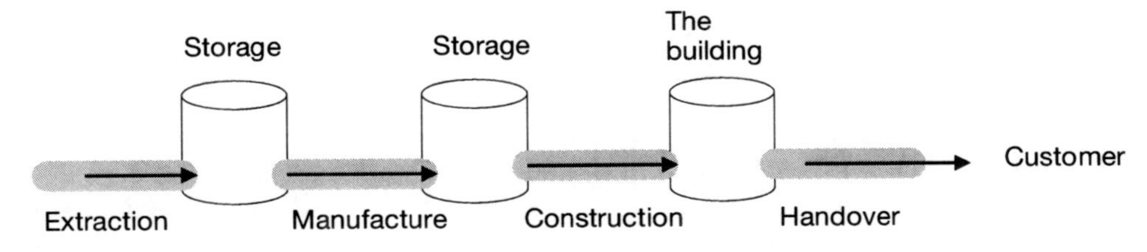

Figure 24.6 Flow of materials in supply chain

start their function, such as a structural element, or may not until commissioned, such as a piece of plant. However, by understanding the flow, decisions can be made to manage the chain. For example investigations of the causes for storage between manufacture and construction, which can take place both at the factory and the site, may reveal excessive storage times, increased risks of damage, double handling at both locations, only part loads being dispatched, and restrictions on the sequence of manufacture, all of which contribute to increased costs. This clearly shows that to manage the chain it is essential that all parties understand each other's business requirements and processes, hence the need for an emphasis on building-materials education to introduce students to manufacturing processes as well as the properties of materials.

Supply chain management in construction is a form of partnering and the same formula for success is required, as laid out in section 11.6, that there are clearly defined lines of communication and statement of roles, procedures for resolving disputes, a statement of working in good faith and openness, especially concerned with costs and delivery problems.

After the strategic decision to engage in supply chain management, the procurement of suppliers and subcontractors follows. Construction purchasing is well established and the significant players in the industry already have in place methods of procuring and selecting preferred suppliers and subcontractors. They look for key performances against certain criteria such as cost, quality, delivery, working to programme, flexibility and reaction time, experience and reputation.

A dilemma for purchasing managers is to decide whether all the work obtained should be given to one supplier or subcontractor taking up a high proportion of their capacity, known as single-sourcing, or whether the work should be spread about to several, known as multi-sourcing. Both have advantages and disadvantages, which are similar to the decision whether to outsource or do the work in-house as argued in 24.4.

In contracting there are limitations to the extent of being reliant on one organisation, because sites can be anywhere within the region of work as discussed in the introduction. So the debate is more likely to be about the extent of multi-sourcing that should be considered, especially subcontracting. Single sourcing for bulk materials is a possibility. For example, a construction company having produced their business plan for the year will have a reasonably accurate prediction of the quantity of reinforced concrete required for the same period and therefore can consider single-sourcing the ready-mixed concrete and reinforcement steel.

The whole aim of supply chain management is to improve performance so that all in the chain, from the lowest level to the customer, derive benefit. One of the major difficulties with the chain is the variability of demand and supply. Construction projects are delayed in commencing or during the construction process, suppliers are in turn let down by their suppliers, or they cannot meet the demand asked of them because other customers have increased their requirements. Any of these can affect the amount of stock suppliers have, with direct consequences on their cash flow. The key to this is communication between one party and the other so that all involved are aware of the issues and can assist each other in sorting the problem out, or can take alternative steps within their own business. This information should be passed rapidly from one to the other. If a supplier knows that delivery is not going to arrive on site when required, the site needs to know. They won't like it, but it means they can re-programme their work for the day and reallocate what would otherwise be idle resources. They also have time to mutually discuss how the lost production can be made up the following day.

There are key issues that need to be addressed in achieving effective supply chain management.

Managing communication: the establishment of direct lines of communications between the main contractor and the subcontractors is essential. This means individuals need to be clearly identified within each organisation and their area of responsibility for decision making specified. This could be one person or several depending upon the size of the organisation and the scope of the work. What must not be allowed to happen is a situation where 'too many cooks spoil the broth' and the 'left hand does not know what the right hand is doing', resulting in a breakdown in communication. If the contractor is involved early in the process they are more likely to take ownership of the project and will involve the subcontractors at an earlier stage with the same benefits accruing.

Managing information flow: a problem in construction is the accuracy and completeness of information and the speed at which it is produced. The quantity, variability and the fact the drawings to a large extent are unique to the contract adds to the difficulties involved. All of this information needs to be progressed, distributed and controlled to be effective. Any form of partnering necessitates openness in financial matters, and many parties are still hesitant in releasing this information. This is part historical, because of the confrontational nature of the industry, and part a cultural in-built puritanical protectionism on matters to do with money – 'We don't talk about it'.

Dispute procedures: even though all parties initially enter with high hopes and aspirations, circumstances will arise where disagreements occur so procedures have to be in place to resolve difficulties quickly and effectively. The key requirements are that all parties know the procedures and action is taken immediately the problem arises to establish the causes and agree a solution. The longer there has been a relationship between parties, the less likely there will be disputes. An environment should be created where those in dispute come to the table with a positive attitude, expecting a solution to be achieved.

Value engineering: each organisation in the supply chain has expertise. If this can be pooled in some way, it will inevitably lead to giving the client better value for money. This is partly a communication problem in getting the interested parties together to discuss and develop solutions, but it is also a timing issue, as to have the greatest impact this must occur at the early stages of the design (see 10.3.8). It is easier to move a column on paper than it is once it has been built. If the parties working in a supply chain are together for a series of contracts then the added value should improve further as all involved more fully understand the client and each other's needs.

Compatible systems: over a period of time organisations develop systems for running their businesses. These are not necessarily compatible with other organisations they are dealing with. This is especially the case with information technology systems. There needs to be compatibility so that the systems can 'talk to each other'. Examples of compatibility helping are in programming, scheduling, payments, computer-aided design, the B of Q and quality systems, and more recently BIM.

Quality: the aim is to continually improve the quality of the completed building. To do this requires a total quality management system that encompasses all parties.

All parties should be aware of what the client requires and each other's needs, and the client in turn should appreciate what is achievable within the budget and programme. As with value engineering, the best place to start is to engage interested parties as early in the process as possible so there is more integrated thinking at the design stage. Not all subcontractors and suppliers need to be or indeed can be involved at this stage, so they need to be engaged during the construction phase by means of discussions and brainstorming the issues. All should be aware of the requirements before they commence work and if problems occur between trades during the work, these should be sorted out quickly.

Long-term relationships: there are some long-term arrangements between contractors and subcontractors, but these tend to be informal often as a result of the subcontractors and suppliers being used on a regular basis. There are more formal relationships between client and main contractors that seem to profit both parties. These are often in the retail and supermarket businesses where the client has a long-term policy of expansion or refurbishment and can see the advantages of having this type of relationship with a contractor. Petrol station contracts have also been procured with long-term contractual relationships.

All parties should understand the importance of relationships between the client and the contractor, the contractor and the subcontractors, and between departments within the organisation as well as within individual departments. This should not be just about 'getting on', but should be focused with the intent of improving performance and adding value to the processes. This value improvement can be achieved by considering the removal of all processes in the chain that do not add value, addressing all aspects of the costs involved in the chain, openly assessing each other's performance and seeking to improve each other's capabilities.

Customer relationships are not just between the client and the contractor. The customers are also the users (both employees and visitors). There should be an understanding of all their needs which should include the neighbours in the vicinity during construction and after handover. In the former case, because of pollution and extra traffic, and in the latter case, because the local environment has been altered due to the presence of the new building and its users. Systems should be in place so that all these stakeholders' satisfaction, or otherwise, can be measured so remedial actions can be taken if required.

Suppliers are often selected purely on price and previous performance without ever asking what can be done to help them. What do they need to make their businesses more effective? Does the contractor complement their good performances or concentrate on their failures? When the main contractor requires changes to be made, is there a full understanding of the impact this might have on the suppliers and subcontractors.

24.8 Sourcing from other countries

When sourcing from one's own country, all the local customs, cultures, religious festivals, holidays and even the weather are almost taken for granted and often not considered in terms of them affecting the supply chain, as they are known. However, these may differ elsewhere in the world and have to be taken into account when managing a supply chain. For example, the Chinese celebrate their new year in a big way, with much closing down during the event as many in the large towns and cities make an exodus to the country to be with their families; Ramadan in Muslim countries is a significant event and like the Chinese New Year is not at the same time each year; respect for age is often reflected in status within a company; and women's position in society and what they are allowed to do can vary considerably from country to country.

Bibliography

Bozarth, C. and Handfield, R. (2016) *Introduction to Operations and Supply Chain Management*, 4th edn. Harlow: Pearson.

Egan, J. (1998) *Rethinking Construction*. London: Construction Taskforce: HMSO.

McGeorge, D. and Palmer, A. (2002) *Construction Management New Directions*, 2nd edn. Oxford: Blackwell.

Slack, N., Chambers, S. and Johnston, R. (2009) *Operations Management*, 6th edn. Harlow: Prentice Hall.

Womack, J.P., Jones, D.T. and Roos, D. (1990) *The Machine that Changed the World*. London: Macmillan.

25 Quality management

25.1 Introduction

Quality has always been an issue for mankind, be it the area of agriculture and food production, the development of tools for hunting and gathering, manufacturing or health. Japanese manufacturers realised that early attempts at quality control were limited and pioneered quality assurance methods and, more recently, total quality management systems. This may explain why their manufacturing base has been so successful relative to the West, which is now playing catch up. The construction industry is primarily concerned with one-off projects and historically, due to having the design and construction processes divorced from each other, has been left behind. However with the advent of new procurement methods, such as design and build, management contracting, construction management and partnering, the industry is rapidly accepting this change and adopting quality management systems.

25.2 Quality control

This was the mechanism traditionally used by organisations concerned with quality. Its basis is that of measurement and inspection. In other words, it occurs after the fact as a checking procedure so action can be taken to improve future activities, rather than being concerned with ensuring quality is always right by having systems in place before the product or service is made or offered.

A typical quality-controlled workplace required inspection and measurement points to be established. A method or procedure for the inspection and measurement would be established, after which data would be collected and analysed so that trends in deterioration of quality could be spotted and extreme problems avoided. For example, in the manufacture of precast concrete components, the inspection points would be checking the quality of the raw materials on arrival, the concrete prior to discharge into the mould, using standard concrete testing methods such as slump and cubes, followed by Schmidt hammer tests to check the concrete was of a sufficient strength to permit de-moulding. The completed components would be dimensionally measured and had to fall between a range of sizes to become acceptable for dispatch. These measurements would be recorded over a period of time to plot trends, so that as the moulds slackened with use, increasing the size of the component, action could be taken before the situation deteriorated beyond the specified tolerances.

In other words, measurements were made to compare against a specified standard. This meant paying for an inspector, process adding no value to the product, reject stock having to be paid for, and in the end all these inherent costs would be paid for by the customer as part of the supplier's overheads. What it did not do was to improve the quality of the product or service, but just reduced the likelihood of incorrect products leaving the factory.

25.3 Quality assurance and management

So was quality control alone satisfactory? What about whether the service or product was delivered when the customer required and was it at the price the customer was able or wishing to pay? These were not considered as quality issues. It became clear inspection in itself is not enough to sort out quality problems. What was required was a measurement of quality that accounts for all the customer's requirements. This was referred to as quality assurance and more recently, quality management. This focuses on compliance with procedures to ensure product or service quality.

The British Standard for this was BS5750 published in 1979. In 1987 the International Organization for Standardization based their ISO 9000 series on the BS, later adopted by the EC and the UK as BS EN ISO 9001, ISO 9002 and ISO 9003. These were subsequently integrated to become BS EN ISO 9001. Note that ISO is the Greek word for equal and not an acronym for the International Organization of Standardization. All of these have been concerned with setting standards for the system of quality assurance rather than solely the product. They set out a framework by which a management system can be implemented so the needs of the customer are fully met. In other words, quality assurance is based upon the principle that it is better to prevent quality problems rather than detecting these problems after the product has been made or the service provided.

Figure 25.1 shows the full ISO 9000 family as produced by the British Standards Institution. The ISO 9000 family is concerned with the way an organisation's quality management meets the customer's quality requirements and any applicable regulatory requirement, with a view to enhancing customer satisfaction and to continually improving the company's performance in pursuit of these objectives.

Quality assurance activities used to be defined to include:

- how the organisation develops policy in respect to quality;
- the allocation of responsibilities for quality within the organisation;

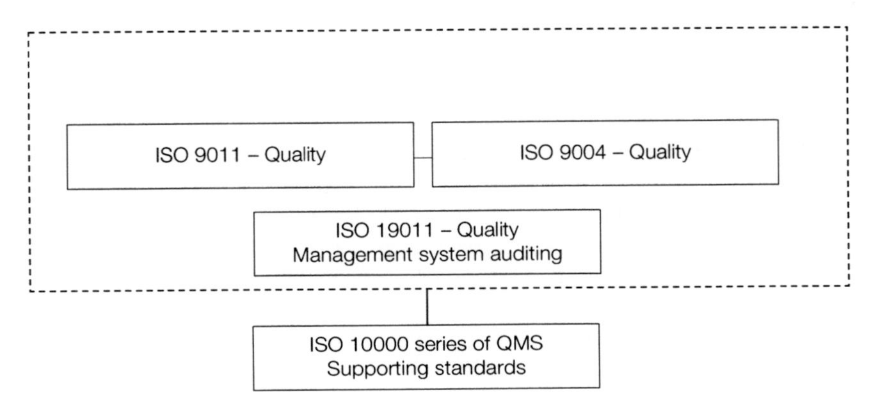

Figure 25.1 The family of ISO 9000 standards

- procedures used to carry out the needs of the business;
- the standards to be attained in the workplace;
- the documentation required to demonstrate both the operation and maintenance of the system and the attainment of quality.

While this still is part of the process, it is very much system orientated rather than people focused. As can be seen in the eight key principles of quality management (25.4) referred to in ISO 9001 and ISO 9004, it has become more a management of quality issues that includes people as described below. This is often referred to as total quality management (TQM) described by the Department of Trade and Industry (2005) as: 'A business philosophy that focuses on quality throughout the organisation. It aims to deliver complete customer satisfaction, benefits to all staff and benefits to society as a whole.'

25.4 The seven key principles of TQM

There have been various suggestions on what are the key principles of TQM, such as from Oakland (2005) and as shown in 25.5, but for discussion purposes the author has chosen the International Organization for Standardization publication, *Quality Management Principles* (2015), which states that the seven key principles of TQM are:

- customer focus
- leadership
- engagement of people
- process approach
- improvement
- evidence-based decision making
- relation management.

25.4.1 Customer focus

All organisations depend upon their customers and therefore should understand their needs, be able to meet their requirements and try to provide a product or service that is better than expected. In reaching this goal, the organisation will find they have become more flexible and speedier in reacting to the market, their own organisation will be better prepared in meeting the customer's needs and this will result in more business as a result of customer loyalty. Historically in construction, the industry has not been customer friendly, resorting to a more adversarial approach adhering strictly to the conditions laid down in the contract documents at the expense of everything else. However, in recent years this attitude has changed considerably with the development of new forms of procurement such as partnering and management contracting.

An organisation focusing on the customer has to research and understand their needs and expectations and ensure that this is communicated throughout the organisation. Performance has to be monitored and deficiencies addressed rapidly. This should not be carried out on an ad hoc basis but systematically managed to make it work. However, satisfying customers must be balanced with other interests such as the shareholders, employees, suppliers, financiers, society and the environment.

25.4.2 Leadership

Without leadership it is unlikely the whole organisation will take the matter seriously. Management has to create an environment that encourages all to become fully involved in meeting the organisation's objectives. Good leadership will ensure the employees both understand the needs and are motivated to achieving the objectives set. It also reduces the likelihood of miscommunication within the business.

The advantage to the business is a clear vision of the future, communicated to all, enhancing motivation. People know what is expected of them, have clearly defined targets to meet and their successes recognised. Levels of trust increase and fear diminishes as fairness prevails. Staff development is improved and encouraged and their needs increasingly satisfied.

25.4.3 Engagement of people

As a result of good leadership, employees at all levels of the organisation are enabled to use their full range of abilities for the business's benefit, especially when the attitude is towards a more participative form of management. Involved personnel tend to be more motivated and committed and contribute ideas to the process, being prepared to be accountable for their actions as they wish to continually improve.

This means they understand the importance of their contribution and bring to the attention of management any constraints affecting their performance. They accept their responsibility and monitor their own performance against their own goals. They look for opportunities to widen their ability and experience and are prepared to discuss problems, sharing their ideas and knowledge with others.

25.4.4 Process approach

The ultimate objective is achieved more efficiently if the activities and resources are managed as a process. The benefits include lower costs and speedier operations, more predictable outcomes in terms of quality and time, leading to more accurate planning, and problems being readily identified, rectified or improved.

By systematically defining what has to be done to achieve the desired result, clear responsibilities and accountabilities are identified. Overlaps and gaps in responsibility between the different functions in the organisation are spotted and their ability to manage the situation clarified. It means, that by looking in detail at the methods of work, how it is resourced and the use of materials, the process will be improved.

Finally, through risk assessments, the impact and consequences of the work on customers, suppliers, employees and others can be evaluated and action taken to improve.

25.4.5 Improvement

There is no point in any organisation setting targets and sitting back satisfied at their achievement. As these are being attained, then higher and improved standards should be set so continuous improvement occurs. There are clearly commercial and competitive advantages achieved by doing this, not just by being cheaper and producing higher quality, but having the ability to react quickly to opportunities with a more motivated workforce.

To achieve this requires continual improvement to occur on all facets of the organisation, to provide employees with appropriate training to be able to improve and to look for ways of continually improving the processes and systems. This requires monitoring procedures to be in place to track and record improvement, and to recognise and acknowledge improvements.

25.4.6 Evidence-based decision making

Informed decisions should only be made after an analysis of the available data and information. The result of any decision should be monitored so the business can learn from the mistakes resulting from incorrect or bad decisions. The organisation should have the freedom to review and challenge decisions and there should be a culture in management to accept that their views may not always be right and they should be prepared to change their stance if a better argument prevails. To enable this to happen, there needs to be a source of reliable and up-to-date data and information that is available to those that need it. Any analysis of the data should be carried out using valid methods, and decisions should be made based upon factual analysis, taking experience and intuition into account.

25.4.7 Relation management

The organisation and its suppliers are independent and in the case of construction, the design team are independent from each other. However, there are mutual benefits in working together as it will lead to all parties creating increased value, through increased flexibility in reacting to situations rather than being tied to traditional contractual relationships.

This means having improved communications, working together with common objectives, sharing information and expertise, and in the case of suppliers in particular, discussing future plans. Based on this, partnering agreements can be developed and the organisation can assist the supplier in improving their processes and systems. With the development of BIM this has become easier to facilitate.

25.5 An alternative view of TQM

Another way of describing TQM is shown in Figure 25.2, giving an overall summary of the key issues relating to TQM in diagrammatic form with some different headings, notably survival. It should be remembered the reasons for carrying out TQM are about providing customer satisfaction without which the business will lose customer support, resulting in decline and eventually ceasing to exist as many well-known companies have experienced.

In summary, some organisations see validation as something that should be done because others in their field are doing the same and therefore do not to wish to be left out. Those with that attitude will not derive benefit, but will find it an expensive and costly process. Validation has to be conducted in the full knowledge that not only is there much work to be done in obtaining validation, but that it may well mean some significant cultural changes within their organisation.

As the standard is international and increasingly companies of all commercial interests throughout the developed world are adopting ISO 9000 in lieu of their own

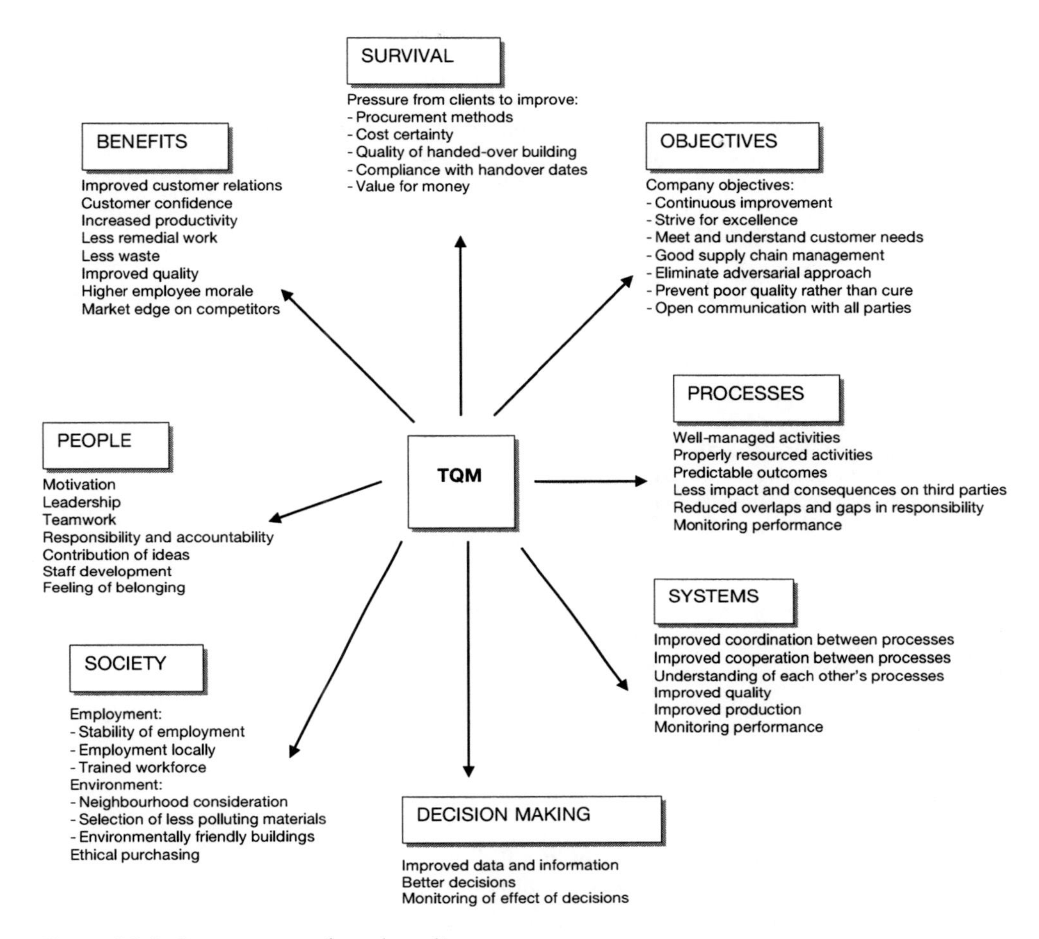

Figure 25.2 Components of total quality management

national standards, harmonisation is occurring. Customers are also increasingly expecting construction companies to be accredited with the ISO certification.

25.6 ISO 9001 and 9004, the process and implication to the construction industry in particular

To be validated, the organisation has to be audited by an independent company accredited to do so. The audit is: 'a systematic and independent examination to determine whether quality activities and related results comply with planned arrangements and whether these arrangements are implemented effectively and are suitable to achieve objectives' (ISO 9001).

In summary the benefits of ISO 9000 implemented properly are:

- improved customer focus;
- increased confidence in the client that the contractor will meet their requirements;

Table 25.1 The main clauses ISO 9001: 2000

4 *Quality Management systems*	6 *Resource management*
4.1 General requirements	6.1 Provision of resources
4.2 Document requirements	6.2 Human resources
– Quality manual	6.3 Facilities
– Control of documents	6.4 Work environment
– Control of quality records	7 *Product realisation*
5. *Management responsibility*	7.1 Planning of product realisation
5.1 Management commitment	7.2 Customer-related processes
5.2 Customer focus	7.3 Design and/or development
5.3 Quality policy	7.4 Purchasing
5.4 Planning	7.5 Product and service provision
5.5 Responsibility, authority and communication	7.6 Control of measuring and monitoring device
– Responsibility and authority	8 *Measurement analysis and improvement*
– Management representative	8.1 General
– Internal communications	8.2 Measurement and monitoring
– Management review	8.3 Control of non-conformity
	8.4 Analysis of data
	8.5 Improvement

- facilitation of continuous improvement;
- establishment of consistent work practices and quality throughout the organisation;
- improved working relationships with suppliers, subcontractors and design organisations;
- improved management decision making;
- embedding of positive attitudes to quality in all employees;
- reducing the dependence on individuals so that when they are absent, leave or are promoted there is not a void;
- plenty of evidence to demonstrate that it adds value to the business.

The main clauses in BS EN ISO 9001are summarised in Table 25.1. If the reader requires a detailed analysis of the clauses, it is recommended the reader refers to David Hoyle (2008).

25.7 Key performance indicators (KPIs)

The idea of using key performance indicators (KPIs) in construction developed as a result of Sir John Egan's (1998) report *Rethinking Construction* which challenged the industry to meet higher standards and measure its performance over a range of activities. KPIs are measures of performance of activities that are vital in the success of an organisation. Supported by the department for business innovation and skills, data is collected to produce national KPIs.

To measure an organisation's performance using KPIs the company needs to identify what its goals actually are. These goals should reflect different areas and activities of the organisation such as: client satisfaction, impact on the environment, whole-life performance of the building, amount of defects both during and after the construction process, quality, productivity, health and safety, adherence to the construction programme time and budget, profitability, amount of waste, employee satisfaction, staff turnover, sickness and absentee rates, and qualifications and skills.

Having established the goals, these need to be quantified, first by establishing current figures and then deciding what the goal figures for each should be, such as how much profit is required, reduction in the number of complaints from clients, reportable accidents, labour and staff turnover, amount of overrun on projects and waste. This should be done in consultation with the affected staff to ensure that they are engaged and supportive and that the goal figures set are realistic. If not then the incentive to succeed will evaporate.

These performance indicators have to be monitored at frequent intervals, the durations of which have to be decided. These will vary depending upon the type of activity. For example, safety performance will be done relatively frequently whereas staff turnover will be measured less often. From the measurement of each, a percentage of improvement can be calculated and this informs future predictions, adjustments that can be made to expected rates of progress, and analysis of why the improvement has, or hasn't, taken place, i.e. benchmarking.

25.8 Benchmarking

While benchmarking has been used for many years the recent interest has been generated by the publication of national KPIs that enable companies to measure their performance against them and then set their own. Its aim is to produce best value for the client. It is a method of improving performance by comparing one's performance against others'. The starting point is to identify the KPIs the company wishes to compare itself against, then the basic questions to be asked are: Who performs better? Why are they better? What actions are required to improve our performance?

Benchmarking involves not only the collection and comparison of data, but will include fact-finding studies to establish the reasons for superior performance of the other businesses. The ways of accomplishing this are: first, within one's own company by comparing one site against another, which for certain indicators will have been happening for a long time, for example, health and safety statistics; second, comparisons with one's competitors, which is more difficult, although knowledge of their tender prices after the contract is let is usually readily available; and third, comparing against similar activities in unrelated industries both at home and abroad.

It is important not to set the target too high otherwise it will probably not be reached and all involved will become demoralised. The target should be set at a level that is achievable and once this is approached or met, the bar can be raised further. Equally important is to select a few areas at first that are important to the business and not easy options.

Having established which areas need to be improved, the data collection methodology has to be decided. The analysis of the data is to discover the difference of performance between what is currently happening and the target and to identify the reasons. A plan of action is then drawn up, implemented and regularly monitored and compared against the target. Finally, the whole process is repeated so that it becomes part of the culture of the organisation.

Bibliography

Ashford, J. (1989) *The Management of Quality in Construction*. London: E & FN Spon.
Egan, J. (1998) *Rethinking Construction*. London: Construction Taskforce: HMSO.
Griffith, A. and Watson, P. (2004) *Construction Management: Principles and Practice*. Basingstoke: Palgrave Macmillan.
Hoyle, D. (2009) *ISO 9000:2008 Quality Development Handbook*, 6th edn. London: Taylor & Francis.
ISO 19011:2011 *Guidelines for Auditing Management Systems*.
ISO 9000:2005 *Quality Management Systems – Fundamentals and Vocabulary*.
ISO 9001:2008 *Quality Management Systems – Requirements*.
ISO 9004:2009 *Quality Management Systems – Guidelines for Performance Improvements*.
Oakland, J.S. (2005) *Oakland on Quality*. London: Butterworth-Heinemann.
Quality Management Principles (2015) International Organization for Standardization.
Sasaki, N. and Hutchins, D. (1984) *The Japanese Approach to Product Quality*. Oxford: Pergamon Press.

Part V

Business management

26 Organisations

26.1 Introduction

The prime purpose of this chapter is to explore some of the basic concepts of organisations and how they come to exist. Organisations are formed from groups of individuals who have different authority and responsibilities that, when working well, allow the business to meet its objectives. The authority and responsibility given to people is the basis for managing the organisation. Some of these relationships are formal and others informal.

Organisations need to group individuals into particular functions with clearly defined tasks and someone in charge, the manager or supervisor. How authoritarian this managerial role appears is dependent upon the company's management style or the type of business. The armed services will have less debate within the group about decisions than a creative advertising team. These groups may also be subdivided further. For example, the overall group may be 'the production activity' but in construction these would be divided into each individual site, and in each site there could be further divisions of say, each block of flats.

The organisation must take account of how communications are best effected up and down the business, but also needs to consider how communication horizontally can take place. In construction, these horizontal lines of communication are very important and often difficult to enact. The very nature of construction is that each project is different and many of the participants in the process may never have worked together before. While the project manager may select certain key staff, the regional or head office senior quantity surveyor and engineer will have selected their representatives. To make matters worse, the majority of the construction work is carried out by subcontractors, whose primary concern is to complete their particular piece of the work. Their prime line of communication is directly to the main contractor. On top of this, the design team can also be appointed separately without any consultation with the contractor, although the newer forms of contract procurement such as management contracting and design and build, have reduced the problem.

26.2 Span of control

Before designing an organisation structure it is necessary to consider for how many people an individual can be responsible. Generally, as organisations expand, more employees are employed to carry out the business. Whereas in a very small business the owner makes all the decisions, in a larger organisation many of these decisions have to be delegated. Thus the manager of a group is given certain responsibilities and authority. The reason for this delegation of power is simply because the owner begins to lose control if too many people are responsible to them. In other words, there is a limit to the number of subordinates a manager can control. This is known as 'span of control'.

So how many people can one manger control? In theory spans can vary from one to many tens of subordinates controlled by one manager. However, it is usually accepted as a general rule the most likely range is between 3 and 20. Some people argue the range should only be from 5 to 7 to make the organisation more efficient, but to be too prescriptive is dangerous. What is significant is that the size of span affects the style of organisation. If the span is large the lines of communication between the higher and lower levels in the organisation are short and if the span is small, the lines are longer. These are known as tall and flat structures (see 26.6).

There are various factors that can influence the size of the spans of control: for example:

- Senior management may take the decision and decide the numbers because they want a particular type and shape of organisation.
- The talent, personality and ability of the manager to cope under pressure can be a determining factor.
- The experience of the manager is very important. If the manager has been in post since the process started, their accumulated knowledge and responsibility can be extremely high and it is often only realised when that person is being replaced. Their replacement can be by more than one appointment.
- If the ability and knowledge of the subordinates is high they will require less supervision, thus the manager can control more; whereas a less able workforce would require more supervision.
- If the work is very complex and there are lots of changes to the work occurring regularly, increased supervision will be required.
- Equally if quality levels have to be high, extra supervision may be required, especially if the workforce is not of the highest calibre.
- If the subordinates are located a long way from base and it takes the supervisor a long time to reach them, the span of control will reduce. This is not unusual in construction, where a subcontractor's workforce may be spread over several contracts in small gangs, each not requiring full-time supervision.
- Closer supervision is required if the task has a higher than normal level of risk.

26.3 Organisation charts

Organisation charts are diagrams showing the interrelationship between individuals or groups of people. They can be used to demonstrate how the organisation, groups its various activities, and the relationships between functions and tasks. The chart demonstrates who is responsible to whom within the organisation, known as the chain of command. It can also show the horizontal links between functions and groups. The horizontal lines are very important and do not necessarily mean that because two people are on the same level they are of the same seniority.

The author has observed that organisation charts tend to be produced in two different ways. The first commences at the top with senior management deciding who should be responsible to whom until the bottom of the chart is reached. The other, which the author believes is the more successful, starts at the bottom of the organisation and asks the question 'what does this person or group need to best complete their task?' This works on the assumption that in manufacturing and construction as the operatives are the producers and it is their level of output that determines the viability of the business, their needs have to be carefully considered (see 26.7).

26.4 Types of organisation

There are many different ways organisation charts can be compiled, many being a combination of different approaches; this section describes the most common.

26.4.1 Executive or line responsibility

The military is probably the closest to this form with the Chief of Staff at the top, and the chain of command descending through the ranks to the private, in the case of the army. In construction the executive function will exist typically as shown in Figure 26.1.

This assumes the company employs all the people involved in the process. In UK construction this would be a rarity, so the chain of command might stop at the general foreman. However, subcontractors often do employ all of these titles, so their chain of command would be similar, although the smaller businesses would have fewer levels of management. In this example the project manager is the person responsible for

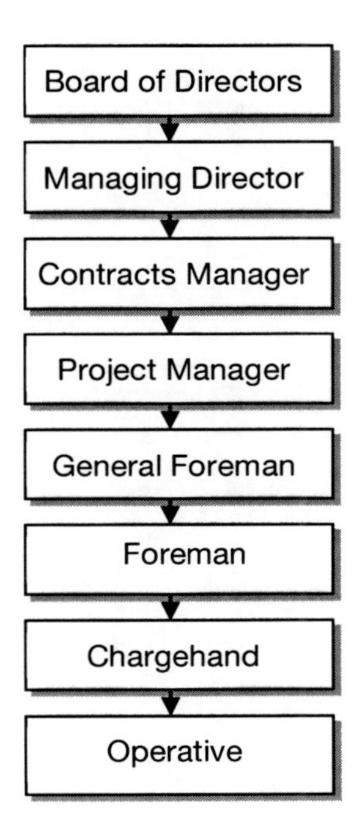

Figure 26.1 The executive function

management of a construction site and the contracts manager is the senior manager responsible for all the current construction projects. This executive line will generally exist in all businesses, but cannot exist in isolation. The people in this chain of command need support to carry out their tasks.

26.4.2 Functional responsibility

It is often stated 'you should only have one master', and on the surface this makes sense, but on further investigation it is found this does not always work. In construction it is quite common for a person to have two masters, but each of the responsibilities for the subordinate is clearly defined.

A typical head office organisation chart might look as shown in Figure 26.2 with the senior quantity surveyor, planner, estimator, buyer, contracts manager and so on reporting to the managing director. On site level the organisational chart might be as shown in Figure 26.3.

Take for example the quantity surveyor (QS) on the site who can have a responsibility to two different people – the chief quantity surveyor in the head office and the contracts manager on the site. This may be seen as a contradiction, but in practice works well. The project manager has to run the project and is responsible for the general behaviour of the staff such as punctuality, appearance and attitude, but may have limited knowledge about the professional competence of the QS. The monthly valuation may be completed on time, but was it done properly? The manager can ask for information by a certain time, but how reasonable is the request? The responsibility of the chief QS is to ensure the site QS is providing a proper and reliable service.

What happens if the site QS takes issue with the project manager and it cannot be resolved at this level? The project manager or QS would normally go to the chief QS and try to resolve the dispute. If this did not result in a solution, the project manager could go to the contracts manager at head office who would then hope to resolve the

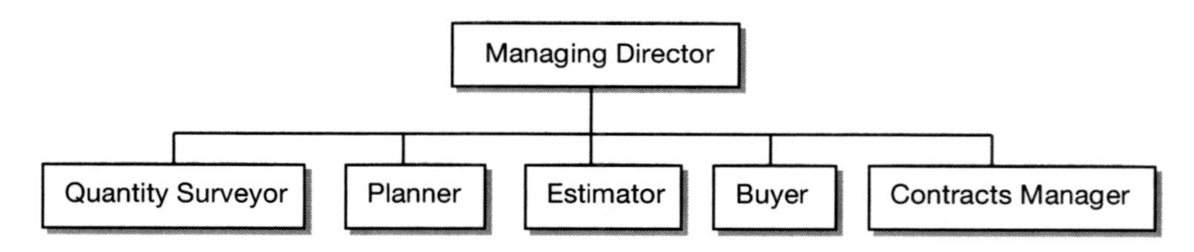

Figure 26.2 Head office organisation chart

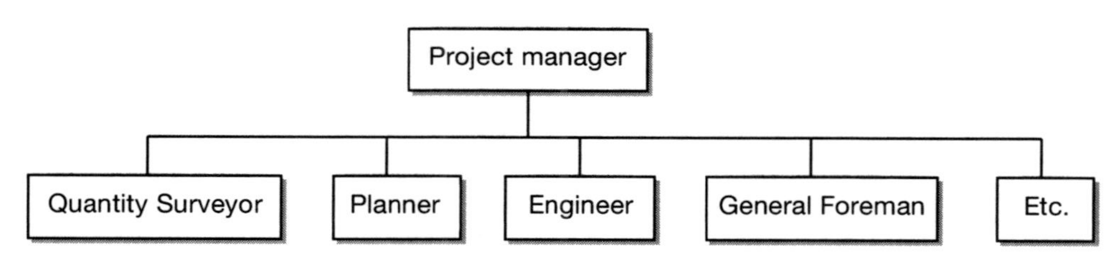

Figure 26.3 Site organisation chart

issue with the chief QS. In the unlikely event of them being unable to sort it out they would have the option of going to the managing director for resolution This identifies an important issue in the construction of an organisation chart in that it can demonstrate the lines of communication in the event of arbitration between two parties being required.

26.4.3 Product structure

Large organisations often divide their structure into product divisions. Schools have science, English and language departments, hospitals divide their work into out-patients; accident and emergency; ear, nose and throat and so on. Construction companies used to divide their business between types of work such as civil engineering, housing and building projects as shown in Figure 26.4, but in recent years have tended to divide between procurement methods such as PFI, design and build, and traditional, as in Figure 26.5, each with their own subdivision reflecting their specific needs.

Although not shown on these there will be certain activities probably carried out centrally on behalf of the group such as human relation management, purchasing, public relations and marketing.

The advantage of adopting a product approach is that attention is focused on each category, which might have different needs and expertise from the others. For example, housing is in a speculative market anticipating customer requirements, where traditional contracting work is obtained by competitive tendering and then constructing what the client has already determined. The contractual differences between the various methods of procurement are significant and PFI work may also include the facilities management element. The disadvantage is there can be duplication of work. This is why certain functions will be carried out centrally.

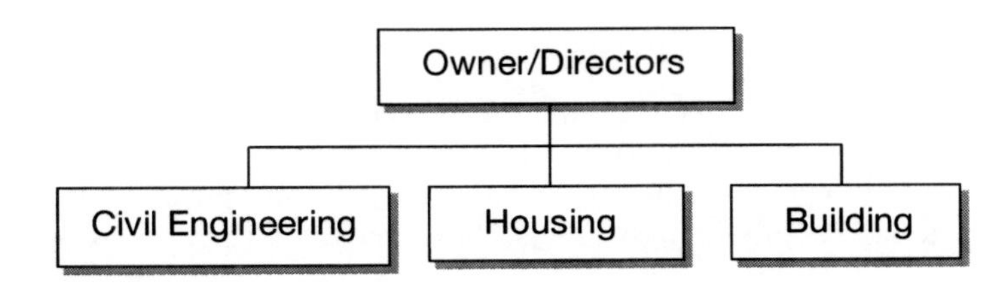

Figure 26.4 Organisation by product

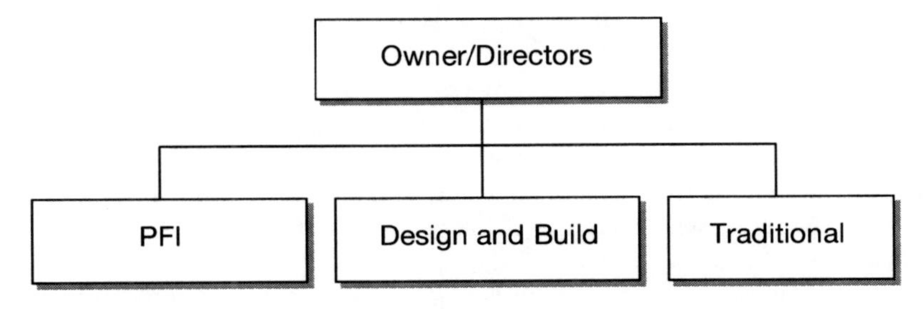

Figure 26.5 Organisation by procurement method

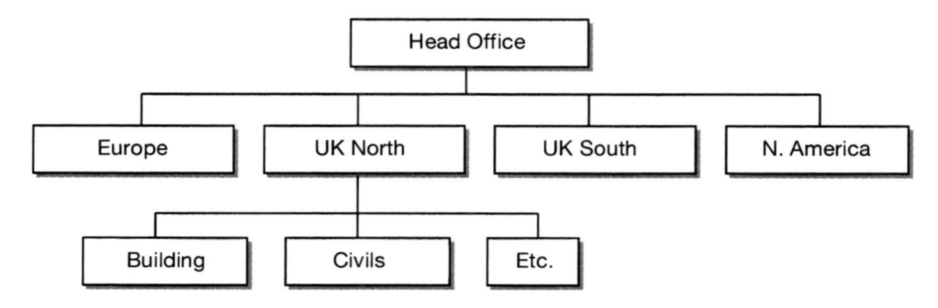

Figure 26.6 Geographical structure

26.4.4 Geographical structure

Geographical structures are produced, especially in large organisations, based on where the company's activities are. If a construction company is working in most parts of the UK there becomes a problem of communications if everything is controlled from one location. Local knowledge is important for many reasons. Senior management builds up a network of local connections and thus potential clients. The local market is better understood, as are the availability of competitive and reliable subcontractors and material suppliers. The same applies internationally, but there are also the issues of time differences, different regulations and laws, and understanding local culture. As with product structures, certain activities will be carried out centrally. Figure 26.6 demonstrates a typical example.

26.5 Centralisation and decentralisation

Centralisation is defined as concentrating the power and authority near or at the top of the organisation; decentralisation is delegating the power and the authority to make decisions to lower levels of the organisation. No organisation is entirely one or the other, unless a one-man business, but will tend towards being primarily one or the other.

What type an organisation is will depend upon a variety of factors:

Management's philosophy. Some managers believe in strong central control and want to be in total control. They build up a close team around them of quality people and make all the major decisions. The lower levels in the organisation are then instructed what to do as with Likert's exploitative and benevolent management styles. Many of these managers also believe the staff below fall into the category of McGregor's Theory X. Other managers tend more towards Likert's participative management style and believe employees fall into the category as described in McGregor's Theory Y. They delegate much of the decision making and accompanying responsibility and accountability to those who have the appropriate information available to make the best decision.

Organisational growth. Organisations that grow and remain centralised do so usually as a result of the way the company was set up in the first place. When organisations merge or expand as a result of further acquisitions they tend to move towards being decentralised. This is for a variety of reasons that include different cultures of the businesses; different products or services; the fact management structures are already in place, although these are likely to be radically changed as part of the merger and takeover process; the businesses will be geographically spread; and the sheer scale of the business.

Geographic diversity. The greater the geographic diversity of the business the more likely the organisation is to be decentralised. This is due to the problems of control from distance. For example, a construction company working nationally will set up regional organisations staffed by personnel who know the local market and supply chain well and can therefore make informed decisions. International organisations have an added problem of the time differences between the different parts of the organisation, not to mention the customs and practices of another country. In construction there may be a limit on the value of a contract the region can tender for, because of the risk involved, and contracts over a certain size may be considered centrally.

Effective controls. Organisations that have difficulty in controlling the lower levels tend to be centralised. This can occur within an organisation that is primarily decentralised, especially if the work occurs at random, such as with a maintenance department of a factory where there is a need to move personnel rapidly from one job to another at a moment's notice depending upon the urgency of the breakdown. Decentralised organisations, by implication, mean if you delegate the work there must be some control to ensure the work delegated is being carried out properly.

Quantity and quality of managers. By definition if the business is decentralised and powers delegated, there have to be sufficient qualified and competent managers to take the responsibility and make sound decisions. Universities are examples of where there are a limited number of good and qualified managers since promotion on the academic side is primarily a function of scholastic ability rather than management ability. This creates a problem for senior management. Should they have many faculties, with the few good managers running them with lots of small departments run by less able managers, or should there be fewer faculties and departments, using the limited management resource.

Diversity of products and services. The contracting industry is very diverse and is changing continually. Traditionally there was the split between civil engineering and building contracting, but over the years the line between the two has become blurred, and qualified civil engineers and builders work on both types of process. Speculative housing has still remained as a separate unit. The traditional form of procurement has given way to others such as design and build, management contracting and PFI, each requiring different skills and knowledge. This means there is a tendency for national companies to decentralise, up to a point, to cope with these diversities.

It is interesting to note many organisations go through cycles of moving from centralised to decentralised organisations and vice versa approximately every two decades. The author is reminded of the following quotation: 'There are those who believe that the best way to meet any new situation is by reorganisation. This is a wonderful method of creating the illusion of progress while producing confusion, inefficiency and demoralisation' (Petronius Arbiter AD 65).

26.6 The effect of tall and short structures

A flat structured organisation has fewer levels in the hierarchy, but with many more subordinates reporting to one manager, a wide span of control. The tall structure has many more levels in the hierarchy, but with fewer people reporting to the manager and with a narrow span of control, as shown in Figures 26.7 and 26.8. In each of the two examples the same number of boxes has been used in the diagram, yet the flat structure has only three levels of management whereas the tall structure has five.

In the tall structure the flow of information vertically is interrupted many times more than in the flat structure, but in the latter there tends to be more discussion and consultation. So there is a balance between, on the one hand, more time for the flow of information to take place, as against more time required to both discuss issues and coordinate many subordinates. The taller structures tend to be more dictatorial than the flatter structures as a result. However, because in the taller structures the span of control is small, the manager has more time to consider issues as less time is spent supervising. The low levels in the hierarchy in the flatter structures usually know the senior managers much better than in the taller structures, which can increase the quality of personal contact. Taller structures are more centralised organisations than flat structures. To manage a flat structure, because of the size of the span of the control, there needs to be more delegation of responsibility and this tends to result in higher morale of the employees. It can be seen from these few arguments it is a complex issue to determine which of the two is the better.

26.7 Case study

Two factories were set up to produce the same product, but the managers created different organisation structures. The factories produced precast concrete structural components for one of the industrialised building systems of the 1960s and 1970s. The equivalent position on a construction site is indicated in brackets in the boxes in the chart where appropriate in Figures 26.9 and 26.10. There were other roles not shown on these charts, but they were the same in both cases. The production manager was responsible for all the production activity, the maintenance engineer to ensure that the plant was working to as close to 100 per cent capacity as possible and the quality controller checked all the components were being made to the correct dimensions with materials of the appropriate performance and quality. In both cases the production manager was the second in command, the planners and maintenance

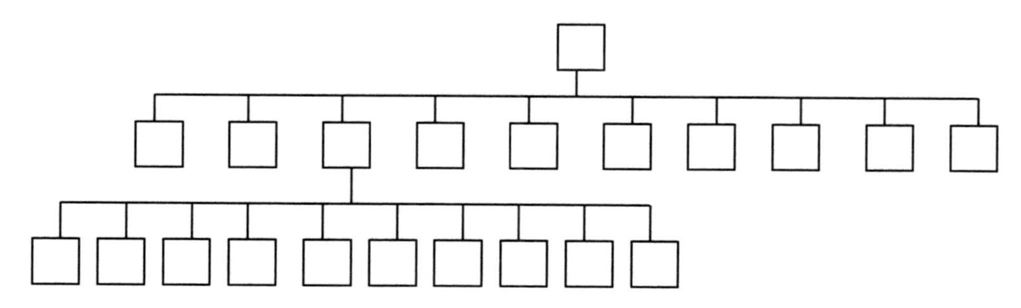

Figure 26.7 Flat structure

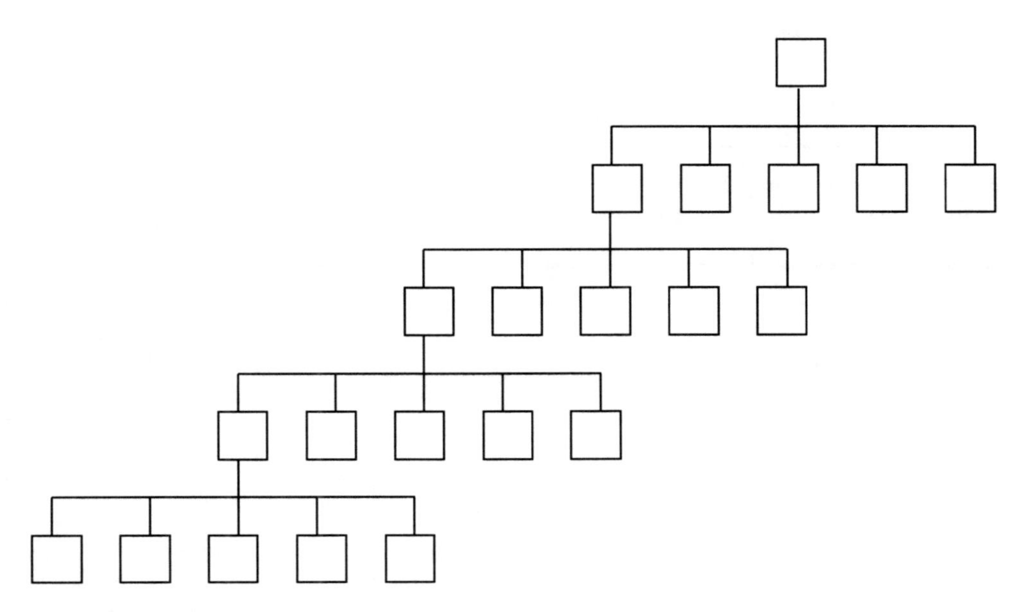

Figure 26.8 Tall structure

engineers were well qualified and the quality controllers had previously been supervisors.

In Factory 1, the factory manager had determined the organisation based upon the following logic. The production manager was responsible for all of the production so it was sensible for him to be responsible for all the departments connected with that activity, i.e. the production controller, maintenance engineer, quality controller and the supervisor. The production manager was also his second in command. The reason for the secretary being responsible to him is perhaps a throwback to the days when it was the norm for the senior managers to have their own secretaries.

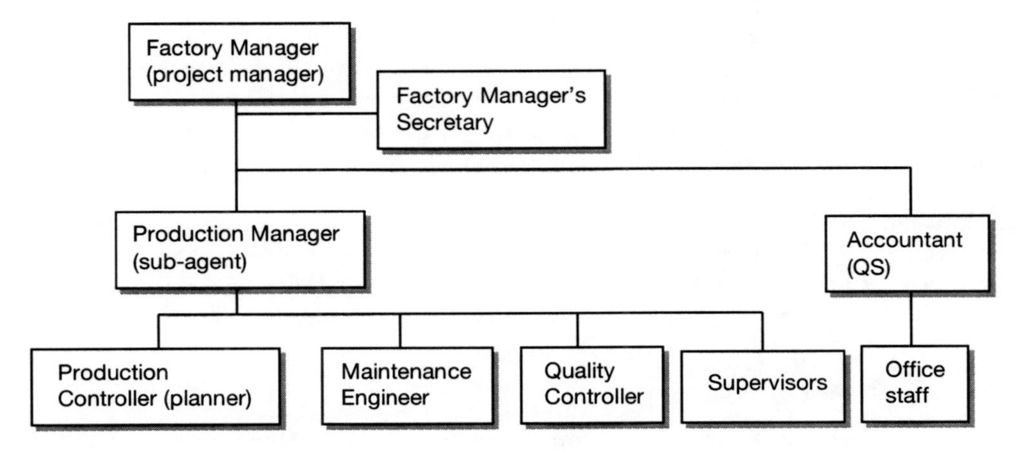

Figure 26.9 Factory 1

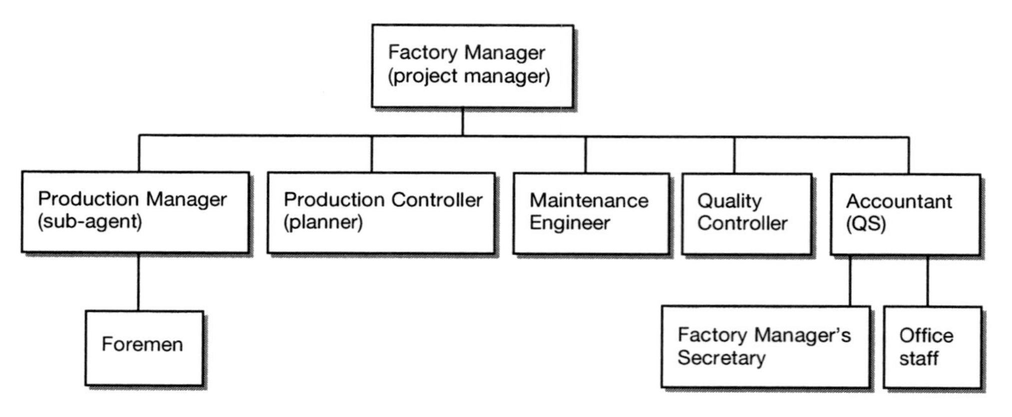

Figure 26.10 Factory 2

In Factory 2 the manager saw the same organisation in a very different way. The job of the production manager was to obtain as much production as possible each working day. If the plant broke down, the maintenance engineer was expected to make a rapid temporary repair and if necessary work on at night to make the repair more substantial and lasting. The engineer on the other hand would want to make the repair properly at the time of the breakdown. The quality controller acted as a brake on the production manager. As production rates increased, then quality standards would deteriorate. The production manager leaned towards being prepared to make substandard components if they could be rectified afterwards and keep production rolling. The production controller had to ensure there were sufficient units of the type required by the site to be ready in store for when the site called for them. Ideally, the production process would prefer to make all of one type then change to the next, because every change would stop production while the changes to the plant were being made. The operatives then would have to go through a learning process that could diminish production outputs. All this meant that there was potential conflict between the production manager and the others.

By putting all these roles on the same level, it was incumbent on them to resolve their differences. If they were unable to, they could go to the factory manager for arbitration. Most people would prefer to solve their problems rather than resort to going to their superior for action. In Factory 1, the production manager could tell the others what to do and, while there would be some discussion, in the end he made the decision, which did not ensure that the best solution was always used.

In Factory 2 the factory manager soon realised there was not enough work to keep the secretary fully occupied and decided to make her responsible to the accountant, who was also the office manager, so any unused time could then be put to good use, but still retaining priority call on her when requiring work to be done.

26.8 The board of directors

It is necessary to have a team at the top of any organisation to manage and direct the company. Although shareholders own limited companies it is not their responsibility to run them. They employ directors to do this on their behalf, can vote them off the board at the annual general or extra-ordinary general meeting. The board is responsible for setting the strategic objectives and making sure there are sufficient resources in place to meet them. They are responsible for all matters concerning the running of the company including health and safety, tax and employment law. They have a duty to ensure the financial accounts are accurate and honest. Indeed, it is a criminal offence to do otherwise. The board members take collective responsibility for the overall performance of the company.

There can be executive directors and non-executive directors. The former are full-time employees of the company and will have specific responsibilities for certain aspects of the business such as director responsible for finance, human resource management, contracts and so on, as well as their board responsibility. The non-executive directors are appointed because of their specialist knowledge and expertise or sometimes their contacts or image and standing within the community. If the business operates overseas then a retired foreign secretary could access influential politicians in another country. A member of the royal extended family gives a feeling of stability and integrity to the business. Their role is also to challenge the ideas put up by the executive directors as they are outside the normal day-to-day thinking within the business and are able to approach matters with a fresh and unbiased point of view.

The board must work within the limits set down in the Memorandum and Articles of Association that may have laid down certain restrictions on how and where the business operates, the type of work to be carried out and whether or not money can be borrowed. They must treat all the shareholders equally, irrespective of the number of shares owned. If they have any conflict of interest this must be declared. This can often occur with non-executive directors who may have interests in other companies. They must obey the law and occasionally can be responsible for the actions of their employees such as in matters of safety where, if they have not set up standards procedures and effective monitoring, they could be accused of negligence. Finally, they must not make personal profits at the company's expense such as selling or buying company shares when they have privileged information.

The board will meet at regular prescribed intervals. The frequency will depend upon the business itself. It would normally be anything from once a month to once a quarter.

26.9 The chairman of the board

The chairman of the board has different responsibilities to those of the chief executive. The chairman or chairwoman, sometimes referred to as the chair, is as the name implies responsible for chairing the board meetings as distinct from running the company. Their role normally includes:

- setting the agenda after consultation with the chief executive;
- ensuring the board runs effectively and all items on the agenda are given appropriate time for discussion;
- concluding each agenda item with a course(s) of action when necessary;
- ensuring through the company secretary all the papers for the meeting are distributed, giving sufficient time for proper consideration, and the subsequent minutes are promptly sent to members;

- ensuring there is proper communication from the executive to the non-executive members about the running of the company;
- ensuring there are appropriate checks and monitoring procedures in place so the board can properly evaluate the success or failure of the business.

26.10 The chief executive

The role of the chief executive or managing director is to run the business on a day-to-day basis on behalf of the shareholders. Their responsibility includes:

- developing and delivering the strategic objectives that have been agreed by the board;
- giving responsibilities to senior management to carry out this work in meeting the objectives and monitoring their performance;
- preparing the annual budget and financial plan for the business and establishing medium-term financial projections;
- recruiting, developing and retaining good quality staff;
- establishing and monitoring risk-management strategies;
- reporting to the board with accurate and timely information so the board can discharge its responsibilities properly;
- consulting with the chairman of the board on all significant matters;
- representing the company on matters pertaining to the business of the company.

26.11 Small and medium enterprises (SMEs)

In smaller organisations, there is no need, or indeed personnel available to fulfil the functions described in 26.8–26.10. Sole traders such as those offering contraction trade services have to register with HM Revenue & Customs and produce accounts that do not have to be made public. Partnerships usually have an agreement or a 'deed of partnership' that is legally binding between all partners describing how the business will be run. They do not have to publish their accounts. As businesses increase in size employing more personnel they can become a limited liability company. To set up a company, called incorporation, they need the Memorandum of Association and Articles of Association. These are submitted with a fee to the Register of Companies at Companies House which then issues a 'Certificate of Incorporation'.

Bibliography

Freeman-Bell, G. and Balkwill, J. (1996) *Management in Engineering: Principles and Practice*, 2nd edn. Englewood Cliffs, NJ: Prentice Hall.

Gray, J.L. and Starke, F.A. (1988) *Organizational Behaviour: Concepts and Applications*, 4th edn. Columbus, OH: Merrill Publishing.

Luthans, F. (2005) *Organisational Behaviour*, 10th edn. New York: McGraw-Hill.

Megginson, L.C, Mosley, D.C. and Peitri, P.H. (1989) *Management Concepts and Applications*, 3rd edn. New York: Harper & Row.

Mintzberg, H. (1979) *The Structure of Organizations*. Englewood Cliffs, NJ: Prentice Hall.

27 Strategic planning

27.1 Introduction

Companies evolve over a period of time, expanding, diversifying and contracting, depending upon the circumstances. There is a need to periodically re-evaluate the business, not just because there is a problem, but because the business world is always changing and the company needs to position itself so as to take full advantage of opportunities that might arise. Sometimes the changes are small, but in other cases the changes are dramatic, moving the business in a different direction with major implications to the employees and their future with the company. Examples of this are when in the 1990s Wimpey and Tarmac decided to exchange parts of their business, Wimpey giving their contracting arm to Tarmac and in return received Tarmac's house building. In 2001 Alfred McAlpine sold their house building and used the proceeds to build up their facilities management and utility services business.

Strategic planning has become increasingly important over the last few decades because both the external and internal environment have become much more complex, and the techniques available to management, and their skills, have become much more sophisticated. It takes place in both the private and public sector. In essence the company is saying 'Where do we want to be in say five years' time and how do we get there?' This requires senior management decisions, the ability to allocate large sums of money if necessary and to understand the company's interaction with the external environment and the implications with the internal environment. It is done to identify and achieve competitive advantage, to give the business a direction all the employees can identify with, which can encourage innovatory ideas to achieve the goals set. Once the strategy has been decided it will fall into one or more of the following: expansion, contraction, diversification or consolidation.

27.2 Levels of strategy

Depending upon the size of the company, there are three levels of strategic development, as shown in Figure 27.1. Single businesses will only have two.

The group corporate-level strategy determines which of the businesses of the company will continue to operate, how resources will be allocated within the group, and whether any of the third level will be coordinated for the group as a whole. For example, research and development may be carried out on behalf of all the businesses. It will also determine the amount of growth and profit expected from each of the businesses and their contribution to the group finances.

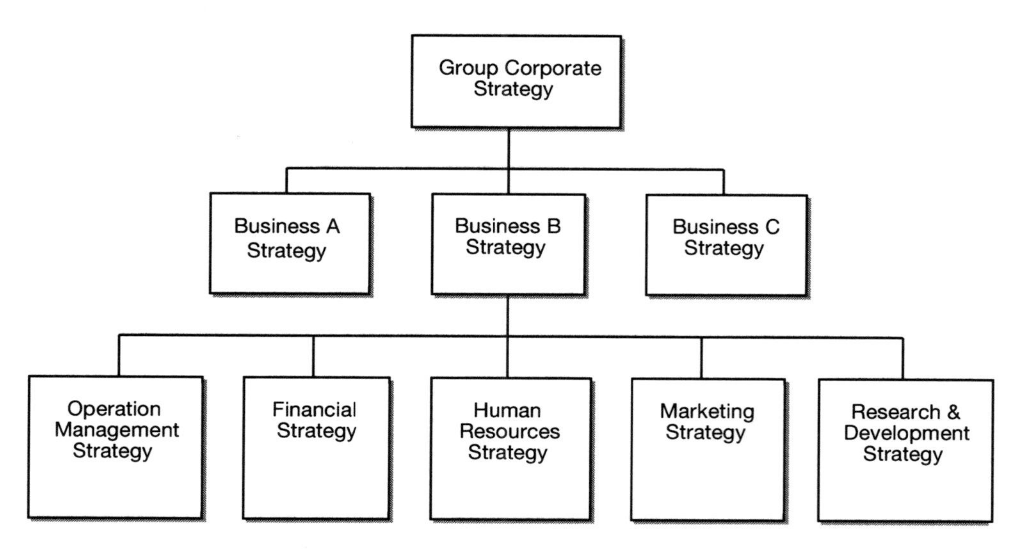

Figure 27.1 Levels of strategy

Business-level strategy looks at the requirements for each of the businesses, assuming each of the businesses is autonomist and has its competitors. In the case of a national and international construction company, this could be a UK operation, a European operation, house building, general contracting or civil engineering. Within the UK it could be a northern and southern operation or cover smaller areas such as the North West, Scotland or the South East. The strategy will consider the amount and level of competition, the opportunities within which the business is operating, and determine its objectives.

Functional-level strategy looks at how each functional area within the business needs to perform to meet the overall strategy for the business. For example, if the business is expected to expand by a given amount, the marketing strategy has to be geared to make that happen. The human resources section has to take steps to recruit and train the required number and type of personnel. If the business is to shrink, they must develop a redundancy strategy to have the number of staff to meet the new demands, or if the business is to diversify, recruit personnel expert in the new work of the business. The group needs to coordinate all the strategies across all the businesses to maximise efficiency and to ensure the overall corporate objective is being achieved.

27.3 The strategic management process

The components of the strategic management process are demonstrated in Figure 27.2. The first stage is to identify the company's current mission, and strategic goals and objectives. The strategy may be modified as a result of the exercise.

The first stage is to identify the company's mission: what is the business the organisation is in, where it is heading, and when does it wish to achieve its objectives? Using the example of higher education, is the university in the business of training students for particular jobs or is it in the business of producing graduates with a well-

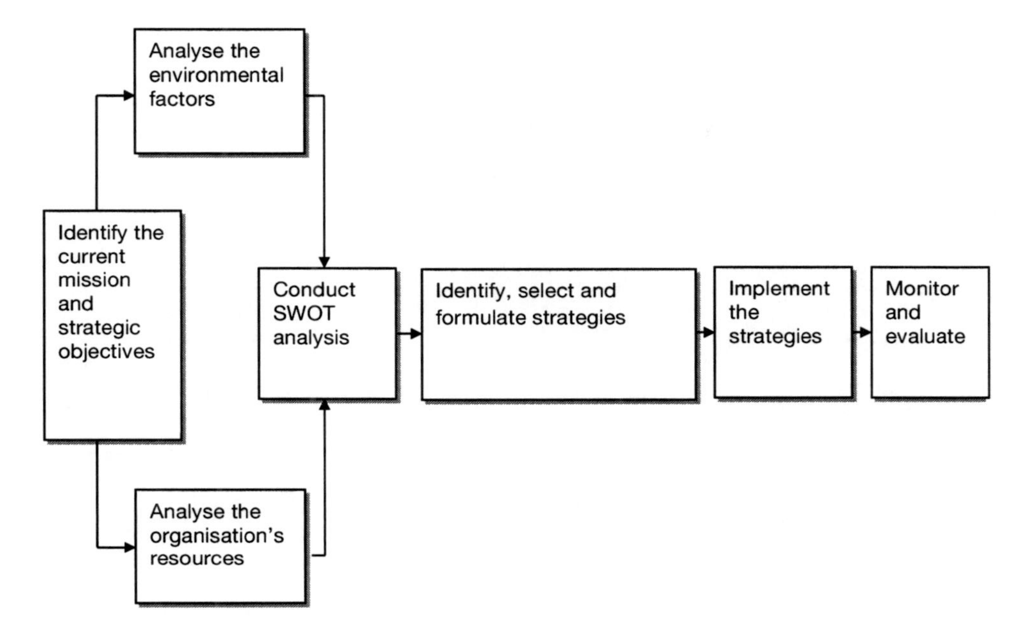

Figure 27.2 The strategic planning process

rounded education so they are adaptable to the work situation? Does the university wish to take the less able students and add value to them by high levels of staff involvement or take the higher ability student and expect them to read for the degree on their own with few lectures and individual tutorials, as at Oxbridge?

In construction, questions could be: Is the business that of constructing speculative housing and/or general and civil engineering contracting or a particular part of the market? Is the business one that reacts to tender enquiries or is it one that is active in generating work? Is the company concerned with providing a quality service or product? Does the company want a zero-accident culture? Does the company wish to move to being known as 'an environmentally friendly' business? Does the company wish to minimise the time spent making good faults after handover?

Many companies lose their way when their mission objectives are unclear or wrong. There are well-known examples in the retail trade where customers drift away because the shops no longer provide them with what they want and they go elsewhere to those that do. The railway business in the USA had a problem because they saw their business as being the railway business rather than transporting people and products.

The second stage is to assess the environment in which the business is to operate and the resources within the organisations, as summarised in Figure 27.3. Without this analysis it would be difficult to plan the future of the company. Customers, present and potential, are the crucial element of any analysis of the external factors affecting the development of a strategy. The status of suppliers of materials and components determines the opportunities for expansion. If they are overstretched, new ones have to be sought out and, if not available, this will affect any decisions made; similarly with subcontractors, can they meet the demand and quality required? The availability

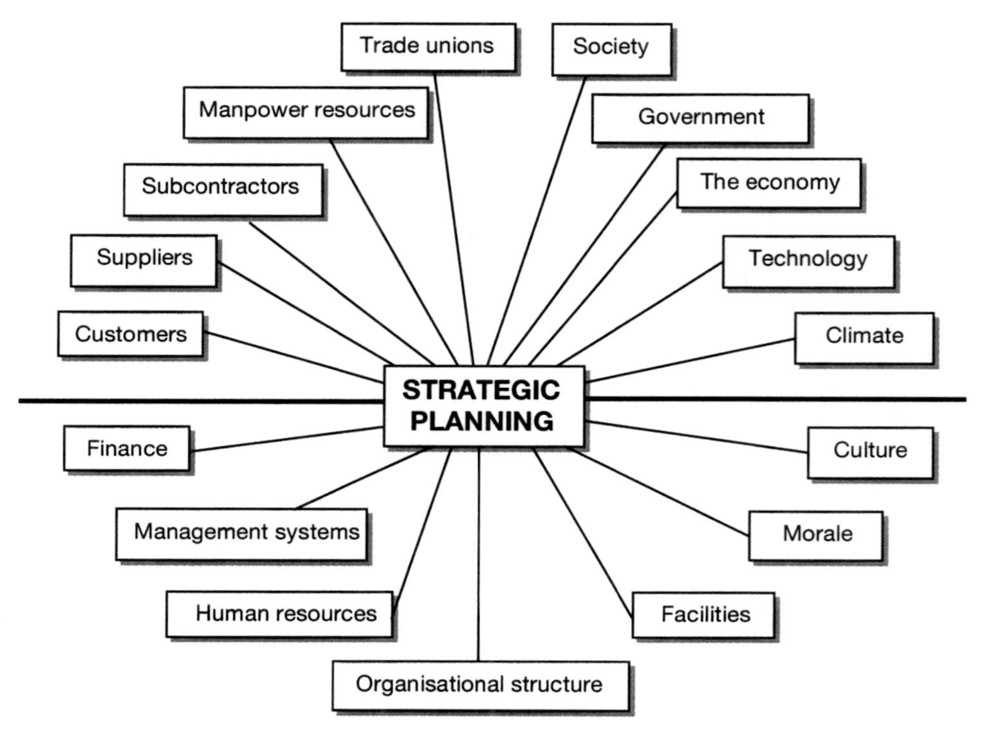

Figure 27.3 The internal and external environment

of adequate trained manpower, both manual and cerebral, can impact significantly on the business, especially if one of the requirements for working in a particular area is that local labour has to be employed. Trade unions have become very pragmatic and work with employers for the good of the industry and their members. The days when, for example, many companies would not work in Liverpool because of the militancy have now gone. There are parts of the world where the way labour is organised and controlled needs to be understood, as it can affect the prospects.

Social changes have an impact on the construction industry; for example, in housing the increase of one-parent families and the need for social housing. Society can also determine other changes in requirements, as did the revulsion against multi-storey housing as a solution to housing shortages in the 1970s; a trend now reversed with many multi-storey flats being constructed or other multi-storey buildings being converted in our inner cities. Environmental pressures on where construction occurs and the types of buildings are other examples of society's influence. Government (local, national and European) determines policies and enacts laws that influence the market place. Nationally, the Public Finance Initiatives are a good example of this and the current push for affordable housing another, as this may result in the relaxing of planning regulations and the availability of previously protected land. The European Community (EC) produces directives affecting employment and free trade, hence public contracts over 5 million euros have to be advertised across the EC. Contractors need to be aware of the economic climate and future trends especially as a result of Brexit. This is affected by national governments and world events since we now exist in a global economy. Technologies are changing at a fast rate, not just impacting on the way construction takes place, but also in the requirements of the clients in the design of buildings both to accommodate the new technologies and to cater for the way their business operations change. Finally, climate can have a significant impact on the construction process, depending on the part of the globe the company decides it wishes to operate in.

Internally, the financial state and ability to borrow money to finance further operations needs to be analysed. Are the management systems in place appropriate to the organisation's needs or should they be modified or improved? The human resources department will assess the type of manpower available within the company and the match with the strategic plan to establish whether or not training and promotion from within will cover the needs, or whether recruitment from without is necessary. Organisation structures can become ineffective and changes must be made. Offices may be inappropriate with poor internal communications as a result. The buildings housing staff may be outdated, the wrong size or in the wrong geographical location. More difficult to establish is the overall morale of the staff. If it is low then radical changes to the organisation may be difficult. The culture needs also to be explored; changing this takes time and needs to be managed carefully. The culture may have been determined by many things, such as the business previously carrying out traditional contracting and it is now having to move into the management contracting arena.

27.4 SWOT analysis

SWOT is the acronym for strengths, weaknesses, opportunities and threats and is used in the early stages of developing a strategy. The strengths and weaknesses are concerned with the internal environment of the company, whereas the opportunities and threats are factors affecting the company from outside. Typical issues that might be explored in a SWOT analysis are shown in Table 27.1.

It is normal practice to quantify the relative merits of each of the four categories and prioritise them. By carrying out a SWOT analysis, management has a much clearer understanding of where the company is currently and the environment in which they operate. This gives the opportunity to address the weaknesses and use the strengths to overcome the threats and explore and take advantage of any opportunities.

While Table 27.1 gives indicative issues that might be raised, in practice the issues evolve from senior management brainstorming or by consulting key staff, asking them to list their views on the four categories. These are then collated and a list produced for each of the categories, prioritising the responses from those consulted.

One of the major issues will always be competition from other companies and looking at this issue can enhance the SWOT analysis. Michael Porter, a strategy expert, developed his 'Five Competitive Forces Model', comprising rivalry, bargaining power of customers, bargaining power of suppliers, threat of new entrants and threat of substitute products or services. Rivalry is the way competitors use varying tactics to obtain work, by offering discounts, reduced construction periods, advertising and

Table 27.1 Typical issues for consideration in SWOT analysis

Strengths and weaknesses	*Opportunities and threats*
Financial status	New markets
Competitive edge	Existing customers
Image	Competition
Strategic direction	National economic revival or decline
Facilities	Labour availability/redundancies
Management expertise	Suppliers' workload
Market leader	Subcontractors' workload
Cost advantages	Expansion of the European Community
Reputation for quality	Inflation levels
Reputation for completion on time	Terrorism and war
Construction expertise	Mergers and takeovers
Staff turnover	Location of the business
Land bank	Employment legislation
Morale of staff	Risk

various incentives in the housing market such as free legal fees, stamp duty paid and carpets throughout. This gives an indication of how much competitors are spending that can affect their profit margins.

The bargaining power of customers is how strong they are in being able to force prices down or obtain an improved service for the same price. This is a function of the market forces of the day or the strength of their order book. The stronger the bargaining power, the lower the profit, except in the case of serial or term contracts, where there are other advantages because of the continuity and familiarity with the work.

The threat of new entrants in construction can come from overseas and from existing contractors who have decided to diversify into the particular type of work. For example when the workload of the industry declined in the 1980s, some of the larger contractors diversified into small works. This inevitably puts pressure on profit margins as companies strive to maintain their market share against the new competition or reducing market.

The threat of substitute products or services can be interpreted in construction as where new forms of contracts are used, requiring a differing approach in attitudes and culture. The changes can be significant, as shown by the significant increase in design and build, management and construction management contracts and the PFI. The threat of substitute products is more of concern to the suppliers to the industry such as with uPVC replacing timber for window frames.

As a result of all these analyses and the SWOT analysis, the mission and objectives of the organisation may have to be amended.

27.5 Identify, select and formulate the strategies

The strategies selected need to be set for the three levels of the business – corporate, business and functional. There are two accepted approaches to this: grand strategies framework and corporate matrix strategy.

27.5.1 The grand strategies framework

Sometimes referred to as the master strategy it comprises three main approaches, growth, consolidation and retrenchment, sometimes referred to as defensive.

Growth strategy

A growth strategy as the name implies means employing strategies that expand their business in some way. The first, concentration, is either to direct effort into a part of the business with the aim of securing a larger market share, or to develop a new product in the existing business, for example, in speculative housing by using timber frame as well as or instead of masonry construction.

Integration is taking on work in the supply chain currently carried out by others, or not producing for a client, but selling directly oneself. This is interesting in relation to construction as the UK industry has done exactly the opposite in the former case, and in the latter, other than in the speculative market, has always built 'a product' for a client. The reason for subcontracting work has been one of cost savings and the ability to call in labour as and when required rather than having to find work and guarantee continuity of employment in the temporary and changing nature of construction projects. There seems little likelihood of this reverting to the former scenario of employing labour directly. However, it would be interesting to explore the possibility of becoming the developer as well as the contractor as a means of growing the business.

The third approach is diversification into new markets. The advantage of this is the business is less exposed if one of the other strands comes under competitive or economic attack. However, there are risks involved when entering any new market as it is unfamiliar territory. In construction, this could mean moving into the PFI market place, offering high-quality exclusive executive speculative homes as distinct from the general market, carrying out the refurbishment and maintenance contracts or starting a facilities management operation to complement the PFI work.

Stability strategies

This is a situation when the company decides it does not want any further growth and will concentrate on consolidating its business. This can occur when a business has grown so fast and feels it is overstretched, resulting in unsatisfactory service to its customers so a period of stability is required. In a declining marketplace, maintaining its place in terms of market share can also be considered a stability strategy.

Retrenchment strategies

Sometimes called a defensive strategy, as some companies dislike the term retrenchment, it is a strategy designed to reduce the company's operations normally by cost-cutting measures and/or by reducing the company's assets.

A harvest strategy is used when the company intends to move out of a market, but in the meantime has to carry on until the existing order book has been completed. This requires as minimal investment in the business as possible, while trying to maximise profits and cash flow. Examples of minimising investment would include not decorating the premises, not investing in new plant and equipment, and ceasing manpower training unless retraining for a new occupation elsewhere in the group.

A turnaround strategy is employed when the company is in decline and the aim is to reverse this trend and restore the once-achieved profitability levels. This will involve attention to all the cost centres and usually involves cost-cutting measures, including redundancies or selling off the non-profitable parts of the business. The latter is known as divestment.

Bankruptcy is applicable to individuals who can no longer pay off their debts and cannot be pursued by their creditors once the courts have allocated any assets they might have. Companies go into liquidation either compulsorily or voluntarily. The former is when a creditor goes to court and the court appoints an official receiver to sell its assets and pay off its debts, the latter is when the company agrees to sell its assets to pay off its debts.

Combination strategies

This is where different parts of the business follow more than one of the above strategies where, for example, one part of the group is pursuing a growth strategy while another is contracting.

27.5.2 Corporate portfolio matrix

The grand strategies approach looks at the company's overall direction, whereas portfolio strategy aims to help managers determine the types of business the company should be involved in. The Boston Consultancy Group (BCG), a well-established consultancy organisation, developed it in the 1970s. It comprises four cells, referred to as stars, question marks, cash cows and dogs in the top left, top right, bottom left and bottom right respectively. The horizontal axis expresses the percentage of the market share and the vertical axis the anticipated growth relative to growth in the economy as a whole. In the matrix squares are used representing the size of the individual business percentage revenue relative to the other businesses in the company as a whole.

The businesses in the star category are high growth and have an anticipated high market share. This is a rapidly expanding market and the business may require a significant investment that is greater than they can earn. Those in the cash cow quadrant have a high market share in a low growth market. They usually generate high returns but have only a limited opportunity for future growth. Those in the question mark have a low market share, but in an anticipated high growth market; in essence there is a question mark over them as they are high-risk ventures that might or might not be successful. They will require substantial investment if it is decided to support these businesses. Dogs are the low growth and anticipated low market growth area. They produce little profit, but do not require much investment. These need either to be revitalised as in harvest and turnaround or divested or liquidated (section 27.5.1).

Construction businesses do not have as many separate companies as some of the large manufacturing groups, but this method can be used to look at the relative performances of the regions of a national and international construction company and may indicate, for example, two or more regions should be merged into a larger region. It should be remembered this is a guide to assist senior management in the making of strategic decisions and should not be taken in isolation. Each separate business should be analysed independently as there may be good reason why the business is shown in a particular quadrant at that point in time.

27.6 Formulating business-level strategy

This is concerned with the way a business within the group competes and operates (see 27.2). In the construction industry this will usually be at regional level or the organisation as a whole. This section will explore two approaches: the first, adaptive strategies, developed by Raymond Miles and Charles Snow; and the second, competitive strategies, developed by Michael Porter.

The adaptive strategy identifies four strategies types referred to as defenders, prospectors, analysers and reactors. Defenders produce only a few products or services aimed at a narrow part of the market, and work very hard to try to keep others out of their 'patch'. If successful they maintain a niche market other competitors find difficult to penetrate. Prospectors look for innovation and believe their success will be founded on finding new ideas and exploiting that market. To achieve this they need to have organisations that are quick to react and prepared to accept and absorb change. Analysers aim to minimise risk and maximise the opportunity for profit. They copy good innovatory ideas prospectors have developed. They are more efficient than prospectors, but because they take less risk, produce smaller profits. The reactors behave inconsistently and tend to perform badly. They are reluctant to follow any strategy aggressively and usually lack focus.

Competitive strategies consist of three alternatives: cost-leadership strategy, differentiation strategy and focus strategy. The common requirements for each of the three strategies are as follows:

- A cost-leadership strategy is where the company sets out to be 'the' cost leader in the business and not just one of the contenders. The product or service provided must be seen as being similar to the rivals'. To be a cost leader, and have cost advantage, the company needs efficient operations, innovation, low labour costs (this does not necessarily mean low wages), economies of scale and access to suppliers who give preferential rates. All has to be achieved without affecting standards of quality, safety and environmental issues. To do all of this requires very tight controls.
- A differentiation strategy is about producing products and services unique to the industry. These differentiation factors could be high-quality, innovatory designs; quality of service; speed and reliability of completion; very low accident rates or a combination of several of them. 'What makes us different?' has to be marketed as a concept. Both the cost-leadership and differentiation strategies are aimed at a wide range of the market.
- A focus strategy is aimed at a very narrow band of the market by specialising, and is probably best suited for small businesses. The segment of the market could be a geographic region, a type of customer or a specific part of the market, for example a structural steel frame erector. The strategy still requires either a cost-leadership or differentiation approach.

27.7 Strategy implementation and monitoring

Once the strategy has been formulated it has to be implemented. It is essential any implementation should be carried out throughout the business in all departments and functions at the same time. Galbraith and Kazanijian (1986) highlighted five principal factors that should be considered: technology, human resources, rewards systems, decision processes and structure.

Technology is defined, in this case, as the tools, equipment, work techniques and knowledge used by the company to deliver its product or service. There has to be a balance between the technologies available in matching the strategic objectives. So, for example, if the strategy is to reduce costs, new technologies may have to be employed to achieve the strategy.

The organisation needs the right human resources with the appropriate skills employed in the correct positions, both geographic and functional, and unless this is in place, the strategy cannot be accomplished. Once it is, there are further advantages that affect the company's competitive edge, because the employees are more likely to be able to be more productive and innovatory in their approach to work.

Reward systems are not just about bonuses, but also concern promotion prospects and job satisfaction, resulting in higher motivation and less staff turnover. However, linking bonuses to achieving the company's strategic objectives is important, especially if designed in a creative way. For example, if one of the objectives is to reduce the number of accidents in a region of a construction company, linking the regional director's annual bonus to the safety record of their patch would certainly focus their attention and that of the subordinates.

Decision processes concern the ability to resolve questions and problems that occur within the business and are, in part, linked to the communications systems in place and the culture of the organisation. A frightened subordinate is less likely to have an open discussion with their supervisor than one who isn't, and a dictatorial supervisor will have little interest in the opinions of those they manage. There is also the issue of having good decision-making processes in place when developing the strategic plan, otherwise how do you know that you have the best plan and have covered all eventualities?

The organisation structure is the formal way in which the various parts of the company interact, be it horizontally or vertically, sometime referred to as the organisation chart. It needs to be designed so the different parts of the business can coordinate their work to meet the strategic objectives.

Finally, once in place management needs to periodically monitor the key factors that affect the achievement of the strategic objectives to ascertain whether or not the implementation is working and then to make adjustments depending upon what is found. Sometimes the strategic change is seismic in proportions and management has to change vary carefully.

27.8 A further thought

A problem often encountered when a new strategy, or indeed any change, is introduced, is that there will always be a percentage of those affected by it who will find reasons why the objective cannot be achieved. The suggestion is that the strategy should be put forward in a different way. Rather than ask the staff to come up with methods to achieve the objectives, why not state that we have already met the objectives and ask the question 'how did we get here?' This tends to place the staff in a positive frame of mind rather than having thoughts of negativity. The author has found this a very useful approach.

Bibliography

Bartol, K.M and Martin, D.C. (1997) *Management*, 3rd edn. New York: McGraw-Hill.
Carnall, C.A. (1997) *Strategic Change*. Oxford: Butterworth Heinemann.
Galbraith, J.R and Kazanijian, R.K. (1986) *Strategic Implementation: Structure, Systems and Process*, 2nd edn. St Paul, MN: West Publishing.
Hussey, D. (1998) *Strategic Management*, 4th edn. Oxford: Butterworth Heinemann.
Megginson, L.C, Mosley, D.C. and Peitri, P.H. (1989) Management Concepts and Applications, 3rd edn. London: Harper & Row.
Miles, R.E. and Snow, C.C. (1978) *Organisational Strategy, Structure and Process*. New York: McGraw-Hill.
Porter, M.E. (1980) *Competitive Strategy*. New York: Free Press.
Robbins, S.P. and Coulter, M. (2015) *Management*, 11th edn. Englewood Cliffs, NJ: Prentice Hall.

28 Marketing

28.1 Introduction

The construction industry is different from most other industries because it provides a bespoke service producing building or infrastructure projects that the client has commissioned that, generally speaking, is different in every case, the main exception being the speculative housing market. The concept of 'selling' is, therefore, alien to many in the industry except to those in the speculative market. At first glance, the industry reacts to clients' wishes by tendering for work and if successful, constructs the building or provides the infrastructure. However, while traditionally this was the case, the construction company had then to contend with getting their name on preferred lists so they could tender, and more recently make presentations to be selected for projects and serial contracts. This has made the industry more conscious of the need to market and sell their services.

The Chartered Institute of Marketing defines marketing 'as the management process responsible for identifying, anticipating and satisfying customer requirements profitably'. Examples of these customers' needs in the speculative marketplace include: what size and location houses should be, the quality of finishes, and so on. In mainstream contracting, as contractors have become more involved in the design processes, they are in discussions with clients to establish what they really need: they apply value engineering techniques; there is greater emphasis on completion on time and less on using the contract clauses as reasons for not doing this to claim extra payments; and they use total quality management to ensure customer satisfaction.

There are various stages of marketing for the construction industry, which can be briefly summarised as follows:

- market intelligence systems and market research
- the customer
- the external environment
- forecasting
- market positioning
- new products or services
- distribution decisions
- advertising
- other forms of promotion
- market planning.

28.2 Market intelligence systems

The success or failure of the marketing process is dependent upon the quality of the data obtained from marketing research and the information it provides. The quality of the management decision is only as good as the facts available at the time. Day-to-day management does not have the time to spend collecting and examining large quantities of information, but rather has to decide very rapidly based on clear information available at the time. However, making decisions about the long term is a different matter and marketing falls into this category. The role of market research is to minimise the uncertainty as far as possible by providing data and information, known as market intelligence. Data is the collection of facts, information is data that has been selected and sorted for a specific purpose, and intelligence is the interpretation of the information after analysis.

Information technology allows marketers to collect, store and organise data. These systems are called market intelligence systems (MIS). The information is obtained from three prime sources as demonstrated in Figure 28.1.

In construction, internal data related to performance is primarily obtained from the estimating, purchasing and contract management departments, giving information on costs, both actual and estimated, success rates in estimating, competitors' successes, completion times, durations of activities, names of the clients and design team members. Further internal data can be derived from written reports, aide-memoires and by questioning company staff.

One should never underestimate the knowledge gained from personal experience. For example, competent site managers and estimators can look at a set of drawings and in relatively short time advise on the approximate cost of the project, its duration and when key stages in its construction are likely to be achieved. Senior managers have an instinct and feel for the market place resulting from their contacts with clients, existing and potential. Purchasing managers are good at predicting how much materials suppliers and subcontractors are likely to reduce their quotations if the contract is awarded.

External data can be collected from the web, providing the person doing so is able to distinguish between useful material and the rubbish. However, data about clients and their businesses is generally sound. It is often possible to discover a lot about other organisations from their websites. Companies have to lodge their annual accounts with Companies House, another source of data. The news media is an excellent source of information. There are the broadsheets, but also the specialist newspapers such as *The Economist*, and the trade press. Libraries, professional institutions and trade associations are excellent sources, especially for directories,

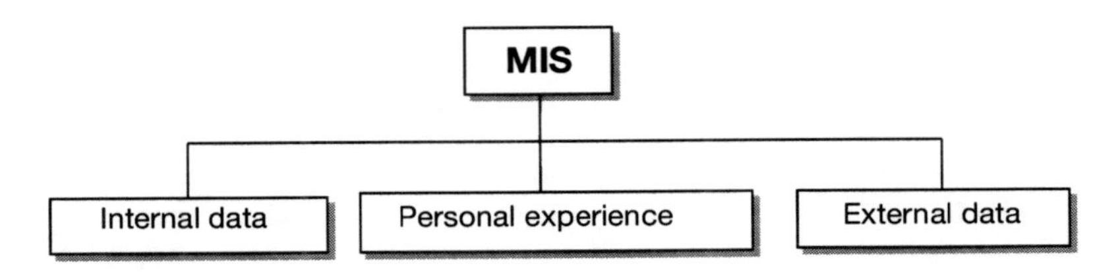

Figure 28.1 Sources of data for market intelligence systems

with their staff often eager to carry out searches. There are an increasing number of computer databases that can be accessed, usually for a fee or subscription.

28.3 Market research

Market research is the process of collecting information from consumers, the users of the building in the case of construction, and the clients. Sampling is normally utilised, because questioning all clients and users would be too large a task. It can be carried out by the company's staff if qualified, or by employing specialist market research firms.

The first stage of the process is to define the objectives of the research: why is the research being done and what do we want to gain? It could be the company wishes to find trends in a type of work, the projected availability of labour, the competition if diversifying, clients' aspirations, or what the users actually require from the building, which may differ from the clients' beliefs. At another level, who are the users? For example, in a hospital there are patients, doctors, nurses, administrators and maintenance staff, all of whom have a view on the best way to carry out their work or be treated.

The second stage is to plan the research. How much depth is the study going to go into, as the deeper the questioning, the more expensive it will be. It can vary from identifying problems and determining possible courses of action, to conducting detailed surveys.

The third stage is to select the method of research, demonstrated in Figure 28.2. The methods can be defined as the qualitative approach and the quantitative approach. Qualitative research discovers people's attitudes, opinions, ideas and work methods. This is achieved by several methods. One is by observation of the people in a building and seeing how they react. For example, watching patients in the outpatients department in a hospital and seeing how they react, how much they move about, and their ability to find the toilets and doctors. In-depth interviews are conducted usually by means of a structured interview face to face, whereby all the questions are thought out before the interview takes place and answers solicited in a planned sequence, or by using semi-structured interviews where the main questions are planned, but are more open-ended to allow more freedom by the respondent. These are more complex

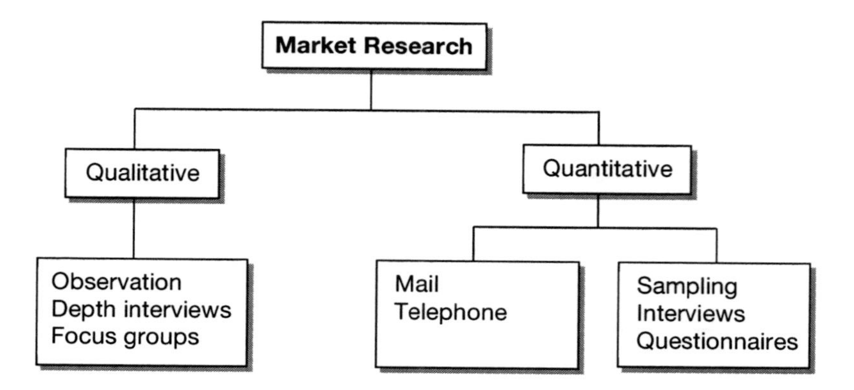

Figure 28.2 The research method selection

to analyse afterwards. Another method is the use of focus groups, comprising 6 to 10 people who are encouraged to discuss and debate the answers to prepared questions. The interviewer is there not just to pose the questions but also to act as the chairperson in encouraging everyone to contribute.

The quantitative approach aims to obtain enough data for a statistical analysis. This is conducted using questionnaires. There is a science in producing questionnaires, which should be studied before composing one, and the reader is advised to look elsewhere for this such as, Hague's *Questionnaire Design* (Hague, 1993). It is important to choose the sample carefully to ensure it is representative. It is important to obtain as high a return response as possible otherwise the statistical validity can be questioned. These are difficult tasks. Once the questionnaire has been formulated it can be sent out to the respondents for completion. Alternatively, it can be used as it is or modified to interview selected people by telephone.

All the data obtained has to be analysed using statistical analysis, which can be done using computer software packages, and then a report produced. This should be written in a form the reader can understand and that tell them what they wish to know. The author of the report needs to be aware of any biases he or she may be prone to and try to eliminate these from the findings. The reader needs to know how reliable and accurate the conclusions are and how relevant the findings are to the original objectives.

28.4 The customer

The construction industry serves a wide range of customers including national government, local government, the hospital trusts, commerce, industry, the retail sector, housing associations, individuals who either commission work or buy from the speculative sector, and users of the building who may not necessarily be the owners, such as visitors and employees. It provides new buildings, maintains and alters existing buildings, provides a complete facilities management service and then demolishes the building after it has served its useful purpose.

Deciding whether or not to have a project constructed or modified is different from purchasing a manufactured product in many ways, depending upon who the customer is. For simplicity, the customer is considered in three broad categories – the public sector, the private sector and the individual. The prime determinates for the public sector are the government and local council, which determine the type and use of the project, both building and civil engineering, and the amount available for spending. In recent years the amount spent has also been aided by the use of Public Finance Initiatives where private sector money has been sourced. The prime marketing position for obtaining this work is being on the approved list for tenders and having the ability to put together bids and presentations. It also requires market research in anticipating trends and changes in political leanings. Good networking is also useful.

The private sector and the individual customers are more complex as it is necessary to understand the nature and requirements of the customer. Figure 28.3 demonstrates some of the factors that can influence customers. When considering these factors, it is important to be aware they are continually changing as society and the world changes.

Economic influences are key issues in any decision to build, usually determined by government policy and the current and perceived future state of the country's economy.

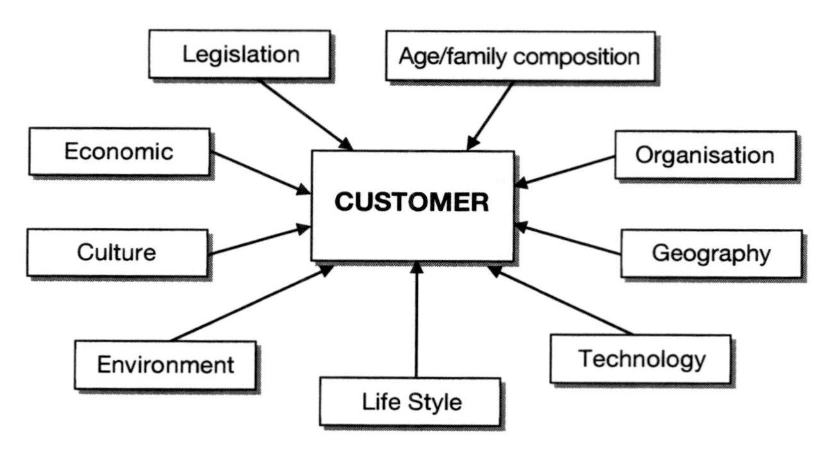

Figure 28.3 Factors that influence customers

How much government is prepared to spend and how much it passes to local authorities is determined by the Chancellor. The public sector takes a longer economic view considering likely trends and the influence of global economics. The individual purchasing a house or having extensions built is concerned with their residual income, interest rate trends, any government policy affecting house purchase such as levels of stamp duty, and the ability to raise finance from the lenders as well as the cost of property. Older couples are using downsizing as a means of releasing capital for their retirement. The decline of company pensions will in the long term have an impact on the housing market.

Legislation from time to time creates construction opportunities, although not always from a continual need to reform. Resulting from the tragedy at Hillsborough in 1989, and the subsequent Taylor report, there was a major change and requirement for the design and construction of sports stadiums. Part M of the Building Regulations, 'access and facilities for disabled people' is another example of how construction work has been generated.

The age of the customers in house purchase is a serious issue as first time buyers are becoming older as property prices increase. The size and composition of families is changing with an increase in single-parent families, fewer children per household and with people living longer, resulting in more elderly couples or single habitation. Younger purchasers have higher expectations than their parents who were prepared to wait until they could afford to purchase things they aspired to. Younger purchasers are more likely to expect the house to have everything they need already in situ when they move into a new house. The choice and style of the design of fitments such as kitchen units can be related to the age of the purchaser. The manner in which a show house is presented could be different depending upon whom the houses are aimed at. For example, if for families with younger children, then the children's bedrooms would be filled with furniture in keeping with how they would wish to live, with attractive up-to-date toys and games.

Organisations have their own way of making purchasing decisions when it comes to buildings and will select from the wide range of procurement options available to them such as traditional contracts, design and build, management contracting and serial tendering. The company needs to establish which method companies are most likely to adopt and, if able, adapt the business to cater for it and promote their ability to carry out work under the specified form of contract.

Geography can exclude certain activity outside the area of current operations. However, if the company makes a strategic decision to extend its area of working, market research could be necessary to break into the new area. Some customers operate nationally and may wish only to have to deal with one office even though the work is in another area of the construction company. For example, much of Marks & Spencer's work was carried out by Bovis Lend Lease's southern office and the northern director watched from his office in Manchester as the new M&S prestige store was being constructed after the IRA bomb, by the southern division. In the housing market, identifying areas people would prefer to live in can be determined by the school in the area, facilities provided and transport access, not to mention attractive areas for holiday homes, although to date this is provided mainly from the existing housing stock. This means there is a need to provide affordable housing in these areas as the less well off are priced out of the housing market.

Cultural factors, while mainly concerned with working overseas, can influence customer choices in the home country. The UK comprises a very diverse population of different races and religions, from which there are many potential customers both in business and as individuals. Their way of life and beliefs has to be understood in developing good customer relationships. Overseas, additionally, there are cultural practices that determine the means of carrying out business. A good example is the Chinese virtue of 'face', a very complex subject, but crucial to have some understanding of when working in this environment. For, example, it can be something as simple as knowing a calling card should be handed out held in both hands, to appreciating what causes a Chinese colleague to lose face and the implications, or that age is revered and often is linked to seniority. In some countries women are not afforded equal status with men, which may have implications in staffing operations there. Working overseas can require different approaches to marketing, sometimes with moral implications, as in parts of the world it would be expected to 'grease the palm' to obtain work or be offered contracts.

Technology development moves on apace and an understanding of the likely impact and opportunities is important in the marketing process, especially when involved in the design process, as accommodation for the future needs to be designed for and built in. A demonstration of the company's awareness and understanding of these issues could assist in obtaining work. The development of information technology and the accompanying communications hardware and software increasingly relies on smaller and wireless technologies. Employees can now take their office with them, work from home or hot desk, affecting the type of office accommodation required. Whereas once there was one television set in the lounge, now small children have one in their room plus computers and other gadgetry. There are many more labour-saving pieces of equipment all of which require servicing.

Life style is continually altering as culture and prosperity change, technology being one of the main drivers. The use of the Internet has changed shopping habits, which in turn affects the retail markets' building requirements. People are choosy when they look for a home, preferably near good transport links and other facilities in the area.

Environment is now centre stage politically. Since buildings are significant contributors to carbon emissions both in terms of embodied energy and that required for heating, cooling and lighting, there is potentially a great marketing opportunity for companies involved in both the design and construction processes. Public knowledge, concern and perception have changed considerably over the last decade and governments are slowly catching up, enacting new laws and regulations. New designs, innovatory use of materials, both currently available and those in development, will be used. Methods of becoming more sustainable and self-reliant will be on the agenda, especially in terms of energy use. How life styles will change as a result is up for debate at the time of writing, but the industry needs to be ready to meet this demand and play its part in marketing the available solutions both in the home market and for industry and commerce.

28.5 Forecasting

Forecasting is about reducing uncertainty. The more accurate the information is, the better the decisions are likely to be. It is also about looking historically to establish trends. Accurate forecasting permits the business to plan for the future and gear up to have the appropriate resources in place to carry out the plan, and it can establish opportunities and likely threats in achieving the plan. Forecasting can be applied for the short, medium and long term.

Short-term forecasting in construction is most prominent in the production of the detailed programme of work for the contract. This is where a month or three months of the contract programme is produced in more detail to assist in controlling progress of the project.

Medium-term forecasting is the production of the master programme for the contract and the annual business plan, which is the control document for the business as a whole for the forthcoming year. This looks at the projected turnover necessary to cover the business overheads and the profit required. Both documents are used as the main control documents by monitoring actual performance against the forecast.

Long-term forecasting is the strategic requirements of the business and can cover a period of up to five years. The company is deciding from the position it is in now, where it wants or expects to be in five years' time, taking account of the issues likely to hinder or assist this goal. Once the strategic direction is decided, steps can be put in place to achieve this by developing the resources and advising the employees and shareholders of the intentions. Employees perform at their best when they are informed about their future so they can engage in its achievement, and in the case of the shareholders so they appreciate how their investments are being protected.

28.6 Forecasting techniques

28.6.1 Expert opinion

Whether it is a small one-person business or a large organisation, expertise is an essential part of forecasting. The individual and the heads of sections have accumulated knowledge of what is realistic based upon their experience of the past, which will be more finely tuned if they have collected data upon which to base their opinion. An experienced project manager can look at a set of drawings and determine how long it should take to erect the structural frame using differing techniques and so on. With data the planner can confirm it is possible. Purchasing managers can predict the likely costs of building materials and subcontractors if the contract is awarded. Readers of this text can budget their own personal finances based upon experience and the known cost outlays for the year. How successful they are is dependent upon the level of optimism or pessimism they apply to the figures. Some of the knowledge can be sought from current and potential customers.

28.6.2 The jury method

Sometimes referred to as the expert panel method, this is a well-established means of forecasting. A panel of experts is brought together to proffer their forecasts and from this an agreed forecast is reached. An example of this is the monthly meeting of senior economists advising the governor of the Bank of England where interest rates should be set to control inflation. The success of this method is dependent upon the quality of members' expertise. Meeting monthly gives more opportunity for their views to be amended than less frequent meetings do, which can cause a higher degree of uncertainty.

28.5.3 The Delphi method

This is also a panel of experts, but in this case they are not brought together under one roof so they cannot influence each other or follow the majority, if they feel it prudent to do so. A questionnaire is sent to each member of the panel asking for their predictions on key issues. The collected replies are then circulated to all members along with further more directed questions. This can occur on several occasions with the questions becoming more specific, until a detailed view is obtained.

28.5.4 Decision tree analysis

This type of forecasting looks at all the possible outcomes to the major factors affecting the business and then predicts the likely result. This can involve the use of statistical analysis. For example, the decision may centre on whether or not to diversify into civil engineering work. The key factors could be the level of management experience needed, the amount of projected work in the firm's area of activity, the level of competition, availability of experienced subcontractors and so on. Taking the example of management experience, the tree would develop the questions: Is it available in-house, and if so what training is required? Can this be done in-house or not, and if not where can it be provided? What will it cost and how long will it take? If the experience is not available in-house, example questions would be: where is it obtained, how long will it take to recruit and what will it cost? From this and other branches of the tree, information comes to light to inform the decision-making process.

28.7 Market positioning

So, what is the construction market? The industry used to divide itself into building and civil engineering, and those companies that did both had two separate sections or companies. Perhaps this was at a time when there were major clearly identifiable civil engineering projects such as the motorway programme. Building projects were smaller than they are today. In recent years in the UK, civil engineering projects are fewer, and larger projects are being carried out with large civil engineering content, often as joint ventures combining a range of expertise. This is significant, as the two marketing functions have also merged. The reality is that the industry covers a very wide range of activity. In terms of type of construction, there is speculative housing, housing, hospitals, offices, educational establishments, factories, supermarkets, out-of-town shopping complexes, theatres, sports stadiums and facilities, fairgrounds and leisure facilities, prisons, petrol stations, fast-food chains, airports, roads and motorways, railways, docks and harbours, sea defences and bridges. Taking the speculative housing market alone, there are multi-storey flats, low-rise flats, maisonettes, executive houses, terraced houses, bungalows, semi-detached and detached houses.

In terms of types of contracts, there are competitive, negotiated and serial tendering, design and build, construction management, management contracting, partnering agreements etc. There are also maintenance and alteration contracts and recently, facilities management contracts. Some are small and others worth several hundred million pounds. The customers are also wide ranging as discussed earlier.

So, the company must ask: 'what part of the market does the company wish to participate in and what penetration of this market does it wish to accomplish?' This is a strategic senior management decision that analyses the availability of resources and the potential market. In essence, though, it covers three main areas: the type of product or service provided, which also includes the type of customer; the geographical area the work is to be carried out in; and, especially in the case of the speculative market, the price range and quality standards.

28.8 New products or services

In construction, generally, a service is provided rather than a product, the obvious exceptions being the speculative housing market or the speculative commercial/industrial unit market. This is because the industry constructs a building the client has determined is required.

The first stage is to conduct a gap analysis. The market research may have indicated there is a gap in the market as a result of shifting requirements. An example of this is the changing composition of the population with the increase of one-parent families and the elderly population. Another is the massive increase in the numbers of people requiring accommodation in the centre of large city areas. Offices and warehouses have been refurbished and new apartments have been constructed. Associated with this is the social infrastructure needed to support this changing in the needs of the population in the inner cities, such as bars, clubs and restaurants. In the 1990s there was a major expansion in the out-of-town shopping complexes such as Trafford Centre in Manchester, Meadowhall in Sheffield and Bluewater in Kent. Planning restrictions brought out-of-town locations to a halt, and the emphasis has changed to inner-city developments such as the Bull Ring in Birmingham. Facilities management, a significant development often linked to PFI contracts is another example of a gap offering opportunities.

It is occasionally possible to create an opportunity. Many of our manufacturers are housed in out-of-date buildings that have been modified over the years to house the new plant and equipment necessary to be competitive. The buildings are often unsuitable for purpose and energy inefficient, but their greatest asset can be the land upon which they are constructed. There are examples where the developer has offered to construct a purpose-made building to the latest building regulations, but on a new site, provided the owner gives them the land currently in use, either as full payment or in part exchange, the developer realising the profit they can make from developing the land for a different purpose.

Once a gap in the market has been established it is necessary to establish what the likely competition is so as to take steps to be ahead of the field if possible. It is interesting to note nearly all the large out-of-town shopping centres were built by Bovis as the managing contractors, primarily because the chief executive, Sir Frank Lampl, spotted the opportunity long before anybody else and put steps in place to ensure the company was positioned to procure the work: an example of vision. If there is competition, then the costs of re-sourcing the work need to be analysed so a decision can be made whether or not there will be an acceptable contribution to the profit of the company.

28.9 Pricing decisions

This is a senior management decision. In the case of the tendering market this decision will be made at the time of tender, when senior management reviews the current state of the market and the business, and makes a decision on what percentage to add to the estimate. When the tender is being put together most bill items will be priced similarly by each contractor. Indeed, it is the estimator's job to establish what the true cost of building will be in order to enable senior management to assess the company's financial position in the market place and decide what profit margin and/or contribution to overheads the bid should include. There is also some flexibility in pricing the items by applying commercial judgement as to what the actual price will be if the tender is successful. In other words, the supplier or subcontractor might be prepared to change the price if the contract is awarded (see 13.22).

The speculative market is different as the building can only be sold on if the price is right. The developer will have done its valuation calculations such as the residual method or discounted cash flow analysis (see 6.3) to establish whether the figures will give the profit level required. There will also be some flexibility depending upon the market at the time of sale. In the case of the speculative market the sums will already have been done prior to the project having been started.

Setting a price is a complex business and there are many influences that affect it. Besides wishing to make a profit, a buyer has to be found who is prepared to purchase at the required price or somewhere near to the asking price. Speculative builders anxious to ensure a sale will often offer a fixed price, providing the purchaser pays a substantial deposit. This means the builder has a guaranteed sale and the purchaser will not have to experience any increases in price. Alternatively, they will ask for a small deposit but expect the purchaser to accept any increases in costs, the downside being if the rises are too much the purchaser may pull out of the contract. On a large retail development, the developer may offer a lower rental price to those that buy-in before the development is completed. All of this is set against a background of supply and demand that at times can be volatile. World events can affect the global and hence the country's economy. Legislation can also impact. The introduction of home information packs in the domestic market in 2007 resulted in sellers putting their homes on the market ahead of this date to avoid the costs of the pack. The result was a slowdown in house inflation, which affected the speculative market as well.

28.10 Promotion and selling

With the exception of speculative builders and developers, construction companies do not have a brand to sell like those involved in manufacturing consumer products. What they do have is a company image to promote. Those images most relevant to the industry are to do with completion on time, on budget, to a prescribed level of quality, rapid reaction time to defects, a good safety record, paying suppliers and subcontractors on time, not hiding behind the small print of the contract to make claims, and above all, looking after and out for the client and other third parties' interests. There are a variety of places this message can be promoted, which include television, radio, newspapers, both general and trade, the Internet and other electronic means, in-house publications, signboards, booklets and leaflets as well as word of mouth. Methods of promotion fall into four main categories: personnel selling, advertising, sales promotion and public relations.

Personnel selling in construction is utilised when making presentations to prospective clients, which can occur when tendering for a project and when applying to become a member of preferred lists. It occurs indirectly when attending conferences, seminars, exhibitions, professional institute functions, sponsored sporting occasions and corporate events where prospective and existing clients may also be attending. These can be important occasions and should not be seen as a 'junket'. They give the opportunity for businesses to meet people, nurture relationships and discover what is likely to be happening in the future. These long-term relationships develop feelings of trust and understanding and enhance the possibilities of obtaining work in the future.

In the speculative market the personnel selling normally occurs on the site, in offices in or adjacent to the show house or houses, depending upon the scale of the development. Here staff, not always construction personnel, are on hand to show prospective customers around the premises and to provide information about the costs, types and availability of property, and choices available for the interior finishes and equipment. They may also be offering financial packages to aid purchasing and be able to discuss the amount of deposit required if the customer is interested.

Large developments, where the infrastructure is in place, such as occurred at Salford Quays and London's Dockland, will also have a sales office on site, but their function will tend to be more commercially orientated because of the nature of prospective clients, who may also be developers.

The first stage of advertising is setting the objectives, which comprise three progressive tasks as demonstrated in Figure 28.4.

Informing is telling prospective customers and the general public that the company is in business for providing the service to construct or it is constructing a building either to sell or rent. Persuading is about creating an attitude in the 'informed' that the product or service being provided is one they want. Reinforcing is ensuring that once the customer has changed to using the company, they will wish to continue and as a result further business will follow. It does this with a message that comprises five progressive stages as demonstrated in Figure 28.5.

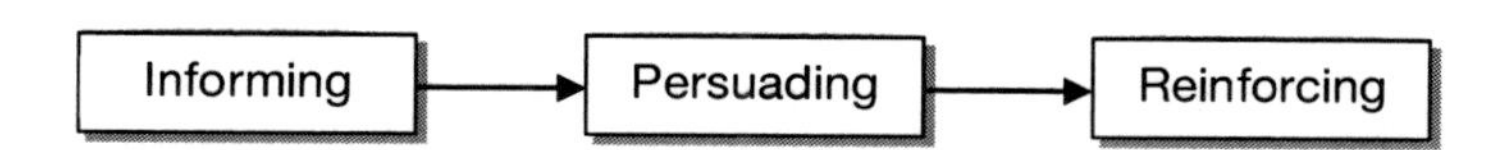

Figure 28.4 Setting the objectives

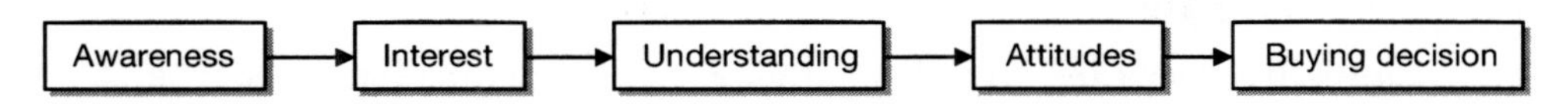

Figure 28.5 Creating the message

Source: reproduced with permission of Wiley-Blackwell from Czinkota *et al.* (1997).

The first stage is to create awareness or 'attention getting'. This can be achieved with words or with pictures such as a logo. After World War II, the name Wimpey was synonymous with construction, and most house buyers will have heard of Barrett. Both Wimpey and Barrett once spent a lot of time and effort promoting their house-building programme on both television and in the press, the former using Paul Daniels, the magician, and the latter the actor Patrick Allen flying a helicopter.

Once the audience is attentive, the next stage is to create interest so the reader or viewer continues to want further information. An example of this is the headlines shown on the board outside the paper shop advertising the evening edition of the local paper such as 'local hero rewarded'. Television and cinema use trailers of a film or programme they wish to encourage people to see. The construction industry sometimes uses artists' impressions of the completed buildings.

Once interested, there needs to be information made available so the prospective client understands what they are going to get for their money, that the company is equipped and able to carry out projects to the client's satisfaction and standards, and that the project is going to suit their needs. In speculative housing this will usually mean providing a show house or houses to demonstrate what the final building will look like, as the majority of purchasers will have difficulty in relating two-dimensional drawings of the building to the final three-dimensional form.

The next stage is to persuade the prospective client the product or service provided is what they need. The most successful show houses are those fully furnished in the style suiting the lifestyle of the prospective customers. Small inner-city flats for young business clients will be different from very expensive four- and five-bedroom detached houses. Placing appropriate toys in the child's bedroom and allowing them to play with then while parents view the rest of the house creates an attitude of desire, as does a well-equipped kitchen to whoever does the cooking. Cleverly designed landscaping gives a feel of what it will be like on a nice summer day in the garden. Similarly, a high-quality presentation to a client by a construction company is done to create a positive attitude. Many companies use a video recording of the key personnel as part of the presentation, rather than allowing them to perform live. This means the personnel can get it right before the presentation and can be filmed with different backgrounds such as on sites of successful previous projects. It is not possible to be specific here as the nature and style of presentations is continually changing as are clients' expectations.

Finally, the buying decision has to be made by the client. In the case of the speculative housing market, this can be quite rapid, much based upon emotion and the ability to pay. However, the constructor needs to capitalise on the fact that by viewing the show houses the customer has shown an interest. Offering other services such as financial and legal advice, discussing alternative finishes and keeping in touch with them all go to reinforce the interest shown. It should be noted that this has to be done carefully as overzealous contact can have a negative effect.

28.11 Choosing the media

The majority of advertising is carried out in the press, the Internet and on television. Less frequently used are radio, billboards and cinema. Television is at its best when showing action, so showing pictures of the house and its interior is not a very stimulating experience for the viewer. Advertising space on television is also expensive, so to have an advert showing a moving view through the property would be very expensive because of the time needed to do it justice. Hence, the use of personalities to talk about the product so there is an association with the celebrity and the building. Another approach is to use stills to show where the site is, with words, both on the screen and spoken, to excite the viewer to go and have a look. This becomes a shorter advertisement and is therefore cheaper. However, the costs are still high and generally only the larger builders can afford it. Unlike a newspaper advertisement, the viewer cannot refer back to it, so it has to be shown repeatedly over a short period of time to obtain the reinforcement mentioned earlier. Where television can be more profitably used is to promote the company as a by-product of a news item such as the opening of a prestigious building.

The press comprises the national newspapers, regional newspapers, magazines and the technical and trade publications. The national press is basically divided into two categories, the tabloids and the broadsheets, each with their own set of readers. Usually at the weekends they produce a colour supplement with featured articles rather than news items. These rely heavily on revenue from advertising. The regional papers are either daily or weekly; some have readerships covering large rural areas, while others cover densely populated areas such as large towns and cities; and others may be aimed at smaller towns and communities. Magazines cover every range of the public's interests, but tend to be very specific by nature, dividing the reading audience by age, gender and interests. Trade and technical papers are specific to an industry, trade or supplier group. Where the contractor believes the audience to be will determine where it places an advertisement. Again, as with television, there will be opportunities to publicise the company as part of the on-going public relations strategy.

It is important to note that while senior management may have a view on the composition and method of presentation, it is a very specialist business with its own expertise, and advice should be sought if it is not available in-house. If the company goes to the agency with a predetermined view as to how the advertising campaign should go, then of course the agency will quote and produce whatever is required. However, if the company goes with an open mind expressing the overall aim of the campaign and leaves the agency to develop ideas freely, a much more successful outcome is likely to ensue.

28.12 Public relations

'Public relations (PR) is being good and getting credit for it' is an important adage as it is a misconception that PR is about successfully promoting irrespective of quality. It has to be of sufficient consistency to engender interest. There will be incidents when a PR exercise will be conducted to limit damages when something has gone wrong such as a serious accident, but this is only effective if it is an occasional event rather than regular and, in any case, is more about controlling the outward flow of information from the company, especially to the media.

Every member of the company who talks to or meets people from outside is indirectly in the process of PR. Their attitude and manner convey signals as to the type of organisation they are representing. Indeed every time an employee picks up the telephone to answer an outside call, a public relations exercise takes place. The manner and way the employee reacts is important. One of the most important people in this chain is the reception telephonist as this is often the first person to whom the client speaks.

The role of PR is to be the interface between the company and its employees, and people and organisations outside the business; Figure 28.6 demonstrates the wide range. PR has several objectives, but in general it is to foster the reputation of the company, to promote the services offered, and in the case of the speculative market, the buildings.

High on the priority list is communication and liaison with the media. The way this is carried out is determined by the overall marketing strategy and the amount of finance made available for it, which in turn will dictate which section of the media is primarily to be used. At one end of the scale large amounts of money can be used for producing advertisements for broadcasting on television while at the other there are newsworthy stories the media will use at no cost to the organisation other than the development and presentation of the information. The media will only use factual stories and if they find the story given is not truthful, the reputation of the public relations department will suffer and they may find difficulty in having stories accepted

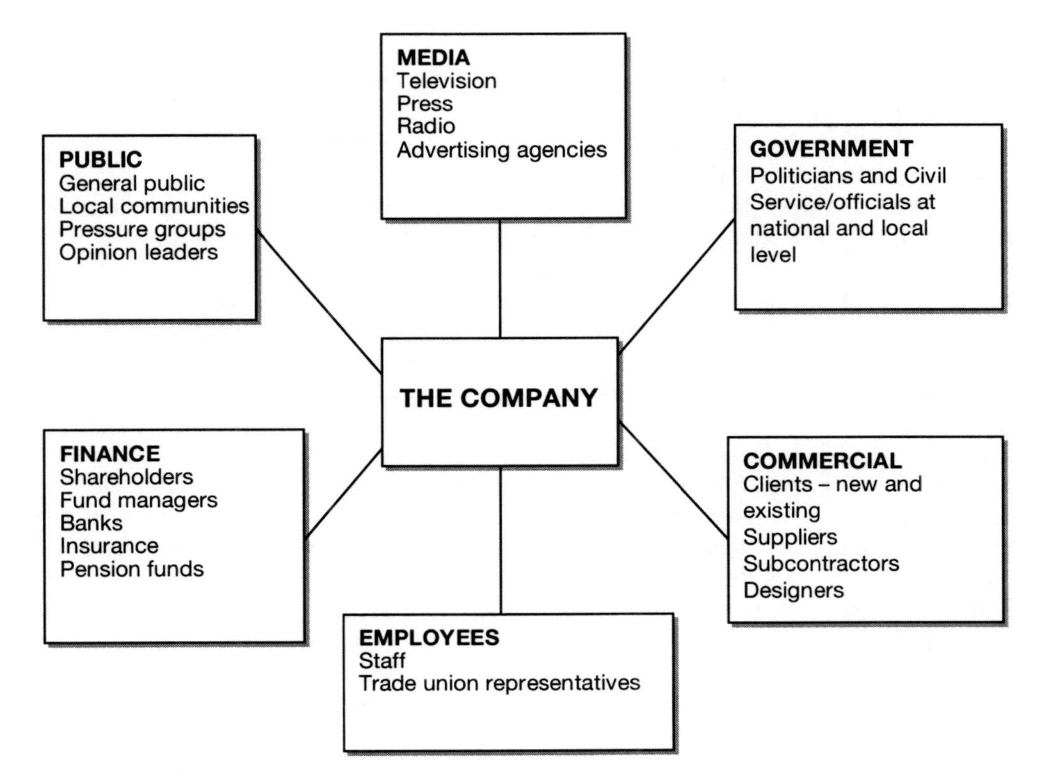

Figure 28.6 The PR interface

in the future. The problem with any story given to the press is that once handed over, the company loses control of what is actually written or broadcast. It is standard practice to produce press releases with a time embargo, which means the media cannot publish until the date or time given. These can be given for the media to use and they can also be supported with a press conference if the story is significant. The press release will be used in this case as encouragement for the press to attend. Press conferences are usually called if the news to be announced is significant, such as a merger, takeover of a major company, the appointment of a significant personality, or resignation.

Writing press releases is a particularly skill, but detailed below are a few key issues that should be addressed. Remember, the media receiving the release needs to be convinced it is a story worth publishing. There should be a headline that grasps the reader's attention that is factual and to the point such as 'The Prince of Wales will officially open the new head office of Winston Construction' or 'Winston Construction is to pay tradesmen a minimum of £50,000 per annum for work on the new shopping centre at Newtown'. The opening paragraph should be a brief summary of the content of the release, followed by the key points in descending order of importance and, as with the headline, should be factual and to the point. The overall length should be as short as possible, ideally no longer than one page. The editor of the newspaper or programme is busy and needs to be excited as quickly as possible, otherwise the document will very rapidly find its way to the wastepaper basket.

Timing is important for the greatest impact, but cannot be guaranteed. The editors will publish information when there is room for it and this is a function of its relative importance to other news that breaks on the same day. This can have advantages and disadvantages. If the company's news is good and it coincides with other more newsworthy items, it may not be published or alternatively used on a future date. Equally, if the same occurs and the news is bad, then it may be buried on the inside pages, much to everybody's relief. A major world event can push the political party's annual conference off the front pages, resulting in less exposure and the likelihood of their points in the opinion polls being affected. The PR staff will spend a lot of time just meeting the press and media at informal meetings to develop relationships so when a story needs to be reported the media are familiar with and trust the source.

Lobbying is used if the organisation wishes to influence decision makers and is part of the public relations function. It can be done in a formal way by belonging to and being active in trade associations and professional institutions. These would include the Institute of Directors, the Chartered Institute of Building, the Institution of Civil Engineers, the Royal Institution of Chartered Surveyors, the Royal Institute of British Architects and the Contractors Confederation. Employing an MP as a parliamentary consultant is another approach. Their role is primarily to advise the company of pending legislation that might affect the company's business, but can also involve making contact with the appropriate minister. Informal routes involve getting to know MPs, local councillors, members of the chamber of commerce, and resident associations.

The sponsorship of sporting events or the arts is a means of associating the company's name with a well-established event or team and, although also advertising the company, demonstrates the company's interest in the wider community. This can be done at national and international level such as the cricket test match series and premier football teams, or at local level. Good causes are also areas of consideration such as sponsoring charity walks, runs and fundraising events. It can be something as simple as providing a lorry for the local carnival parade or supporting national campaigns for fighting cancer.

Having a presence at trade shows can be a means of bringing the company to the attention of others, especially if involved in a specialist market or wishing to demonstrate one's credentials in a specific area of expertise. This is relevant to house builders who may wish, for example, to have a presence at the Ideal Homes Show.

Many large companies now have a corporate social responsibility programme that includes environmental aims, involvement of staff in community groups and charities, ethical policies, sustainable development and a clear code of business conduct, and these should be communicated to the various stakeholders.

PR is also concerned with controlling information if the news is potentially harmful to the business such as a serious accident or protest. It is normal practice when an accident takes place to instruct all staff not to talk to the press but to direct to them to a named person, who will deal with the situation. Reporters will phone the site seeking information and these can be readily transferred to the person charged with communicating with the press, but they will also stand outside the site and accost personnel as they leave. Some of the media stalk the accident and emergency departments of hospitals looking for newsworthy items. Even when the accident occurs elsewhere to a different company, it can have an impact. For example, the author was instructed not to talk to the press, but to refer them to the public relations department, when the progressive collapse of the multi-storey flats occurred at Ronan Point. Another company using a different system constructed these. However, the joint design between the floors and the wall was similar to that used by the author's company, the one used at Ronan point being an earlier version. Protests at the construction of buildings are either because of the inconvenience caused to locals or due to pressure groups not wanting the project to have started in the first place. In the former case, much can be anticipated and steps taken in advance, not just to reduce the inconvenience, but also to contact and liaise with local residents and those most affected, before and during the construction process. The latter is more difficult, as the protesters are already in place and will obtain high levels of publicity when the contractor starts on site. The problem for the contractor, besides losing time, is that the press see the people dealing with the situation adorned with the company's name, not to mention site signage.

It is not always possible to control information and the author, on a job interview, arriving early went for lunch at a local public house that was full of workers from the factory who were all on strike. Overhearing the conversation was quite enlightening. Any event the press are invited to needs to be stage managed. It is essential all safety issues be covered, especially if the event necessitates walking on the site itself. Signs should be strategically placed, and cameramen advised of them and other interesting things, because when a film is made for a television report, the director needs 'link' footage to move from one issue to another if there is a break in the filming. All personnel should be wearing safety helmets with the company logo.

Internally, PR can be applied by informing the staff about the company. This is normally accomplished by using an in-house magazine or leaflets. Staff are interested in the future prospects of the company, so new contracts awarded and their value demonstrate what future prospects there are. Rumours also occur within any organisation, especially about re-organisation, mergers and takeovers. The magazine is an ideal place to put the record straight, although this can also be done using internal memoranda. There is always an interest in knowing who has been promoted, who has left the company. Staff profiles can be used; for example, a senior person joining the company or an achievement such as a sporting medal, receiving an honour from the Queen or raising money for charity. For the technically minded, items on new technology used on a site are of interest as are topping out or high-profile opening ceremonies of the completed project. All of these are aimed at making the reader feel included in the organisation.

Finally, the amount of exposure of key personnel needs to be controlled. On the surface it seems odd that there can be too much exposure. However, after a time the public can lose interest in the message being given or become bored or irritated by seeing the same person over and over again.

28.13 Public speaking and presentations

The industry has increasingly moved in recent years to construction professionals making presentations to potential clients in an effort to obtain work or commissions.

28.13.1 Public speaking

There are four stages of oral public presentations – planning, preparing, practising and presenting.

Planning comprises three main parts:

- *Introduction*: issues that need to be considered are thanking the audience for allowing the company to present; stating who you are, who the presenters are and which sections they are to cover; the reasons for the presentation; explanations of any supporting documentation; when questions are permitted; and deciding upon the duration of the presentation.
- *The main body*: the points to be made should be listed and then sorted into a logical structure: this means making a decision as to what the key points are at each stage; deciding the amount of time to be given to each; and determining what supporting documentation or visual aids is required.
- *Summary*: the key points you wish to remind the audience of; and giving the audience an opportunity to question.

The next stage is to prepare. Trying to memorise the presentation is dangerous, especially in the event of stage fright on the day of the presentation. It is acceptable to use small note cards or similar as prompts. Any visuals used should be attractive, brief and clear; stress the important points; be simple and uncluttered; have large enough letters to be readily read by the audience with a selected font; and show the company logo.

Having established the duration of the presentation, practising will establish problems as well as giving the presenters confidence in themselves and their colleagues; in particular, you will see whether it can be done within the time or needs to be adjusted accordingly, and have an opportunity to check all visuals are working well. Ask colleagues to criticise each other's presentation both in terms of content and stance, oral skills and so on.

Finally, giving the presentation: speak clearly in a voice all can hear; use pauses for effect and to allow the audience to digest; speak with enthusiasm and excitement; keep good eye contact with the audience; and smile and demonstrate you are pleased to be present.

Appearance is extremely important at any presentation. It is both a demonstration of the speaker's standing and attitude, but also of the respect given to the audience. There is a code of dress or convention appropriate to the occasion. If addressing a group of fellow construction professionals or the client, the dress code might be different from addressing a group of small children.

The body language of the presenter indicates much to the audience: make sure hands are not in pockets, make eye contact with different members of the audience, use the hands to make a point where appropriate, and stand up straight. These all give a positive image to the audience of confidence and enthusiasm. If it is a group presentation then those not involved at the time should show interest in what the speaker is saying, even though they are only too aware of the content. They should never do anything to detract from the main event of the moment: the speaker.

Bibliography

Czinkota, M.R, Kotabe, M. and Mercer, D. (1997) *Marketing Management: Text and Cases*. Cambridge: Blackwell Business.

Davies, J.W. (2001) *Communication Skills: A Guide for Engineering and Applied Science Students*. Englewood Cliffs, NJ: Prentice Hall.

Hague, P. (1993) *Questionnaire Design*. London: Kogan Page.

Hague, P., Hague, N. and Morgan, C.A. (2004) *Market Research in Practice: A Guide to the Basics*. Sterling: Kogan Page.

Jobber, D. (2009) *Principles and Practice of Marketing*, 6th edn. New York: McGraw-Hill.

Mercer, D. (1996) *Marketing*, 2nd edn. Oxford: Wiley Blackwell.

Stokes, D. (1997) *Marketing: A Case Study Approach*, 2nd edn. London: Letts Educational.

29 Human resources management

29.1 Introduction

Personnel management, as it was once referred to, was in the 1950s, the 1960s and early 1970s primarily concerned with ensuring sufficient appropriately qualified employees were recruited into the company to satisfy the needs of management. Apprenticeships took care of skill training needs and the larger companies organised a certain amount of in-house training for line and service managers. They would also have a responsibility for safety issues and sometimes were involved in developing industrial relations strategies. It was not unusual for personnel managers to be recruited from the military. Ranks such as major and group captain were commonplace because of their experience in administration and their availability, as they often retired from the services aged around 40 years old.

However, in the late 1960s and early 1970s there were significant changes in legislation, encouraging companies to take a much more detailed look at the ways they recruited, what work conditions were provided and notably, how employees were dismissed. Examples of these Acts of legislation included the Industrial Relations Act 1971, Sex Discrimination Act 1974, and Health and Safety at Work etc. Act 1974. All of a sudden companies found themselves attending tribunals for breaches of industrial law, often losing the case because they had not procedures in place to ensure employees were treated fairly.

Whether or not this legislation coincided with a change in society's attitude to how labour should be treated can be debated. The fact remains, companies started to take much more seriously the importance of the type of labour they employed and how they used and looked after it. It became important to 'beef up' the personnel function, give it a voice on the Board of Directors, and introduce many more systems of control into the organisation. It is perhaps significant that towards the end of the 1980s and early 1990s the term human resources management (HRM) started to be used rather than personnel management. An appreciation, perhaps, of the changing attitude and the realisation that labour is a resource, and a very complex one at that. Companies aim to obtain 95 per cent plus efficiency from a piece of plant, increasingly we try to keep the wastage of material down to a minimum and these days recycle. It is, therefore, reasonable to attempt to obtain the highest output from employees, be it manual or cerebral, to ensure the company is competitive and sustainable.

As a demonstration of the difficulty of the task ahead, consider drawing up a shortlist of 12 from Table 29.1 that the reader would consider the most important issues and conditions an employee might want from their employment. This is not a definitive

Table 29.1 Possible employment requirements

Possible employment requirements		
Adequate wages	Responsibility	A chance to contribute ideas
Security	Safe working conditions	Satisfaction
No Saturday work	Promotion prospects	Long hours
Work away from home	Variety of work	Short hours
Long holidays	Work close to home	Regular average bonus payments
Large wages	Sports club	Recognition of one's work
Music while you work	Good supervision	Transport to and from work
Job satisfaction	Acceptable workmates	Quality output in one's work
Work requiring no thought	Clean working conditions	Canteen
Work requiring thought	Fair treatment	Bonus payments
No responsibility	Friendly work colleagues	

list. The selection made could be for a manual worker or for a manager. If the list is different for both, consider why this is and whether they should be.

After completing this, ask the question 'what are the four most common grievances you have experienced in your workplace or have read about or heard of in the media?' If these do not coincide with those in the shortlist of 12, then why not? Surely they should. Although it would be too simplistic to draw any conclusions from this brief exercise, it is worthwhile noting that when asking students to carry out these two exercises in groups they are able to answer the latter question quickly, with little dispute and generally all coming to the same conclusions. Whereas in trying to answer the former, the groups have lengthy debates and their conclusions usually are at variance with other groups'.

However, the function of the HRM unit is not just about providing an appropriate working environment for employees, but also about planning for and meeting the staffing needs of the corporate plan, which can include laying off personnel as well as recruiting and retraining, ensuring the organisation complies with and is aware of the current legislation, has training plans in place, recruits the best personnel for the job, advises on pay structures and pay rates, anticipates and deals with industrial relations problems, and provides a safe and healthy working environment.

29.2 Strategy and organisation – human resource planning

Human resource planning is part of the corporate business planning process. It is all very well senior management deciding to expand, contract or diversify the business, but if it is impossible to provide the appropriate people to cope, the plan will fail. If the business is to contract this will mean redundancy and there are all sorts of implications here to the reputation of the business and to industrial relations. There are also financial implications. It is therefore essential the HRM function of the organisation is consulted and has an input into this decision-making process.

Human resource planning is influenced by exterior and internal factors, all of which need to be understood in both carrying out the board's requirements in the corporate business plan and advising the board when developing the plan.

29.3 External factors affecting human resource planning

Increasingly, clients, especially local authorities, are requiring that as part of the contract a relatively high percentage of the labour employed on the contract must be recruited from within the area. Other clients require as part of the contract, that the contractor should train the staff that the client is to employ. This has both cost and resource implications for the HRM department. When working in another part of the world there are further complications, mainly cultural. Other interested parties may control labour, and this may involve the criminal fraternity, and there is the issue of paying cash to others for services to be provided. In the UK this may be classed as corruption, but can be normal practice elsewhere. It is important the HRM department is aware of the differences in the way staff and labour are employed in these situations. This is one of the reasons why companies working in other countries partner a local contracting organisation. Wimpey, for example, partnered with John Lok, a local contractor in Hong Kong, when building the Hong Kong and Shanghai Bank's head office.

Legislation is constantly changing and this has a direct effect on the company who need to be aware of current and proposed changes initiated both at Westminster and from Europe. This can be concerned with hours of work, flexible working, health and safety, maternity and paternity leave, the minimum wage, holiday arrangements and trade union membership. Legislation in other countries may be different also.

The demographic changes need to be understood. While the total population is increasing slowly and the split between male and female is changing marginally from year to year, the age profile is altering more rapidly with an increasing older population and a decreasing younger one. This has an impact on recruitment. Add to this the fact that since the government has planned for 50 per cent of the school population to be able to go on to university and higher education, there is a shrinking pool of young less qualified people from where trade and labourers can be recruited. The image and reality of the industry working outside in often inclement weather, in dirty conditions and with less than ideal welfare facilities do not encourage many to consider the industry as a career, adding to the difficulty in recruitment. It is noted the number of women entering the industry is small and this is a potential recruitment possibility.

The availability and quality of labour, both professional and trade, impacts on the business, especially if expanding or moving into other areas to work. This can be affected by the economic cycle and the amount of construction work being conducted now and in the future. In recent years there has been a shortage of qualified site operatives and many of these vacancies have been filled from Europe and beyond. This has brought further issues to the fore such as the need to have visa checking, as not all applicants are legal immigrants, and problems of communication because of inadequate language skills. The latter is extremely important when dealing with safety issues.

Construction has been and still is being used by governments as an economic regulator as procurement of public works can be stopped as a means of controlling their expenditure and vice versa. This is particularly problematic if a boom in construction is permitted rather than a steady increase. There is no doubt that there has been a reduction in the availability of properly trained tradesmen although the flow of labour from the extended European Community has alleviated this to some extent. While much of this labour will not be directly employed but subcontracted, the HRM department in conjunction with the procurement department needs to understand what the labour market is likely to be. Qualified construction staffing is likely to be a great cause for concern as it relies on the continued output from our universities. The number entering certain disciplines is variable, much reflecting the public's interest. However a further significant issue is the age profile of the current lecturing staff, which is closer to retirement age than graduate entry, meaning there is a likelihood there will be insufficient qualified staff with industrial experience to educate the construction professionals in the future.

Design trends can also have an effect. For example, as has happened before, a trend towards building steel-frame multi-storey structures results in skill shortages in this trade. The effect of this is excessive wage demands. There are continually changing methods of construction, making certain operations quicker, requiring less human resource and in some cases less skill. There is also a trend to move site production into a factory, where quality and output can be controlled in a secure weather-protected environment. How far this will develop is still to be seen. Historically, as in the industrialised building systems of the 1960s, prefabrication has had a limited success for only a short period, but the indicators are that this time it is more likely to succeed.

The wages offered have to be competitive with other industries and other companies. While there is a national wage agreement settled for each year between the employers and the trade unions, this does not mean every company pays that amount. They may well offer more by increased bonus earning opportunities, plus rates to attract labour working in the same area of the country and accommodation and travel expenses. This can become more complicated if the increases and incentives offered attract people from one region of operation to another. The Heathrow Terminal Five project was an example of this where tradesmen were offered reportedly in excess of £50,000 a year.

Currently, the trade unions are nowhere near as powerful as they were. However it should be noted there have been considerable problems in the past in certain areas, notably Liverpool and London, both then hotbeds of militancy, much of it very politically motivated. This resulted in countless strike actions to the extent some companies declined offers of work in these areas. While at the time of writing there is moderate calm, it should be remembered that a few miscalculations by government can create the right environment for unrest to be fostered and there will be always some ready to capitalise on these failures.

29.4 Internal factors affecting human resource planning

The HRM department has to take note of the impact on staffing levels when senior management restructures the organisation. Periodically organisations can be centralised or decentralised, departments merged or divided, work functions moved from one location to another or the business expanded or contracted. There can also be takeovers and mergers. Whenever significant changes occur in the organisation, it is not unusual to find staff are in the wrong place and/or in the wrong numbers. Relocation of staff is not always easy as many have domestic criteria important to them such as children being at a good school or in their GCSE year, and spouses having their own career.

The skill profile of the workforce needs to coincide with the work to be carried out. Achieving this balance may mean making staff redundant, because there is an

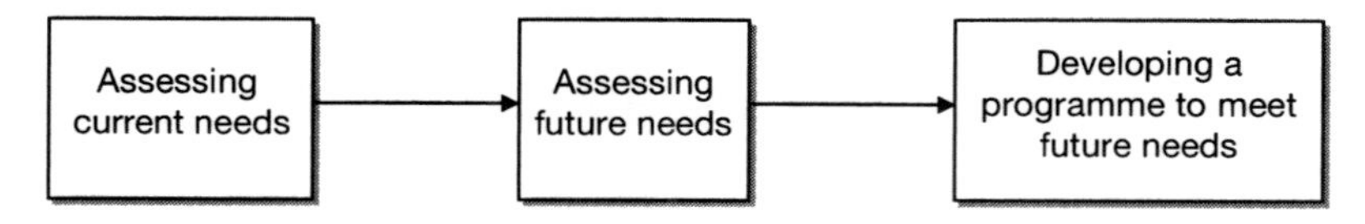

Figure 29.1 The three stages of human resource planning

excess of that skill, or retraining to adapt to the current needs of the business. If retraining is not an option or will not fill all the requirements, then recruitment will be needed.

The business plan may have an impact on the organisational structure of the company. It may not be able to effectively cope with the changes. This means a new structure has to be developed and this can affect the relationships between staff, lines of communication and levels of command, responsibilities and authority. The impact for the HRM department will then be in resolving unrest, morale issues, retraining, new pay structures, redundancy and recruitment. Further, if the numbers of staff change either up or down, or new offices are opened, there are support functions to be considered, such as health and safety. These may not be located at the new premises, but the existing support teams have to be refocused to deal with the new situation.

One of the problems associated with rapid expansion is the amount of money it is necessary to pay new staff. If there is a shortage of them this will almost inevitably mean they may be offered more than staff already in post. However hard one attempts to keep salaries a secret, the truth will out eventually. This can cause great resentment for obvious reasons. Equally, in times of a glut of applicants, paying a salary below that of staff in post will also cause friction.

In summary, Figure 29.1 takes account of the external and internal factors, demonstrating the three stages of human resource planning.

29.5 Sources of potential personnel

Before investigating the selection procedures, it is worth considering where job applicants are likely to come from. Table 29.2, adapted from Robbins (1994), gives a broad indication of these sources.

29.6 Recruitment and selection of employees

Recruitment and selection occurs at all levels of an organisation from the site labourer to the chief executive. Each will require a different approach, but in essence the same question has to be answered. Can they do the job? In all cases it is important to decide precisely what is required. Sometimes this is relatively simple, but it can be quite complex. It is not enough to advertise for a labourer. What skills do they need? Do they have experience in laying paving, drains or concreting? Are the joiners experienced in formwork, first fix, or second fix, as each requires a different kind of skill? While it would not be necessary to write a job description in these cases, simple questioning on the type of work they have recently carried out would usually suffice, but would not be for foremen, office staff and all those in middle and senior management.

Table 29.2 Sources of potential personnel

Source	*Advantages*	*Disadvantages*
Internal search	Low cost; improves morale of staff if they believe there to be opportunities; candidates familiar with company and vice versa	Limited supply; may not recruit best available candidates; danger of perpetuating the status quo and not bringing new ideas from outside
Advertisements	Can be targeted at specific groups;can be placed to cover wide catchment areas including overseas	Often many applicants are unqualified for the post
Employee recommendations	Current employee gives candidate knowledge about organisation; can generate strong candidates; employees know whether or not candidate will fit in	Tends to attract people of similar type to the person recommending; danger of nepotism; may not increase the overall mix
Private employment agencies	Wide contacts; they carry out the screening	Can be expensive; agency may require up to six months' salary as payment of fee
Job centres	Free service	Candidates tend to be unskilled and untrained; often lower calibre, the better candidates not using the service
Temporary agencies	Fills temporary needs; fills peaks and can be released when workload drops.	Expensive; usually limited to a specific role
Universities	Academic ability is measured; interviews can often be conducted at university; expected salaries are usually known before hand	Academic ability is not necessarily a measure of employability; expectations often differ from reality and what is on offer
Schools	Large number of candidates	Since so many go on to university, the majority have only limited qualifications and aspirations

Source: Adapted from Robbins (1994).

29.7 Job analysis, description and specification

Until an analysis, description and specification of the job is produced, there is nothing to compare the applicants against. It is not necessarily the same as that used for the post just vacated. Over a period of time jobs can change, either because the job itself has changed or, as is often the case, the person in post has developed the job to carry out more functions as they become more experienced. It is not unusual to discover on analysis, especially if the person has been in the job for a long time because of their knowledge and understanding, that it may be necessary to divide the role into two or more jobs, or reallocate some of the work to others. Equally it may be found others in the team could carry out all the work and the job has in effect become redundant.

The first stage is to analyse the job to establish its main duties, the contacts with others as part of the normal routine, the knowledge, skills and abilities needed to carry out the role, and the working environment in which the job is conducted including the type of plant or equipment to be used. It is also necessary to establish where in the organisation the post is, and for whom the post is responsible. This enables the HRM department to produce a job description that is a statement of the duties, specific requirements needed to carry out the job and the working conditions. From this a job specification or a statement of the qualifications required can be written, stating the skills, abilities, previous work experience and the minimum education standards required.

The job specification is divided into two broad categories: the tasks to be carried out and the skills and aptitudes needed to accomplish them These are divided into essential and desirable categories – essential being those standards prospective candidates have to meet so they can function in the job and desirable being those that would be an advantage or could be taught when the candidate is in post. This document is invaluable at the interview stage as it permits all members of the panel to be working in a consistent manner when coming to a conclusion on whom to offer the job to.

The factors making up these would normally include the following:

- *Experience and past performance*. This is probably the best measure of the candidate's ability as they can demonstrate a track record for the suitability for the post. It would always be subject to taking up of references to confirm what the candidate has written is correct, although this can normally be teased out during the interview phase. The level of experience needed should be clarified.
- *Technical skills*. Most posts require some level of technical skill from using a computer to operating machinery. The levels of requirement are usually easy to measure and specify, and can be tested either from past experience, certification or tests on the day of the interview.
- *Education*. The candidate producing the original certificates of their achievement demonstrates this. It is important to note that photocopies are unacceptable as these can easily be doctored, although even the originals can be modified and not spotted except by the experienced. The levels required can readily be specified. However, there is a danger of asking for higher qualifications than are needed. The risk is by appointing an over-qualified person, they will soon become disillusioned and look for another post elsewhere, or just use it as a staging post until a more suitable appointment turns up.
- *Communication skills*. This covers a wide range of abilities and skills. In writing the specification, the writer should be very clear what is required. Testing for it is complex. Certain aspects are covered in past experience and technical skills, but the more personable skills have to be tested at interview.
- *Personality traits*. These are important for two reasons. First, will the candidate be able to work with the others in the group and second, how will people outside of the organisation view them? Measuring them will be a subjective issue, but the interviewers must have a clear understanding of what they are looking for.

- *Health, energy and stamina.* Is the candidate physically and mentally up to the job? At one extreme, footballers are given strenuous medicals before a transfer is agreed, whereas in many jobs the company might consider no form of medical check is required. Many companies do ask for a statement from the candidate of their health, not least of all to see whether they have any special needs that will have to be catered for. This is important because of disability legislation. Further, as stress becomes more understood and accepted as a health issue, the candidate's ability to cope with the pressures of the job needs to be ascertained.
- *Interests.* While not always necessary or essential, depending upon the post, knowledge of the candidate's interests is often an indicator of their overall personality, attitude and aptitude. They can give a picture of how rounded the person is and whether or not they are likely to want to be relocated in the future, which could be important.

It is rare for a candidate to have all the attributes required for the post so, as the specification is being developed, questions must be asked about which are essential for the job and which are desirable. Linked to this and a determining factor as to which are essential or desirable is to decide which requirements can be learned on the job and which will require special off-the-job training. Training costs money and the decision will be informed by how much it will cost, whether or not the organisation has the resources internally to carry it out and how long it might take – the latter being important, depending on how quickly the person in the post needs to fully cope.

29.8 The selection procedure

29.8.1 Advertising

It is sometimes said the ideal advertisement should be compiled in such a way it attracts but one candidate who is a perfect match for the job. This is a slight exaggeration as it is important to be able to select at interview from several candidates so comparisons can be made. However, a too loosely worded advertisement can attract many applications from people who are totally unsuitable to fill the post, resulting in wasted time by those sifting through the applications, not to mention the time spent by those seeking employment. The basis of the advertisement is constructed from the job specification discussed previously.

Where the advertisement is placed has to be carefully considered as it needs to be read by potential candidates, and it is also expensive. It will depend upon the type of post: a secretary will be recruited locally whereas the managing director might be found nationally or globally. Construction professionals are likely to read the journals and newsletters of their institutes/institutions or the construction press. Those in education will tend to look in the *Times Educational Supplement* or *The Guardian*'s weekly education section. For example, who are the readers of *The Big Issue*? These days there is increasing use of the Internet. Due to this, if the advertisement is to be placed in more than one newspaper or journal it is important to have a reference code on each one, requesting that the prospective candidate quotes it in their application. This is important, as a later review of the applications will indicate where they saw the advertisement. This indicates whether or not money spent in placing adverts was effective.

The advertisement should include instructions on how the applicant should apply. This could be to send a letter of application and curriculum vitae, write in for details and an application form, or complete the application form provided on the Internet.

29.8.2 Head hunting and agencies

An alternative decision is to go to a recruitment agency. The main reason for doing this is to obtain temporary personnel. It is a well-established practice in many industries, such as warehousing, where the labour force required fluctuates considerably during the year. For example pre-Christmas and the January sales often require a greater throughput of goods. Subcontractors can avail themselves of this service, but increasingly main contractors are using agencies for the temporary recruitment of site engineers. When recruiting senior managers, it is sometimes of benefit to use agencies to head hunt for this expertise. They are in a specialist area and know the appropriate people that could be recruited this way. They will approach prospective personnel directly and then put them forward to the client for interview. This is an expensive method as the agency fee can be quite high. The advantage is they have been personally selected and are more likely to fit the job profile than results through advertising.

29.8.3 Application forms

The purpose of an application form is to have standardised information provided by the candidates in the same sequence. This enables those selecting candidates for interview to make comparisons fairly and reduces the chances of litigation by candidates who believe they have been discriminated against. It also improves consistency if more then one person is involved in this stage of the selection process.

It is advisable to have an application form for all posts. However, the composition of the forms can vary depending upon the post being advertised, but in truth for most of the junior, middle management and administrative roles in construction, the same form can be used. There are standard application forms for graduate vacancies. Depending upon the role, a letter of application may also be appropriate. The application form can either be hardcopy or electronic, the candidate completing it on the Internet and sending it back to the employer.

The information to be solicited in the application form will include personal details about the person such as age, address and contact details; dates of education and qualifications, both academic and professional; industrial experience in either chronological order, or the reverse, to include dates, roles and brief experience details; a general section to permit the candidate to indicate outside work interests and any other details the candidate would wish to bring to the employer's attention; and finally names, positions and roles of referees. There is still reluctance on behalf of many referees to act upon electronic requests for references. This is because of concern as to legitimacy of the request in handing out personal information about third parties.

29.8.4 Selecting for interview

The application forms have arrived and the next stage is to select from these the candidates to be shortlisted for interview. If the response is inadequate or the applicants unsatisfactory, steps most be taken re-advertise. This may require revisiting the composition and content of the advertisement, the places it is posted and even the salary being offered.

It is important to realise personal prejudice can affect this selection if steps are not taken to overcome it. This can be largely overcome if two people select independently of each other using a third to assist if, when they come together, they cannot agree. There are various ways of going about this, depending upon the numbers of applications received. A typical method is for each of the two independent selectors to mark and separate the forms into R for reject, P for probable and Q for maybe. Within the latter two the selector should then place them in order of priority. Their decision should be based on comparison of the completed application forms against the job specification. The two then get together and make the final selection for interview.

29.8.5 Preparing the interview

It is important to remember the interview process is not a one-way process, where the employer is selecting a candidate to fill the post. The applicant, unless desperate for employment, is also selecting the company. While the prime objective is to select the best person for the job this can only be achieved if the candidates are put at ease as much as is possible so that their true selves are exposed. Information about the post and the company culture should be provided to allow the candidate to make an informed decision as to whether or not they wish to accept the job. There is little point in going through this process only to find out after the candidate has been working for a short time that the job is not as they expected.

The first stage of this process is to get the candidate to the interview itself and to deal with other housekeeping issues. For example, taking account of the distances candidates have to travel. If the candidate has distance to travel it may be appropriate to interview the candidate later rather than first thing in the morning, or offer to provide overnight accommodation and subsistence, the former booked for them if necessary. Clear instructions should be sent on how to get to the company offices and where within the organisation they have to report. Remember the candidate may be coming by rail, bus, air or private car and, if the latter, advising where to park the car should be addressed. Candidates should be made aware of what expenses they are entitled to and ideally provision made to either collect them on the day or have an expenses form sent with the interview details so that it can be rapidly processed on completion of the interview. The amount involved can be a significant amount of money especially if they have to travel long distances and may be an issue if they are currently out of work.

The receptionist and security should be made aware the interviews are taking place so the candidates can be taken care of on arrival such as offering light refreshments if appropriate, indicating where the toilets are and taking and storing their coats. They need to be provided with a list of those expected and the times they are to be interviewed. A space should also be allocated for waiting candidates.

If members of staff are required as part of the process such as escorting them around the premises, giving talks about the organisation and its work and ethos, they should be well briefed in advance as to their responsibilities on the day.

29.8.6 Preparation of the room

The layout of the room is important in creating the right atmosphere for the interview. Providing an informal 'sit round the coffee table' approach will make the candidate feel ill at ease if they would normally expect to be sat on a seat in front of a table facing the interviewers on the other side. This could result in some of the objectives of the interview being lost.

A decision needs to be made about where the interviewers are to sit relative to the controlling interviewer, and the use of name plates in front of each can assist in communication between them and the interviewee. The candidates should not be positioned so they have direct sunlight in their eye and the room should be clean, tidy and fit for purpose. The seat provided should be functional. One too comfortable will make the candidate feel uneasy, and an uncomfortable one distracts them.

It is essential instructions be given out that there must be no interruptions during the interview. Providing cups of tea or coffee are also ill advised as the candidate may find difficulty in finding times to drink and will almost invariably not imbibe at all. It is an unnecessary distraction. Refreshments, if provided, should be in the collecting room for applicants.

If as part of the interview the candidates are expected to make a presentation, then the appropriate equipment should be provided and technical support staff available in case anything goes wrong.

29.8.7 Types of interview

Interviewing is a skill that has to be learned. Many organisations insist that any staff member asked to take part in the interviewing process should have been on a staff development course prior to being put in a real situation. There are a number of different ways the interviewing and selection process can take place. These are the interview panel, selection boards and group testing. Some form of testing might supplement all of these.

The interview panel is the most common in the majority of organisations. The composition of the panel will vary depending upon the nature of the employment, the level and complexity of the post to be filled. The panel tends to be small in number. For middle management it would typically comprise someone from personnel, the head of the section the post is in and perhaps their immediate supervisor. If it were a technical post (accountant, planner, quantity surveyor) it would be normal for someone at a senior level in that discipline to be in attendance as well.

The selection boards are much more formally structured and found in public institutions such as the civil service, universities and the health service. The composition may be laid down in the organisation's procedures. These boards tend to be larger than the interview panels, which does bring with it the disadvantage of a committee making the decision rather than those closely involved with the role to be filled. The members will usually also include personnel from outside of the immediate department to ensure a fair decision is made. So, for example, a board set up for a university lecturing post will include professors from another department. The danger is they will ask questions of insufficient depth if not controlled by the chair of the board. The author well remembers being asked by a professor of computing about his domestic plumbing, somewhat off target from the debate that had previously taken place about the differences between the educational needs of civil and building undergraduates. It was suggested he looked in the yellow pages for a plumber.

Both of these types of interview are in depth and should be designed to encourage a two-way flow of information between the candidates and the interviewers. In both situations the interviewers should have been briefed on what is required and the areas should each carry out their questioning so that there is a balanced approach. While it may be appropriate to have a series of common questions each applicant is asked, there should be flexibility to allow the interview to go in any direction the chair feels is appropriate. However, this should not detract from the reason for the interview in assessing whether candidates meet the criteria of the job specification.

Group selection is both expensive and takes considerable time, not just in the duration of the event, but also in the preparation. It is a method used by some large organisations and especially the armed services in officer recruitment and selection. The basis is to bring together six to eight candidates in a group (there could be more than one group in the case of the armed services) and put them through various exercises, scenarios and tests, working together as a team. These will be supplemented by personal interviews, and candidates may be asked to make presentations. It is a mechanism to observe the levels of leadership, communication, group working and interaction skills, initiative, logical thinking, confidence, listening ability, inventiveness and ability to accept criticism, all often under pressure. Observers who rate and rank them against the criteria just mentioned continually assess the candidates.

Testing can take varying forms whose use is in part a function of the job requirements. These forms include:

- physical testing – used when the fitness and health of a person is paramount to their contribution to the organisation. A professional sportsman would fall into this category where the testing would be vigorous and thorough. However in today's workplace, the physical and mental health of the candidate is important to stand up to the rigours of the workplace, and medicals are sometimes required as part of the process. This might appear to discriminate against disability, but legislation covers this and protects against this, providing the candidate has declared their disability at the time of application;
- ability tests – used to test mechanical skills such as working a machine or driving a vehicle. It would be normal to test a typist for their ability to use computers and various software packages, both for speed and accuracy. They can be used to test mathematical, linguistic and spatial skills, as well as sensory skills such as listening and vision;
- personality testing – increasingly used by larger organisations to measure personal characteristics such as their ability to work with others, whether they are very detailed in their work or more cavalier in wanting to complete the job as quickly as possible. These tests usually ask the candidate to select from a series of two statements and assess which they consider they are most like. For example, 'I am a hard worker', or 'I am a fast worker'. From the responses the assessor produces a personality profile. It should be noted there are varying views as to the validity of these types of test and they should always be read in conjunction with the findings from the formal in-depth interview. Depending upon the test, the person making the assessment may have to be qualified and be a member of an appropriate professional body such as the Institute of Psychology. It should

be remembered there is an ethical dimension in assessing the personality of an individual.

Finally, there is the one-to-one interview usually utilised in employing tradesmen and general operatives on the site. The person carrying out this should be qualified because there are many pitfalls in not complying with current legislation, especially if the applicant believes they have been rejected on the grounds of gender, age, race, trade union membership or disability.

29.8.8 The conduct of the interview

The chairperson should agree with all the interviewers how the interview is to be conducted, the sequence of questions, the areas to be examined, the approximate duration of the interview and ensure that all have the correct paperwork with them; this includes the candidate's application form and the job specification. They should be reminded they should be testing the candidate's abilities against the categories listed in the job specification and they should look beyond these and be prepared to make a judgement about the personality of the candidate. This includes their appearance, humour, confidence, accuracy of response and so on.

When the candidate sits down all the members of the panel should be introduced by name and position in the company. It is usual to open the interview with a few pleasant remarks and ice-breaking questions such as 'Was your trip here all right?' and minor personal questions so as to relax the candidate and give a reminder of the purpose of the interview. It also means the candidate gets used to the style, pace and approach of the interview. They should be told when the decision to appoint is to be made. Sometimes this is on the day, in other cases it will be within a few days.

Other issues the chairperson should address when controlling the interview are the direction and standard of the questions so the purpose of the interview is maintained and its pace is within the timescale allotted. It should be remembered the candidate will probably be nervous and allowances made for that fact. The questions posed should be clear, concise and direct and open ended. They should give the opportunity for a considered response and allow the candidate to open up and elaborate. Generally the more they speak the better, providing the length of the response is controlled by the chairperson if it is in danger of going on too long. If there is any doubt about the clarity or accuracy of a response this should be addressed. The panel should assess whether or not the candidate wants the job. This can be done based upon the enthusiasm shown or by simply asking the question 'If you were offered this post would you accept it?' The candidate should be given the opportunity of asking any questions they have. The interview should be firmly but politely brought to a conclusion, thanking the candidate for their attendance and, if appropriate, resolving the issue of their expenses. Above all, all members of the panel should remember the most important and most difficult thing is to listen!

There are actions members of the panel should not do when interviewing. They should not use the interview as a platform upon which to air their views and knowledge, nor should they dominate the conversation, lose their temper or argue. Debating an issue, though, is acceptable. They should not allow their personal liking or disliking of the candidate to cloud their impartiality and hence influence their judgement, nor should they jump to conclusions as to the appropriateness or otherwise of the candidate for filling the post until the analysis is done at the end of the interview. The candidate should not be interrupted during their response unless they are clearly missing the point of the question or are dragging it out. The questions asked should not be loaded so as to solicit the answer required. If it becomes clear the candidate is totally inappropriate for the job, then even though a given amount of time has been allocated for the interview, it should be brought to a conclusion earlier. Promises that cannot be honoured should not be made. The interview should be brought to a conclusion and not be allowed to fade out or be inconclusive.

It is perhaps appropriate to say a few words about personal prejudices. Often when the word is mentioned people immediately think of racial and gender prejudices, but everybody is probably prejudiced against some thing or another. It can be about the way people dress, that they smell, their hairstyle, their politics, their accents, where they live, the kind of car they drive, their spouse or partner, facial hair, that they are vegetarian, where they go on holiday, their weight, that they smoke and so on; the list is endless. You cannot change a prejudice overnight and sometimes never. Often they have developed as a result of one's cultural background, bringing up, or lack of knowledge and understanding. What is important about any prejudice is being aware of it. Knowing one's prejudices means one is less likely to be biased against the candidate, although there is a danger of over-compensating.

The decision to appoint should be carried out as soon as possible after the interviews have taken place while the candidates are still fresh in the minds of the interviewers. Some chairpersons ask for comments after each candidate has been seen, while others wait until all have been through the process. Each candidate should be considered in turn and compared against the job specification to see if they satisfy the criteria and a list drawn up of the best candidates. From those who could fill the role, further discussions should take place about their overall personality and performance before candidates are chosen as the first, second and third choice, providing there are enough satisfactory candidates. If the decision is close, then the board may decide to reconvene the following day to make a decision after having slept on it.

Once the candidate has been chosen, issues such as the salary to be offered should be agreed. All those who clearly are not suitable should be advised immediately. It is extremely frustrating for candidates hoping for employment to be told several weeks later they have been unsuccessful. It also does the company image no good. The first choice should be notified as soon as possible and given a reasonable amount of time to make the decision, even though they may have indicated at interview they would accept the job if offered. They have a right to make a considered judgement as it may mean moving and disrupting their family who need to be involved in the candidate's decision-making process. It is no good pressuring the candidate to make a hasty decision only to find a few months later they are not happy and resign. If they accept, the reserve candidates can be notified, and if they turn it down, the second choice can be offered the job.

29.9 Training and staff development

No organisation can afford to stand still; there will always be change as new opportunities and challenges arise. To cope with this staff will continually need to be trained or retrained to meet these needs. The amount of training required will depend upon the role of the individual, but it is a reasonable assumption that staff will have training throughout their career. However, it must be done for some purpose that will benefit the company and the employee. The Construction Industry Training Board (CITB) was formed in the 1960s as one of 29 Industrial Training Boards (ITBs) set up as a result of the 1964 Industrial Training Act, but by the 1980s this had been reduced to only two, one for the engineering industry and the other the CITB, the latter as a result of strong lobbying by the major construction companies.

The reasons for training are to introduce new members of staff to the company, to make visitors to construction sites aware of the dangers and their responsibilities, to meet technical advances, to bring employees up to speed with new ideas and legislation, to give the employees greater job satisfaction, and to retrain for new work or on promotion.

29.9.1 Induction courses

When a new employee starts they arrive somewhat bewildered, perhaps a little frightened of what's ahead of them and maybe questioning their own confidence. The purpose of the induction course is to resolve these matters. The composition of the course will vary depending upon circumstances. In all cases, the new employee needs to be introduced to those they will be working with and those they will have contact with, especially their immediate supervisor and subordinates. This is not just useful for the employee, but also existing staff that need to know who this new face is and what their role is. They need to know where they are working, where the basic facilities are, such as toilets and canteens, where the stationery office and stores are. Larger organisations may wish to run a more formal course to discuss the overall ethos and objectives of the company, introduce them to safety, quality and environmental policies, explain what to do in the event of fire and to meet senior staff, spending a day or more in the process.

On construction sites it is not unusual to give all visitors an induction course related to safety issues before they are allowed on site to ensure they are wearing the correct protection clothing and know the evacuation procedures and where not to go. Some of these courses are specially developed for specific risks such as working in tunnels and sewers and how to use and understand the safety equipment they have to carry, such as methane detectors.

29.9.2 Training and development needs

There appears to be a link between the ability to hold on to staff and the attention given to staff development especially from senior management. On analysis this is not surprising, since senior management is more sensitive to the direction the company is taking and hence the future needs for staff to fulfil this ambition. The training needs of the company are determined by the overall corporate plan for the future. It is usually beneficial to the company if the human resource needs can be met within the company rather than recruitment from outside, although new blood from time to time bringing in new ideas can help to invigorate the business. The human resource department has to analyse the company's needs based upon the corporate plan and assess the talent available that can be developed within the organisation and the gaps in their ability to cope with the work to be carried out. Alongside this, they must consider the changing work environment as new technologies and legislative requirements come into play as well as staff expectations and aspirations, which can be established through staff annual appraisals.

The next stage is to design the training and implement it in practice. The training can be carried out in a variety of ways, the most common being on the job. Here the employee learns or is coached while carrying out their role usually with the help of a mentor who could be the supervisor or an experienced colleague. The mentor will assess at set intervals the progress of the employee and this will be used as a basis for discussion in the annual staff appraisal. Job rotation through different parts of the organisation may be another method used. Often trainee graduates are expected to spend a certain amount of time in various departments such as planning, estimating, purchasing and quantity surveying as part of their training package. This is not just to learn what happens in different functions of the organisation and how they relate to each other, but also to give the employee the opportunity to discover which types of work they best relate to and might wish to make, or not make, a career in. Another way is to have a planned progression of promotion within the organisation. Job shadowing is sometimes used where the employee stays with the senior manager for a period of time seeing what they do and how they carry out their jobs. Employees are sometimes given temporary assignments for a short period of time. This gives the company the opportunity to see how they are able to cope, as well as probably enhancing the employee's job satisfaction and motivation.

Simulation techniques are also available where the employee is subjected to simulated exercises where they can do no harm but their performance can be closely monitored and feedback given. The most sophisticated building simulation model has been developed at the BMSC Leeuwarden, in the Netherlands. Each trainee is placed in a 'site cabin', connected to the outside world by phone and computer to the central control and the instructors. In each cabin they have the complete set of information for construction of the project such as drawings, specification, the B of Q, the construction programme, and names of suppliers, subcontractors and members of the design team. They are then set a series of tasks to accomplish within a set period of time. Outside the huts and only visible when exiting the cabins is a large screen showing the site at any point in time during its construction that is relevant to the exercise being set. The trainee can use a joystick to go to any part of the site and building. There are also simulated background site noises. When in the cabin, the trainee can access information from the documentation or by contacting suppliers, subcontractors and members of the design team who are played by the instructors. However, this process is two-way and the instructors can contact the trainees and put problems in their way. To make matters worse, various members of the site team with their problems interrupt the trainees at intervals. Actors play these roles. At the end of the sessions the trainees are debriefed. This is an excellent method of developing managers' skills and also of establishing whether or not a prospective manager has the ability to cope under such pressure.

The company can run courses to update employees about changes in legislation, amendments to the forms of contract, the impact of revised or new company policy and systems, and so on. It can also put on courses to develop IT skills and learn to operate new machines. Sending personnel on courses organised by others, or bringing experts into the company premises, as well as being provided in-house, can teach some of these. Sending executives on university courses, outside activities courses and conferences can provide other off-the-job training and development programmes that focus on management skills.

It is important in all these training scenarios that the quality and success of the training is evaluated against the criteria developed when deciding the training needs. Obtaining feedback from both the trainee and course controller/instructor, testing levels of performance and assessing any behavioural changes, can do this. The latter by observations made by the staff member's immediate supervisor.

Retirement needs to be prepared for and many companies run retirement training courses for their employees. It is very important because the change from full-time employment to retirement can be traumatic, resulting in premature death. The main issues are finance, health and what do with one's time. On the surface it appears great to suddenly have no worries and responsibilities, but the truth can be very different. On the financial level, there will be usually a significant reduction in earnings even if receiving a company pension and with many of these no longer being offered the situation could become much worse. Many schemes provide a lump sum and this needs to be invested and used wisely. Assistance in knowing what to do is invaluable. Over the years, many people in full employment, especially if they have had a senior post, have not kept up with body maintenance and this can result in health problems on retirement. Regimes of the kinds of exercise and diet need to be explored to combat the change of life style. Finally the executive used to delegating work and organising others moves into an environment where there is nobody to organise except a partner or spouse. The latter may have been for years looking after the home, supporting the working member of the family and then suddenly there is significant change in both their lives. Equally, if both have been working similar problems occur.

29.9.3 Construction Industry Training Board (CITB)

The construction industry has changed in the last 50 years from one directly employing most of the main skills, such as bricklayers, joiners, scaffolders, steel fixers, crane and machine drivers, and the labouring skills such as drain layers, concreters and flag or pavement layers. The mechanical and electrical services, plasterers and painters were usually subcontracted. The result was that construction companies used the apprenticeship schemes then in place and a lot of on-the-job training, especially for the labouring roles, provided much of the industry's training needs in-house. Since then there has been a steady move towards employing very few site operatives and utilising subcontract labour instead. The result has been a significant deterioration in the amount of apprentices trained directly by the construction companies, except in isolated pockets of the country. The CITB has become the main supply of trained operatives as a result.

In the early days of the CITB the emphasis was on training a skilled labour force including plant operators, but in recent years there has been a movement towards management training. Hosted and partially funded by the CITB, the Constructionarium project gives students the opportunity to get a flavour of what is involved in working on major projects, allowing them to apply their academic knowledge in a practical and safe way.

The CITB produces annually a publication itemising the grants available, their purpose and value. Construction businesses can then apply for them, but it should be noted these are generally paid after the training has been completed unless the training spans more than one year.

29.10 Welfare, health and safety and employee well-being

Employers have both a statutory and moral responsibility to look after the welfare, safety and health of all those they employ and to third parties who either visit the premises or interact in some way with the work process such as pedestrians walking past the entrance to the site, neighbours and other road users. This section is primarily concerned with welfare, and well-being, as discourse on health and safety can be read in 21.14.

The statutory requirements can be found in the CDM regulations (2015) and are primarily concerned with the physical provision of welfare services and are summarised in 21.9.4.

While these are minimum requirements, there is nothing to stop more being provided than is required especially in the support offered to employees both on site and in the office, other than restrictions placed on the employer in terms of the cost of providing them. Indeed there are good arguments to suggest that, by providing high standards of welfare, both motivation and loyalty are increased, so it may be more beneficial to the well-being of the company as a whole to offer more.

An example could be the provision of subsidised canteens. It would be expected the wages of the staff preparing and cooking the food and the food itself be covered in the selling price, but the subsidy could be in the accommodation, fuel, heating, lighting and general overheads. Consideration should be given to the provision and identification of healthy food, although there tends to be a tradition on site to eat considerable helpings of the traditional English breakfasts, chips and so on as a means of employees obtaining a high-calorie intake to balance the energy used in the type of work and cold conditions experienced in the winter months.

Car parking is always problematic in terms of space and is often a cause of unrest. While in the office environment the amount of space is already determined, there should be a transparent policy in place as to the means of allocation and proper provision made for those with disabilities. On site the same should apply, but if there is sufficient space available, steps should be taken to ensure cars using it are not damaged. Some sites require a substantial car parking facility for the completed project and consideration should be made to programme this work in earlier rather than later if possible so site personnel can utilise it.

Personal support for employees is important. This can be on various levels, and some individuals provide it because of their style of management; for it to be universal, the ethos of the company must be partial towards it and this can only happen if determined at board level. While not suggesting a 'nanny state' approach, the fact is employees can have problems at work, at home and in personal health and do not always know how to deal with them. Supervisors should try to develop relationships within their group so employees feel able to speak in confidence to either their immediate supervisor or someone else within the organisation, so supervisors will need training on how to do this. If employees are on long-term sick leave or have recently suffered the bereavement of a close relative, the organisation should be aware of this and take an active interest in the employee's situation. The support given can take many forms, including paid leave, visits from other staff and advice about benefits and other support available within the local area if appropriate. These days the provision of crèche facilities may act as an added attraction to prospective employees who might find the working hours problematic.

Social activities such as sporting events, theatre evenings, visits and trips can be encouraged to engender team spirit and involve family members. With ever-increasing pressure at work, people commuting longer distances to and from work and other interests, there is a limit to their success. Some organisations have had their own sports facilities for staff to use, but these have to a large extent disappeared. They are expensive to maintain, the land may be a useful asset releasing funds or land to build speculatively on, and there has been a decline in interest by employees.

29.11 Construction-related health issues

It is not proposed to go into detail about the illnesses that can occur as a result of working in the construction industry, except that of stress which is covered in section 29.12. The reader should be made aware that not only is the industry dangerous, but the processing of various construction materials and physical exertions can result in workers and management becoming ill.

During the 1950s and 1960s it was not unusual to hear people talk about occupational illness and accidents as almost acceptable as an occupational hazard. Times have changed and it is now totally unacceptable to put workers into a work environment that is dangerous, or if this can not be avoided, to provide appropriate protection.

Examples of potential problems are as follows:

- Asbestos was used until the 1990s in a wide range of products such as sheet materials, pipes and water tanks, rainwater goods and textured walls and ceilings. Serious risks occur when it has to be removed or during maintenance work.
- Timber preservatives are used extensively. The name is misleading as they do not preserve the timber, but kill living organisms that destroy the timber. They are by definition poisons and need to be handled and used with care.
- Much of the work in construction generates dusts and fibres. The nature of these needs to be assessed and precautions put in place.
- Many materials give off gas when used such as paints and solvents, and others continue to do so when built into the building for in excess of 30 years.
- Contact with certain materials may cause immediate irritation (irritant contact dermatitis) and if having long-term exposure, can result in allergic responses (allergic contact dermatitis).
- Physical injury, especially to the spine and upper limbs, often caused by incorrect lifting or repetitive processes such as bricklaying and plastering. Regular use of vibration equipment is another cause of injury, and repetitive strain injury, while not restricted to this, is often a result of long spells on the keyboard.

More detail of potential problems connected with ill-health from materials can be seen in Curwell *et al.* (2000).

29.12 Managing stress

29.12.1 Introduction

Until recently stress as a work-related illness has been considered by many to be macho and part of the job and yet all are aware it has the potential to cause serious illness and even death. Increasingly, there has been a change in attitude and management is becoming aware that stress can be self-inflicted by an individual on themselves for a variety of reasons, or can be imposed by management often without them being aware they are doing it. Besides any ethical reasons for monitoring and controlling levels of stress there are good pragmatic reasons for doing this as the amount of work days lost, the reduction in performance, reduced morale, the implications of a fatality, and staff looking for employment elsewhere caused by stress are significant for the successful running and profitability of the company. As long ago as 1995 the HSE estimated that 6.5 million working days were lost due to stress-related illnesses costing employers £376 million and society £3.75 billion. By 2004 this had increased to 13 million working days, employers' costs to £700 million and the costs to society to £7 billion. If a person is ill and off work for any length of time, the hole left has to be filled and this can be difficult and disruptive. Clearly, if this illness is stress related and could have been avoided, it makes sense to take precautions to stop this happening in the first place.

Further there are legal reasons also that notably include:

- Health and Safety at Work etc. Act 1974 where employers have a duty to ensure, so far as it is reasonably practical, the health of their employees at work;
- Management of Health and Safety at Work Regulations 1999 which includes provisions for stress assessment, prevention and training;
- Working Time Regulations 1998;
- Protection from Harassment Act 1997.

Stress should not be confused with pressure, the effect of which might be stress. Many people enjoy reasonable pressure as it can be stimulating and motivating, but it is when this reaches the stage when the individual feels they can no longer cope with the demands placed upon them it becomes work-related stress. The effect can be either physical such as heart disease or minor psychological illnesses such as anxiety and depression.

29.11.2 Is there a problem?

The first stage in stress management is to establish facts. These can be found by using both quantitative and qualitative methods. Qualitative measurements can be achieved by checking the amount of sickness that has been taken as a percentage of the whole organisation, narrowing it down to specific areas of work if possible and individuals or small groups. In the latter case this information has to be used sensitively, carefully and analytically, as it could either be work related, due to other health problems or just taking time off for personal reasons. The turnover of employees is another potential indicator of stress in the workplace, but could be due to the overall morale of the company and levels of pay. In any case this knowledge would be useful to management. Finally, inspecting trends in productivity where it is measurable may also be an indicator of stress, but should be, as with turnover, equated with other potential problems such as the quality of management and morale.

Quantitative measurements are more a function of the working relationships between supervisors and subordinates. There are formal and informal ways of obtaining these facts on stress. The informal approach of daily contact – listening to, talking to, and observing individuals' performance – can be the first indicator there is a problem. However, if the relationship is strained or very formal this will probably be ineffective. Interviewing employees on return from sick leave can also highlight potential problems providing they are prepared to confide. The annual staff appraisal interview is another opportunity to discover how the employee is reacting to the workloads set and the interviewer should be trained to pick up signals of impending problems.

29.11.3 How is the risk of stress-related problems evaluated and who is most at risk?

Not everybody stands up to or thrives on pressure. This may be genetic or one's environmental conditioning, so there is no norm against which to make comparisons. Further, other factors outside the workplace may have an effect on the vulnerability of the employee. For example, if they are going through a divorce, moving home, a member of their family is ill or has died, they have been diagnosed with a serious health problem or they a have financial problems. There are several factors that impact on the levels of stress individuals are subjected to, as set out below.

Organisational culture. Positive cultures take work-related stress seriously which involves having good communication between employees and management, responding to any concerns, allowing staff to contribute to decisions that affect them, giving responsibility and job satisfaction. All of these reflect good management practice. Added to these are recognising there is a problem and encouraging staff to talk about it, offering support, discouraging long working hours and dissuading staff from taking work home. Finally, creating a culture that permits staff to express their concerns confidentially about their colleagues, without being seen as 'telling tales'.

Excessive demands. There is always the risk of overloading employees with more work than can be accomplished within a normal working period or expecting them to carry out work that is beyond their current capability or knowledge, because it is too difficult, or they have not been trained. Sometimes employees are also asked to carry out tasks that are impossible. In these cases the conscientious employee is likely to work excessive hours, curtail lunch breaks or take work home, worry about it and suffer stress as a consequence. The exception to this is when there is a mini-crisis within the workplace, and 'all hands to the pump' are required. Stress can also occur when employees are under-stretched and bored.

The working environment. It is not possible to be prescriptive here in terms of what are acceptable ranges of noise, vibration, temperature, humidity, ventilation, lighting and cleanliness in the workplace as these can vary depending upon the type of work. For example, a construction site environment is different from a hospital, library or classroom. However, within each of these work environments there are acceptable and unacceptable limits that need to be considered, because if they are not it affects not just the stress people are working under but, as a result, productivity. Before the new airport was built in Hong Kong a large number of the schools were situated directly under the flight path to Kai Tak airport. Every time a plane came in to land, approximately at three-minute intervals, the children on hearing the approaching aircraft would cover their ears until after the noise peak had been reached. Clearly the learning process was severely affected. In some workplaces the fear of violence is also of concern such as with the staff in accident and emergency departments in hospitals, which necessitates steps to be taken, not just to protect staff and prevent occurrences from happening, but also to give staff peace of mind and reduce their already high levels of stress caused by the very nature of their work.

Control of work. To be told what to do without being asked for an opinion on how the task might be accomplished can be frustrating and unfulfilling and result in a deterioration of mental health. Equally, to give employees full control of their work without providing some form of lifeline in terms of either monitoring or advice can have detrimental effects. There would also appear to be possible problems associated with control of the work environment, such as the inability to open windows, but this may also be due to ineffective ventilation systems rather than not being able to have control.

Relationships with managers and colleagues. The quality of the relationships with others is fundamental to satisfaction at work. Without this, the workplace soon becomes a stressful place to be. Problems usually manifest in one of two ways: bullying and harassment. Both forms are unacceptable in an organisation. Harassment is usually defined as 'unwanted conduct' based upon sex, race, colour, religion, nationality, ethnic origin or disability, whereas bullying is defined as 'persistent unacceptable behaviour by one or more individuals in the organisation against one or more employees' (HSE 2001). The latter includes acts or threats of violent abuse, verbal abuse, insubordination, victimisation, ridicule, humiliation, libel, slander, malicious gossip, prying into personal matters, pestering or spying. It also includes such management practices as setting impossible deadlines, excessive supervision, deliberate and unjustified fault finding, withholding information, ignoring, refusing reasonable requests for leave or training, and preventing career development. Sometimes the perpetrator is unaware of the effect of their actions especially with forms of harassment, such as physical contact between employees of different sex, poking fun and patronising disabled colleagues.

Managing change. Few like change and when it occurs employees usually find it stressful mainly because of the change to the comfort of the status quo and uncertainty of the future. Management need to clearly communicate and identify the reasons for the change to allow employees to question and comment, and involve them in the decision-making process where it affects them. This should happen as quickly as possible to stem any rumours that will rapidly spread, causing even more uncertainly and stress. In any reorganisation, job roles may change, and the employee may be working with different and sometimes unknown personnel, all of which should be taken into account.

Role definition. It is unsettling for employees not to have clearly defined roles and lines of responsibility to those above and below. Giving written terms can assist this, with the proviso there can be some flexibility in the interpretation, and also discussing with them so as to ensure they fully comprehend their role, the priorities and what is expected of them. This is an ongoing process so if their role changes for any reason they are still clear. It is important there are no ambiguities in the interpretation of the role and there is no conflict of responsibilities. For example, a site quantity surveyor may be responsible to the site manager to provide the necessary information for the manager to act upon and for their attendance, but the regional or head office senior quantity surveyor may be responsible to ensure the quality of their work is to a high enough technical and professional standard.

Staff development and support. Staff should receive sufficient training to carry out their job properly and this should be monitored regularly as the job may change either as a result of modifications imposed by organisational change, or natural change, which occurs in any organisation as employees take on more responsibility as they grow in confidence and knowledge. New employees should be carefully selected so their abilities match the job closely or, if not, the differences can be filled by training. On arrival they should go on an induction course. This is very important because, if designed properly, it eases the transition from bewilderment on arriving in new employment to being assimilated into the organisation.

Employees welcome being congratulated for having done a good job and this acknowledgement by management helps in relieving work-related stress. When work is not up to the standard expected, it is more profitable to offer support and guidance so performance can be improved, although there does come a time when official warnings and eventual dismissal may have to be considered. The assumption that all employees in a given job have the same abilities and talents is dangerous. It is important to establish their strengths and weaknesses so they can be allocated jobs that play to their strengths, although some weaknesses can be overcome with further training, as they may be knowledge based.

One of the more complex issues to address is support for employees that are going through stress resulting from their personal life. This is because they may not wish to talk about it even if the company is sympathetic to these types of problems. The company needs to develop a culture whereby employees are aware that help is available, management is trained to both spot difficulties and offer, or be equipped to offer, assistance or referral to another party. A word of caution though, referral can also be interpreted by the person seeking help as rejection. Awareness of an employee's personal requirements, such as being a single parent family, so that work patterns can be adopted to suit their needs, reduces stress levels considerably and at the same time the company benefits from a more satisfied and committed employee.

29.11.4 Spotting signs of stress and anxiety

Identifying stress in personnel is very difficult to do as two people working in identical situations may find different levels of stress in the job, one perhaps getting a buzz from it, the other being unable to cope. Thus it is not possible to define the job and say this is an unsatisfactory level of stress, other than engaging in good management practices as outlined in the previous section.

However, there are certain indicators and symptoms of stress management to beware of, all of which are based on deviations from the colleague's normal behaviour. Examples include:

- *lack of concentration* – we all have off days when it is difficult to concentrate, whether we feel ill, something has upset us at home or we just didn't get a good night's sleep. The problem occurs when the lack of concentration continues over a period of time;
- *loss of motivation and commitment* – this is relative to the employee's norm as some are more motivated and committed than others. Similar to lack of concentration, but where the employee no longer comes up with ideas, enthusiasm dwindles and there is an overall lack of interest in the goings on in the business;
- *irritability* – the employee apparently overreacts to minor annoyances whereas before they would take it in their stride or laugh it off;

- *quality of decision making* – dithering and being indecisive, delaying or just making obviously incorrect decisions, compared with previous performance;
- *increase in the numbers of errors* – linked with the quality of decision making, but also in the accuracy of the rest of the employee's work;
- *deterioration in organisation of work* – each individual has to organise their short-term and some their medium-term workload as part of their responsibility. If the level of organisation changes this is yet another indicator of loss of concentration or motivation;
- *inability to control personnel* – if the employee is responsible for the management of others, then excess stress levels can have an effect on their effectiveness in organising and controlling others;
- *reduction of output* – all of the above may be contributory factors to an overall reduction of output and performance. In some types of work, especially manual, this will be easily measured in terms of the amount of work done by day. In other types, it manifests itself in the speed of responses to requests for information or decisions;
- *withdrawal symptoms* – an extreme example would be when a very gregarious personality starts to withdraw into themselves, taking little part in the general office chatter, social events and a loss of humour;
- *alcoholism* – sudden excessive drinking is another indication all might not be well, not to mention its effect on the employee's work output and their own and others' safety;
- *increase in sickness* – stress can cause both physical and mental problems. This can result in increases in absenteeism, behavioural changes as outlined above or noticeable changes in physical appearance, such as loss of weight, drawn features, complexion changes and difficulty in breathing.

The above should not be seen as a definitive list nor should it be used as a checklist for diagnosis; stress-related illnesses are ailments and only a qualified physician or physiologist is equipped to diagnose. These are meant only as indicators for the manager to suspect all is not well and to take further action if it is felt appropriate.

29.11.5 How to help

If it is noted by a colleague and brought to the attention of or is noticed by management that an employee appears to have the symptoms of stress, it is important to remain calm and objective. There are various actions the manager may consider. First, to look at the activities the employee is carrying out, prioritise them, and remove the less important. Unless the situation is very bad, it is suggested taking the important and often more difficult tasks away would further inflame the situation, as the employee would suffer a loss of self-esteem. Prioritising can be done casually without the employee necessarily being made aware of the action being taken, or preferably, with consultation to agree the revised workload.

Second, preventative action should be considered by advising staff on how to identify stress in themselves, suggesting how they might manage it and, above all, creating an organisational culture that encourages them to seek help if they are unable to cope. If any of these symptoms persist, appropriate medical advice should be sought.

Table 29.3 Symptoms of stress

Physical symptoms		
Headaches	Nausea	Indigestion
Cramps or muscle spasms	Tearfulness	Muscular aches and pains
Pins and needles	Susceptibility to infections such as colds	Chest pains
Sleep disturbances	Constant tiredness	Restlessness
Dizziness	Fainting spells	High blood pressure
Craving for food	Lack of appetite	Constipation or diarrhoea
Behavioural and psychological symptoms		
Poor concentration	Memory loss	Irritability
Increased smoking if a smoker	Increased drinking of alcohol/tea/coffee	Lateness for work
More accident prone	Withdrawal from social contact	Reduced work performance
Indecisive	Inflexibility	Fear
Depression	Apathy	Lack of motivation
Low self-esteem	Fear of failure	Aggression

When stressed, the body produces more of the so-called 'fight' chemicals preparing the body for an emergency. These include adrenaline and noradrenalin that raise blood pressure, increase the heartbeat, make one sweat more and sometimes reduce stomach activity, and cortisol which releases fat and sugar into the system. This was all right when it was necessary to fight an opponent, run away as fast as possible or carry out some super physical act, but in a sedentary office environment these chemicals act in a different way. The result of early symptoms that might occur are shown in Table 29.3.

29.12 Salaries, wages and pensions

The difference normally used to distinguish between wages and salaries is that wages are paid weekly and salaries monthly, and traditionally wages have been paid to manual workers and salaries to others. The method of payment does impact on how each view balancing their family budget. Most people have monthly outgoings such as mortgages, insurances and loans. So for salary earners, February is a good month with a fixed income and outgoings, but fewer days, leaving more residual income. For a wage earner it is a bad month as they only receive four wages as distinct from five in some other months.

Setting salary and wage levels is a careful balance between what the company can afford and retaining and attracting staff. What the company is able to pay is a function of the income, expenditure other than wages, investment in research, development and new equipment needs, and producing profit at a level that satisfies the shareholders, if a limited company.

American texts refer to this subject as compensation. You get compensated for your work and effort by receiving wages and benefits. Wages are remuneration for the time the employee has worked, such as the basic pay, overtime and bonuses, whereas benefits are such things as health and life insurance, sick pay, holidays, flexible working, pensions, retirement age and stocks and share schemes. The total package of wages and benefits should be considered as compensation for work.

The management perspective

Contrary to popular belief, generally management would like to pay good wages, but in return they want a high level of output: 'A fair day's work for a fair day's pay.' This is achieved by offering incentives to increase productivity, improving management performance to motivate and increase work rates, and investment in state-of-the-art technologies and equipment.

There can be restrictions on what can be paid. Governments will usually determine the level of pay and annual increases for public employees based upon inflation rates, and can apply pressure to the private sector to keep within similar limits when negotiating annual increases. They also have to pay the minimum statutory wage and pressure is being applied to pay the living wage. Employers also have to take account of supply and demand of certain skills, sometimes paying well over the norm. They can only pay from what the company earns and if profit levels are low this will impact on salaries and wages.

Other factors affecting pay levels result from the collective bargaining at national level between the trade unions and the construction employers that determines the minimum wages and conditions for the construction employees. However, it should be noted, since bonuses are also paid on top of this, the employer does have some flexibility to increase pay. Those employed in office and managerial roles are not covered by any such agreements as, unlike in the public sector, there are no laid-down pay scales. Pay therefore is more likely to be determined by comparisons with competitors within the industry.

The employee perspective

Equity is important, as individuals will compare their income and personal effort against their colleagues in the same section, others carrying out similar work in other organisations, and colleagues in the same organisation apparently at the same level, but carrying out different functions as with a planner, engineer and quantity surveyor on site. Unfortunately, supply and demand can make this difficult to achieve if, for example, there is a shortage of quantity surveyors and they are paid more than the other apparently equivalent roles. Interestingly, employees are less concerned about the pay other professions get compared to themselves, but heaven help the employer who pays more to someone of the same rank if it is perceived they do not work as hard as the others. Differentials within an organisation are a sensitive issue and can be eroded over time if not continually inspected. Much has been written and said about the differential between men and women and there is plenty of evidence to demonstrate there still is discrimination for comparable work.

The perspective on pensions has changed in recent years as a result of scandals associated with company pensions, schemes being closed and a bar on new entrants as employers take stock of the financial implications as well as more public awareness of the volatility of the stock market. (A typical pension scheme may involve the employee putting in 6 per cent of their income and this being matched by 12 per cent or more by the employer.) Those left intact are primarily in the public sector and it is suggested that the implications of having a guaranteed pension or not having one, and needing to make alternative arrangements for retirement, are having an effect on the way employees assess pay offers. Since working patterns have changed and many more women are at work, flexibility and holiday issues become an issue, especially if there are small children to be looked after.

29.12.1 Pay structures

There are several ways in which pay structures can be designed. First, and most predominant in the public sector, such as in the NHS and university estates departments, is fixed published scales, usually set and negotiated nationally, where the employee receives annual increments until reaching the top of the scale. There is usually some overlap between the top of the lower scale and bottom of the next higher scale as demonstrated in Table 29.4. In this example, each annual increment increase is calculated as 4 per cent higher than the one below. This maintains the differential between the grades as they progress up the scale.

This gives the employer a certain amount of flexibility when recruiting staff within a scale as they can offer the job at different points depending upon supply and demand. This can cause a problem if a person appointed earlier has been offered a position lower down the scale. The main problem is there is a limit to the ability of the employer to pay a person on merit to either encourage or hold on to a member of staff if they are at the top of the scale unless there is a promotion opportunity. This can be overcome in part by having a system in place that permits acceleration up the scale so an employee can be given more than one increment. Alternatively, there can be merit increments, shown in highlighted italics at the top of scales A and B, that the employer can use in special cases to reward for extra responsibility or outstanding work. In spite of these difficulties, employees take some comfort from the fact that annual salary increases are negotiated nationally by the unions who have a perceived strength in the negotiation process, because of the number of people they represent

Table 29.4 Incremental salary scales

Increment point	*Scale A (£)*	*Scale B (£)*	*Scale C (£)*
1	10,000		
2	10,400		
3	10,816		
4	11,248		
5	11,699	11,699	
6	12.167	12.167	
7	*12.653*	12.653	
8	*13,159*	13,159	
9		13,686	13,686
10		14,233	14,233
11		*14,802*	14,802
12		*15,395*	15,395
13			16,010
14			16,650
15			17,317

and their experience, with any suggestion of locally negotiated pay deals being treated with caution.

Another method is the use of performance appraisals where an employee and those concerned with their performance, such as their line manager, discuss and set performance targets for the coming year and identify any support that might be required to achieve these objectives. The employee is encouraged to be candid in pointing out what their aspirations are and problems they see in the organisation that are hindering them in both advancement and success in achieving the targets set. At each annual meeting their performance is compared against that which was set the previous year and an analysis of why they did or did not achieve the objectives. This can then become the basis for any pay rise offered or possible promotion. Since the interviewer is often the line manager it is essential that they are properly trained to carry out this task.

The second method of designing pay structures most commonly found in the private sector is based upon job evaluation. This approach is increasingly being used within the public sector as well. Job evaluation is the process of establishing the relative merit of each job within an organisation with the aim of creating the correct differentials between them. There are various methods available, the most common referred to as the points rating method. It selects criteria common to all jobs and rates them on a scale allocating points accordingly. Typical criteria could be experience, supervision, number and importance of decisions, responsibility, effort required, skill required, working conditions, education and training, supervision and complexity of work.

Once the criteria have been selected, a scale of points has to be devised for each. For example, experience could be divided as follows:

Level 1	0–6 months	10 points
Level 2	6 months–1 year	20 points
Level 3	1–2 years	40 points
Level 4	More than 5 years	60 points

In this case 60 points is the highest to be allocated, but another of the criteria, such as a post with a high level of responsibility, may be considered by management to merit more points. How exactly the points are decided is as a result of discussion between senior management and the expert putting the scheme together. Once the scale has been devised, then each job within the organisation is compared against this scale and the points calculated. This is usually supplemented by interviewing staff, since their current role may not reflect what was in the original job specification as jobs often evolve. This then acts as the basis in determining the par rates throughout the organisation. Before using the system, a few experimental trial runs of some of the posts are carried out to see whether the outcome appears to make sense and the pay being suggested appears rational.

This exercise should be carried out at periodic intervals, because circumstances and needs alter as work methods change, products and services are developed and the organisation structure is amended, giving and taking away responsibilities and decision-making responsibilities. The problem arises when differences are found, resulting in some employees now being overpaid. The question is, should they be able to keep their current salary with annual cost-of-living increases, keep it without cost-of-living increases, or have it reduced to the new scheme?

The other aspect is to consider whether there is equity between what has resulted from the exercise with what is being paid in other comparable organisations. This process is called a pay survey or wage and salary survey to benchmark a cross-section of the company's jobs (usually 25–30 per cent) against those in other organisations. This is normally carried out using mailed or telephone questionnaires with companies who are prepared to share their information. It might appear strange competitors are prepared to share this apparently sensitive information, but in practice most of the information is available as a result of personnel changing companies, advertisements and other sources. It is in everybody's interest to be open with this information. The information gleaned from this process is then compared with the outcome from the points rating method so the organisation can develop its wage structure.

29.12.2 Incentive schemes

It is too simplistic, and incorrect, to believe by just offering financial incentives improved performance will result: improved performance meaning higher productivity, improved quality, well-maintained plant and equipment and/or safer working. As has been discussed in Chapter 1, many other forces generate motivation. However, there is no doubt that financial incentives can succeed especially if supported by good management practices.

Incentive schemes for site operatives fall into two categories: piece work, where operatives are paid a rate for the work done with no basic wage; and those that are paid in addition to the nationally agreed hourly wage rate. The former has in the past been applied to work that is simple to measure such as brickwork where the bricklayers would be paid so much per thousand bricks laid.

There are some basic rules that should be applied to any bonus scheme for operatives. These are:

1 Bonus should be paid in direct proportion to their performance where it can be measured, normally with no limit to earnings. That proportion could increase once performance has reached a stated level to act as a further incentive, or decrease to discourage any further increase if it is felt quality and/or safety standards might be compromised.
2 The method of calculation should be fair and transparent and readily understood by all. This is not just good practice, but goes a long way to eliminate suspicion by the operatives about the integrity of the calculation. In the past some schemes have been highly controversial and resulted in industrial action. For example in the post-war years, dock workers could be paid by several employers during the week depending upon the ship they were working on, would have different hourly rates depending upon the cargo and different bonus rates. It was difficult to check the calculation and even if it was done and there was a dispute, the ship would have probably left port so it was impossible to recheck.
3 The nature of the work to be done and the times allowed should be clearly specified. The author remembers a case when the time for cutting a hole through a stud partition was given without a clear specification of what the work entailed. At the end of the week when the work was measured the joiners wanted twice as much because there was a hole on both sides.

4 The times allowed for work should be given before it commences otherwise there could be suspicion the times were decided afterwards and the earnings manipulated.
5 The times should not be altered at a later stage unless both parties agree. Situations do arise when the times given are too tight for the operatives to make a reasonable bonus and need to be relaxed. Equally there are occurrences when the times given allow the operatives to make excessive money. This is more difficult as the operatives are naturally more reluctant to take a cut. One mechanism for doing this is to trade these off against those that need to be relaxed.
6 Allowances should be made for situations when the operatives cannot work because of reasons beyond their control such as being held up by others, waiting for materials, and inclement weather. How much these allowances should be is a moot point, as operatives would argue they should be paid at the same rate as they would have earned if they had been able to work. The employers are more likely to argue it should be a fixed rate or none at all.

The calculation can be produced in several ways. The most comprehensive and therefore most expensive to administer is the unit rate when each separate operation is given a target time. For example, collect tools and materials, fix 19mm × 100mm skirting using plugs and screws and cart away excess material @10metre run per hour; or it can be a fixed amount of time for concreting a complete floor of a multi-storey block. An alternative approach is to offer a fixed bonus if the work a person or gang is charged with carrying out is completed by a stated time. In both cases the number of hours the job is worth is based upon work-study measurement. The difference between the hours earned and the hours worked forms the basis of the bonus calculation, referred to as the *hours saved* method. Another approach is to use the labour element of the rates from the B of Q as the basis for the bonus scheme.

A decision has to be made on how much bonus should be paid to operatives and at what point in their performance they should commence earning bonus. These are referred to as geared schemes. Figure 29.2 demonstrates some examples of a 50 per cent and 100 per cent scheme. In the case of the former, when the operative commences earning bonus it is paid at half the hourly rate and in the latter at the full hourly rate.

This calculation is based upon the assumption that when operatives are working at a standard level of performance of 100 rating, defined in BS 3138, they should earn the equivalent of a third of their hourly wage in bonus. The vertical scale represents the rating performance. In this example it is assumed the hourly wage is £8 per hour. It can be seen that the 100 per cent scheme only commences to pay out bonus when the operative has achieved a 75 rating, whereas those on the 50 per cent scheme receive bonus at a lower rate. However, once achieving a 100 rating performance, those on the 100 per cent scheme earn more than those on the 50 per cent scheme, management benefiting as a result, but this acts as compensation against when the operative is performing poorly for whatever reason. To permit this to happen the standard times used (the bonus rate) have to be adjusted accordingly. In the case of the 100 per cent, the standard times are increased by a third and in the case of the 50 per cent, by two-thirds.

The problem with the 100 per cent scheme is that unless the time sheets are properly supervised and filled in accurately, and the standard times and the targets are set accurately, the system becomes unworkable as either the employer or the operatives

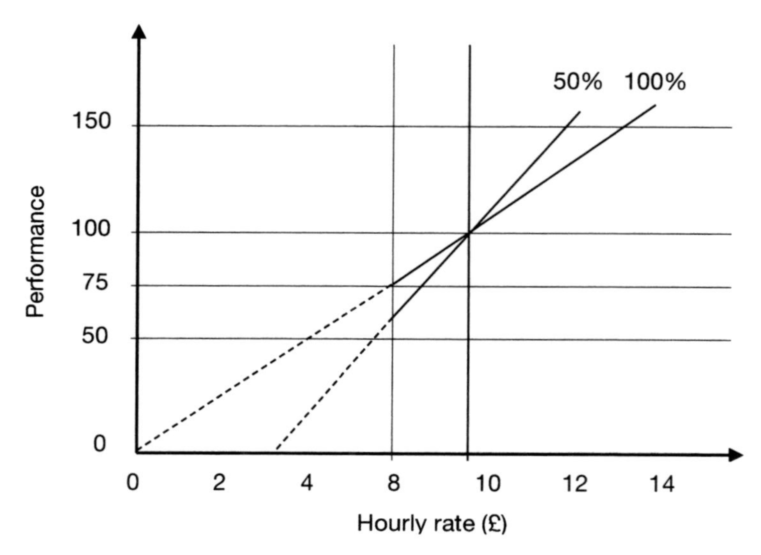

Figure 29.2 Geared incentive schemes

could lose out financially. Due to this the 50 per cent scheme has generally been the most popular in the construction industry.

It is difficult to produce a bonus scheme based upon productivity for operatives such as crane drivers who are servicing many operations. One of the ways of overcoming this is for them to be paid an average of all the various gangs of operatives they are servicing. Those employed on maintaining plant and equipment can be rewarded this way, but another approach is to base their bonus earnings on the percentage utilisation of the plant. In other words their bonus increases, the fewer breakdowns and down-time there is. However, the problem with this is that once they have achieved 100 per cent efficiency and the plant is available for work at all times, they cannot earn any extra bonus.

When decided it is not practical to have a system as above, because of the costs and experience in running it, lack of data to produce it, or because it was felt to be inappropriate, then plus rates can be paid. This is an addition to the hourly rate irrespective of output. It is used to attract labour if there is a shortage, when the work is very complex, when it is difficult to measure and devise a sensible scheme, or to overcome potential industrial problems. The latter was demonstrated in Liverpool in the 1960s when, because of the militancy of the trade unions, bonus disputes were closing construction sites down as a result of strike action, and some companies resorted to paying plus rates in lieu of bonus.

Profit sharing has been given a high profile in the news media over the last few years because of the amounts paid to some senior executives often referred to as 'fat cats'. To put it in perspective, many people are involved in profit sharing and do not receive anything like the amounts quoted. It should also be put in the context of the levels of responsibility a person has and the number of people they employ. Businesses with multi-million/billion-pound turnovers employ large numbers of personnel, not to mention those employed in the supply chain. There are only a few capable of successfully managing such empires or capable of turning round one that is floundering.

Unfortunately, since the payouts are usually based on the success of the previous year's trading, by the time they receive it, the business profits may have taken a dip, temporary or otherwise, and the executive's bonus tends to be compared with that fact rather than the successful year or years (some are based on more than one year's trading). The basic premise in profit sharing is employees, not necessarily senior management, receive a prearranged percentage of the profit the company makes. A word of caution: those further down the hierarchy have no control of the bigger picture, nor do they know how the company is doing until the amount is given to them. If over the past few years having always received a similar amount of monies as a result of the company's success and suddenly they don't, then they can become demoralised, especially if they have come to expect it.

At least one major construction company has been working on the concept that senior management's annual bonus should be related to the safety performance of their region on the grounds that if their income is affected by the performance of those below them in this area, they will be even more determined to have an excellent safety record in their business.

Finally, there is the incentive paid as commission. This is normally associated with personnel who sell and will more likely be found in the speculative housing market. However, there are other potential opportunities for commission to be paid to those whose job it is to obtain work, but since in construction this is normally senior management, they are more likely to be signed up to a profit-sharing scheme.

29.13 Dealing with staff reductions

There are various options when having to reduce the labour and staff in the organisation:

- Redundancy is when the employee is dismissed on the grounds there is insufficient work to justify their continued employment. This can be either compulsory or voluntary. The danger of the latter is good employees see this as an opportunity to obtain a lump sum payment knowing they will be able to start work quickly elsewhere leaving behind the less able. There are procedures laid down in law, which may be enhanced by the employer, that give a given period of notice and so many weeks' or months' wages/salaries depending upon length of service. It should be noted it is the job that is redundant, so it cannot be refilled as soon as the employee has left and thereby used as a mechanism of getting rid of someone the company does not like.
- Early retirement is when an employee is asked to leave employment before reaching retirement age. The employee is usually offered an enhanced pension as an incentive to go. They may also be entitled to redundancy pay. They may be allowed to be re-employed on a part-time basis, usually providing their earnings do not exceed the difference between their salary prior to retirement and the amount of pension being received.
- Temporary layoff is when the employee is laid off for a period of time and then re-employed at a later stage if still available for work. This can be for a few days or longer. There may be some incentive paid for them to remain available for work, but much will depend upon the implications of the impact on state benefits being received while out of work.

- Natural wastage is when personnel leave and their posts are not filled. This is as a result of staff retiring, leaving to find work elsewhere, being promoted, or taking voluntary redundancy.
- Re-deployment is when staff are moved elsewhere within the organisation as their post becomes redundant. They can be moved laterally to another post, which may require retraining, be promoted, or in some cases, demoted. The latter may mean the employee has to take a reduction in wages, or it can be 'ring fenced' until such time as the employee moves out of that post and the new person is employed at the lower rate.
- Reduced working is when the working week or the number of days worked in the week is reduced. In the former, this may be accomplished by stopping or reducing overtime. The latter is more drastic, but is preferable both to the employer and employee to reducing the daily hours below the normal working day. There can be legal implications in reducing the working week because of the contract of employment.

29.14 Industrial relations

Having good industrial relations in the workplace is one of the main cornerstones for having a productive workforce. Management has the responsibility to create such an environment, but from time to time needs reminding of their obligation. The trade unions are the main means of providing this. The term trade union often brings up the spectre of the 1950s and 1960s, when there was a combination of very poor management in many industries and a politicised section of some trade unions, resulting in much of the industrial strife of this period. Much has happened since then, partially due to the employment and trade union legislation enacted in the late 1970s and 1980s by the Conservative government, but due also to a change of attitude by both employers and the unions as they adapted, not just to the legislation, but to the needs of society as a whole. While there is still opposition to unions in some quarters, many see advantages in their existence, not least because it is a much more convenient way of negotiating terms and conditions nationally for their industry, and, at site level, dealing with one representative rather than many is more effective and efficient. A good steward will only bring legitimate grievances to management's notice. The unions on site have the right to elect stewards and a safety representative, and to represent members' interest.

The trade unions represent the interests of their members not just in wage negotiation, but also in the welfare of the members and their families, and much of their good work is not reported as it is not newsworthy. While there appears to be a conflict between trade union pay demands and employers looking after their own and shareholders' interests, both parties realise the need for a satisfactory outcome that suits both parties.

The Union of Construction allied Trades and Technicians (UCATT) is the principal union in the construction industry with a membership of over 84,000 (2012). It was formed in 1971 bringing together the Amalgamated Society of Woodworkers, the Amalgamated Society of Painters and Decorators, the Association of Building Technicians and the Amalgamated Union of Building Trade Workers. It comprises ten regions, each of which meets monthly to discuss regional matters.

UNITE formed as a merger between the Transport and General Workers' Union (TGWU) and Amicus in 2007. Before the merger the TGWU was involved in organising building workers for a very long time, but chose not to merge with UCATT, no doubt due to its much wider involvement in other industries. Similarly the GMB has members working in the construction industry as well as in other industries.

The contractors are represented at the negotiating table by nine trade associations, they are: National Federation of Builders; Civil Engineering Contractors Association; House Builders Federation; Scottish Building, Painting and Decorating Association; National Federation of Roofing Contractors; National Access and Scaffolding Confederation, National Association of Shopfitters and Build UK. The Construction Industry Joint Council – Working Rule Agreement, replaced the Building and Civil Binding Agreements in January 1998. Besides Acts of Parliament, this governs conditions of employment within the construction industry. Amendments to it are agreed at intervals, currently every three years, between the Construction Confederation and the main trade unions representing construction workers, namely UCATT, UNITE and the GMB. It is also the forum for other discussions and negotiations.

The Construction Industry Joint Council comprises 22 seats, 11 belong to the confederation the other 11 to UCATT (six members), UNITE (three members) and the GMB (two members). UCATT acts as the lead negotiator for the trade unions. There is a disputes system in place that is designed to deal with differences over the application of the WRA.

Tactics are a crucial part of the wage negotiation process. Both sides of the negotiation assess their own and the other side's position. The contractors are looking at their ability to pay an increase, taking account of the current and projected market and their profit requirements and those of their shareholders. They look at the trend in wage demands on all employers both in and out of the industry in the public and private sector, and the rates they have been settling. Government signals are also noted. From this they can build up a picture as to what is the range of settlement that might be struck with the unions. The unions are doing exactly the same, taking also note of pressures from members in the union and what their demands and expectations are. From this both sides calculate what is the best settlement hoped for (BS) would be and what they believe to be a realistic settlement (RS). The contractors decide what is the maximum payout they are prepared to make and the unions the lowest offer they are prepared to concede. These positions are called the fallback position (FBP).

Figure 29.3 illustrates the position the two sides might take. The employers will lay down their best settlement terms which the unions reject, and the unions put forward their ideal settlement terms. As can be seen in the figure these are far apart and obviously unacceptable to both sides. There will be some discussion giving the reasoning behind their position and the sides will withdraw to 'consult their members', agreeing to meet on another occasion. It may appear on the surface to be a waste of time, but to the skilled negotiating team it indicates the kind of deal the other party might settle for. They meet again and gradually the two sides move towards each other until hopefully a deal is struck, usually somewhere in the grey shaded area.

29.15 Disciplinary and grievance procedures

Organisations should have published disciplinary and grievance procedures. These outline the procedures both the employer and employee should follow in the event of a problem. In the former, much has been determined by the various pieces of

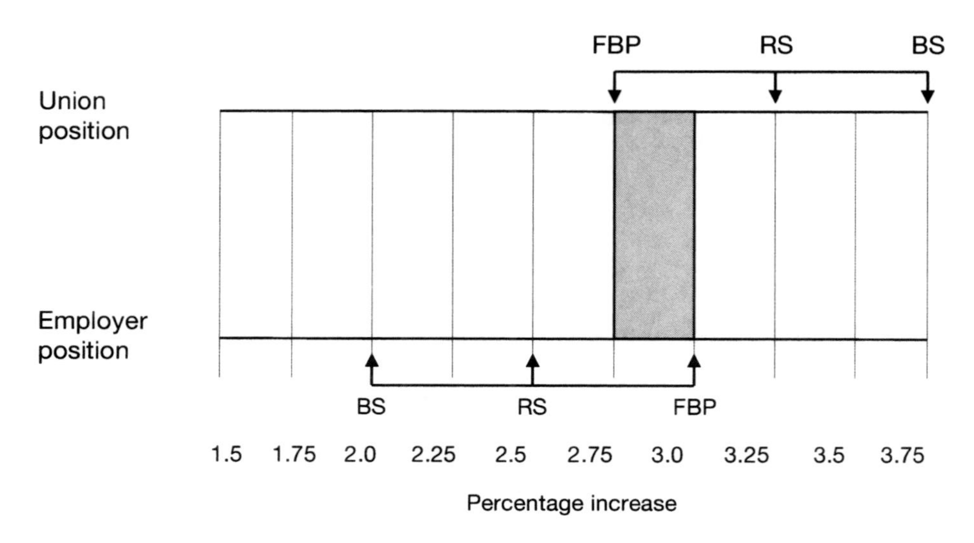

Figure 29.3 Negotiating range
Source: Adapted from Pratt and Bennet (1989).

employment legislation and case studies in industrial tribunals. A typical procedure would firstly define what was unacceptable behaviour, such as unacceptable performance and gross misconduct, the latter usually meaning fraud, theft, deliberate damage to property, violent behaviour to others, drunkenness, accessing pornography on the web, and flagrant disregard of standard procedures and regulations. Employees can be dismissed instantly for gross misconduct. Unacceptable performance can include a wide range of under-performance, such as poor quality work, bad time keeping, use of the web for personal reasons, and excessive sick leave, although it is necessary to tread carefully in this instance. In these cases there will be a series of warnings issued. These are normally first a verbal warning, then a written warning, and then possibly a disciplinary hearing at which the employee may be accompanied by either a colleague or union representative. Warnings issued will have a timescale for improvement associated with it, the duration depending upon the offence. There is much case law demonstrating employees must be given adequate time to improve. This can be as long as 12 months. It is normal practice for warning letters to be removed from the employee's personal file after a stated period of time, often when the warning duration has expired. The disciplinary hearings can also highlight the employee is under-performing because of lack of training or other deficiencies in management that can subsequently be rectified. In dismissal procedures a specific time limit during which an appeal can be made should be given.

Grievance procedures can be either against dismissal or if an employee believes they are been treated unsatisfactorily. If the employee believes they have been unfairly dismissed, before going to an industrial tribunal they can follow the appeal route as outlined above. If they feel they are being unfairly treated, they can follow the standard grievance procedure. A typical procedure is to first discuss the matter with the immediate supervisor. If no satisfactory outcome occurs, the issue can be put to the departmental manager and if no response is made, a written complaint is made to the human resource management department when a meeting is set up with senior management for the matter to be resolved. The employee would expect to have union representation at this stage, although some organisations may permit this at the meeting with the head of department. If the company cannot resolve the matter, the union could ballot their members on strike action after which it would go into the regional and then national stages of dispute resolution.

29.16 Factors influencing industrial relations

There are many factors that can impact on industrial relations. Some are generated within the company and others are outside its control. Figure 29.4 demonstrates the external and internal influences on industrial relations.

If management does not organise and plan the sequence of work to give continuity, and have the correct amount and quality of materials as required for good productivity, employees not only cannot earn reasonable bonuses but become demoralised as they cannot carry out their work sensibly. It is frustrating to be taken off work before it is completed or have to hang about waiting for materials and instructions. Management's attitude in the way they treat employees, their disregard for or lack of interest in safety and quality go to create animosity or apathy from the employees. If facilities such as the toilets and rooms for eating are dirty and unhygienic it is a demonstration of lack of care and interest by management. Perceived or actual low levels of pay and reward for effort have a significant impact, but also disputing the minutiae on expense claims has a long-term low impact on relationships. Whenever there are rumours and then enactment of redundancies, morale inevitably takes a tumble, but how it is handled can mitigate these effects.

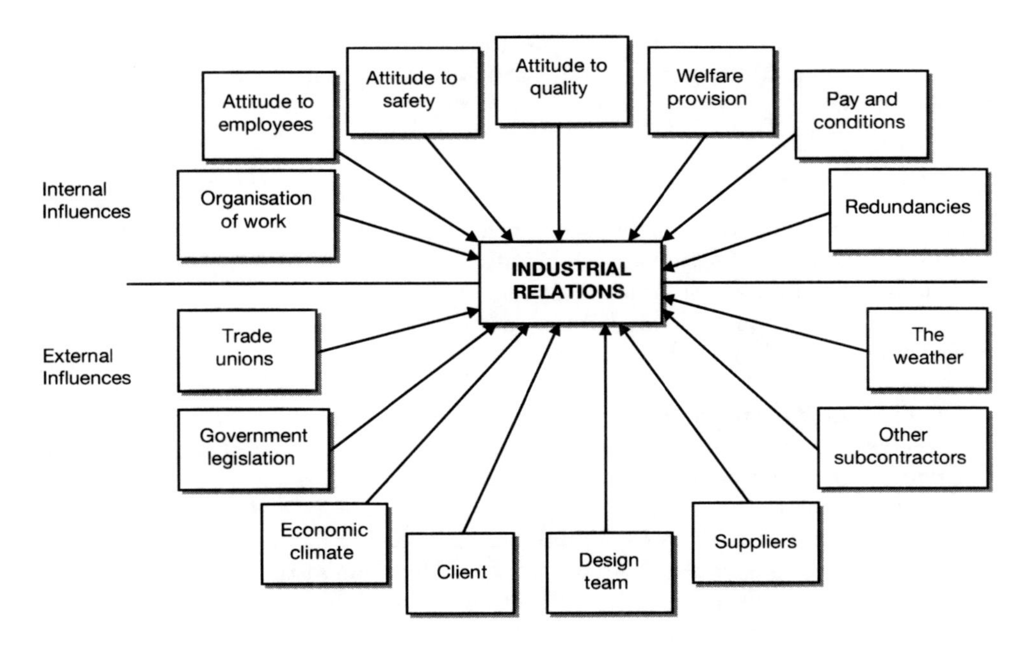

Figure 29.4 The external and internal influences on industrial relations

The trade unions generally wish to resolve issues rather than have confrontation. The problem they have is with a rogue steward on site who can have a strong influence on fellow members especially over the bonus paid. This is not the problem it was when the contractor employed a high percentage of the workforce and the whole contract would be stopped as a result of industrial action. These days, strike action is more likely to be threatened as a result of breakdowns of national negotiations and the risk of this happening is low. Government employment legislation can cause industrial unrest as was shown in the 1970s and 1980s, but the more serious issues have now been addressed, so it is suggested the risk here has receded. Government decisions can have a profound impact on the economic well-being of the country and hence the amount of work being carried out. Uncertainty of employment impacts on industrial relationships. However, a more likely impact on the state of the industry can come from influences outside the country such as other countries' economies, e.g. the USA, China and India, the cost and availability of fuel, war and terrorism.

If a client changes their mind during the contract and makes significant changes to the project, this causes disruption and frustration to employees, especially if they have to demolish work they have already completed. Equally, if the design team make errors that require remedial work, the same occurs. A site in the 1970s in London had a major dispute, the main catalyst being the structural engineers had under-designed a floor that had already been constructed, causing delays when being demolished. Additionally, the service holes through thick reinforced concrete walls had been missed off the drawings, resulting in many drills having to be brought in to rectify the situation. If suppliers are late or subcontractors' work either holds up or conflicts with others', upset occurs. This is in part an internal matter and might have been anticipated and resolved by better management. Inclement weather is normally accepted as part of the job and causes no problem, but excessive situations can be different; for example, the winter of 1962/63 when nearly all construction sites were stopped for three months because of sub-zero temperatures.

29.17 Case study

This case study aims to highlight some of the issues that should be considered when making management decisions to avoid disputes from occurring. Below are some questions the reader might consider at each stage if they wish to participate in the exercise and if so should pause after the questions posed or instructions given, before proceeding; names have been changed:

> Late Thursday afternoon, the precast concrete factory production manager received an urgent message by telephone from the construction site that an extra two loads of components were required that evening. He confirmed they would be delivered as requested. At this point in time there were eight men working in the stockyard, it is almost 4.30 pm and no overtime had been planned for. The production manager contacted the stockyard foreman instructing him to get these deliveries loaded immediately and sent out. Ten minutes later the foreman came back and said that of the gang size of three required, two had agreed to work, but the third had refused.

What advice you would give to the production manager?

> The production manager asked the foreman which man had refused to work. 'Lever – I thought that this was the chance to get rid of him once and for all. When I told him to work on, he swore at me and said that he wouldn't, so I told him if that was his attitude then there was no more work for him here and he would have to pick up his cards on Friday [a euphemism for 'you're fired' based on the fact that National Insurance stamps were physically stuck on a card that employees took with them from job to job]. Did I do the right thing?'

The production manager thought this over and said:

> Well it's not exactly the way I would have handled it, but it's done and we must give him his cards on Friday. However, we've still got the problem of loading these trailers. Go and ask Chadwick to stay on. I think that he'll be agreeable as he always seems to need a bit of extra money.

The foreman returned five minutes later to say Chadwick also had refused to work. He also said Lever had contacted the union steward about his dismissal and the steward was 'kicking up hell outside' and was waiting to see the production manager. Chadwick and Lever were also outside.

Write down your answers to the following:

On the evidence that you have heard so far, and on that alone, what advice would you give to the production manager now?

What is the real problem of the moment? Is there more than one problem and who, if anybody, is at fault?

What are the irrefutable facts so far?

What other facts would you like to have?

For the reader's information some additional facts to assist in the decision making and recommendations are:

- Lever has been with the company for nearly five years. He is an extremely versatile operative and is able to work all types of cranes in the stockyard to a very high level of proficiency.
- The stockyard foreman has never liked Lever, because he finds him a difficult man to control.
- It takes three months to train a competent crane driver.
- Lever does not like working overtime and occasionally takes time off.
- Chadwick is thought to be a first-class worker, willing, 'works all hours'.
- The stockyard hours tend to be erratic because of the type of work.
- It is pay-day.
- Chadwick lives 7 miles from the works and brings two men in with him in his car each day.
- The production manager's opinion of the steward is that he is a reasonable man.

Answer the following questions now:

Write down the irrefutable facts as you now see them.

What are the problems now?

Who is to blame?
What action would you advise now?

The original problem was and still is, the fact that loads have been promised to the site and are now not going to be delivered as promised. The production manager had made a hasty decision without fully thinking out the consequences and should have told the site he would ring them back, checked with the stockyard to see if they could comply and then advised the site he could not comply with their request. To be fair, this was not an unusual case, and the decision to send loads under such circumstances had been made many times before. The result of this poor decision was the site was angry because they had personnel waiting for a load that did not arrive, Lever had been dismissed unfairly, Chadwick had also refused to work but had not been dismissed though treated differently, there was an angry union steward outside, a foreman who had used the occasion incorrectly to dismiss someone he did not like, and a production manager whose fault it was in the first place having all this to resolve. The author is well aware of the situation that had been created, because he was the production manager!

The solution was that the site was informed and fortunately was very understanding. It was agreed Lever should work elsewhere in the factory. It was discovered in the course of the conversation he had always wanted to work in that part of the plant. The foreman was taken to one side by the factory manager and rebuked, as was the production manager, who in turn learned a major lesson in management.

29.18 How to deal with a problem

It would seem at this juncture appropriate to write a few words on problem solving. Although in the human resource management section, these basic rules apply to any problem-solving situation.

The most important thing is to have in one's mind a clear understanding of what the reason is for solving the problem and what is expected as a result, i.e. determining the objective.

Stage one is to obtain the facts, reviewing the record as it is seen; discovering any customs and rules that might apply, as these may act as a restriction in developing solutions, or have to be changed in some way; and then talking with all the parties concerned to get feelings and opinions to obtain the full story. Remember people can be reluctant to tell the full story, either because it might cause trouble for another, or they wish to tell what they think the interviewer wishes to hear.

Stage two is to analyse the facts and information, fitting the information together and considering how they interact with each other. From this, several actions might be possible, but before taking a decision, look at any restrictions that could affect choice. These could be company policy and the requirements of good management practice such as safety, quality and the environment. The options available could also create further problems and have an effect on the labour force, in terms of redundancies, moving personnel elsewhere and retraining, as well as the production process, because looking at one scenario may have a knock-on effect on another part of the production process. Finally, is the action decided upon going to achieve the original objective? It is important in this thinking process, that while possible outcomes may come to mind, it is wrong to jump to conclusions until all the facts and possibilities have been examined.

The third stage is to take action. Is the originator of the decision going to take the action, is help required in enacting it or is it to be delegated to others, and if so, what information do they require to carry it out? Does anybody else need to be informed of the decision such as the immediate supervisor? The timing of the enactment can also be important. Making people redundant the week before Christmas or trying to change a production process at the peak period may not be appropriate. If the action to be taken is unpleasant, the buck should not be passed to others, or others blamed for the decision.

Stage four is to periodically monitor the success or failure of the action taken. A decision needs to be made on how soon after action has been taken the effect will be monitored and on the frequency thereafter. This could include changes in people's attitudes and relationships. Above all, has the action taken achieved the initial objective and if not why not, and can any modifications be made to make it work?

29.19 Causes of unrest

As a final thought, included is a list of causes of unrest advanced by employers and employees back in 1914 abstracted from the Commission on Industrial Relations, First Annual Report (1914, pp. 19–22). While some of the points are now dated, many of these attitudes still persist.

Causes of unrest advanced by employers

1 Normal and healthy desire for better living conditions.
2 Misunderstanding and prejudice. Lack of conception that interests of labour and capital are identical.
3 Agitation by politicians and irresponsible agitators.
4 Unemployment.
5 Unreasonable demands arising from strength of union.
6 Labour leaders who stir up trouble to keep themselves in office.
7 Inefficiency of workers, resulting in ever-increasing cost of living.
8 Rapidly increasing complexity of industry.
9 Sudden transition of large numbers of immigrants from repression to freedom, which makes them easy prey to labour agitators.
10 Universal craze to get rich quick.
11 Decay of old ideas of honesty and thrift.
12 Misinformation in newspapers/Internet.
13 Too many organisations for combative purposes instead of for cooperation.
14 Violence in labour troubles.
15 Sympathetic strikes and jurisdictional disputes.
16 Boycotting and picketing.
17 Meddlesome and burdensome legislation.
18 The close shop, which makes for labour monopoly.
19 Financial irresponsibility of unions.

Causes of unrest advanced by employees and their representatives

1 Normal and healthy desire for better living conditions.
2 Protest against low wages, long hours, unsanitary and dangerous conditions.

3 Demand for industrial democracy, and revolt against the suppression of organisation.
4 Unemployment, and the insecurity, which the wage earner feels at all times.
5 There is one law for the rich, another for the poor.
6 Immigration and fear of over-supply of labour.
7 Existence of double standards, which sanctions only a poor living in return for the hardest manual labour, and at the same time luxury for those performing no useful service.
8 Disregard of grievances of individual employees and lack of machinery for redressing same.
9 Control by Big Business over both industry and the state.
10 Fear of being driven to poverty by sickness, accident or loss of employment.
11 Inefficiency of workers on account of lack of proper training.
12 Unfair competition from exploited labour.
13 The rapid pace of modern industry resulting in accidents and premature old age.
14 Lack of attention to sickness and accidents and the delays in securing compensation.
15 Arbitrary discharge of employees.
16 Blacklisting of individual employees.
17 Exploitation of women.
18 Ignorance of social economics on the part of employees.

Bibliography

Bartol, K.B. and Martin, D.C. (1997) *Management*, 3rd edn. New York: McGraw-Hill.
Commission on Industrial Relations, First Annual Report (1914). Chicago, IL: Barnard & Miller Print, pp. 19–22.
Curwell, S.R., Fox, R., Greenberg, M. and March, C.G. (2002) *Hazardous Building Materials: A Guide to Environmentally Responsible Alternatives*. London: E & FN Spon.
Dainty, A. and Loosemore, M. (2012) *Human Resource Management in Construction*. London: Routledge.
Health and Safety at Work etc. Act 1974.
Health and Safety Executive (2001) *Tackling Work-Related Stress: A Manager's Guide to Improving and Maintaining Employee Health and Well-Being*. London: HSE.
Health and Safety Executive (2007) *Managing the Causes of Work-Related Stress: A Step-by-Step Approach Using the Management Standards*. London: HSE Books.
Kelly, J., Morledge, R. and Wilkinson, S. (2002) *Best Value in Construction*. Oxford: Blackwell Science.
Megginson, L.C., Mosley, D.C. and Peitri, P.H. (1989) *Management Concepts and Applications*, 3rd edn. London: Harper & Row.
Pratt, K.J. and Bennet, S.G. (1989) *Elements of Personnel Management*, 2nd edn. London: Van Nostrand Reinhold.
Protection from Harassment Act 1997.
Robbins, S.P. (1994) *Management*, 4th edn. Englewood Cliffs, NJ: Prentice Hall.
Robbins, S.P. and Coulter, M. (2015) *Management*, 11th edn. Englewood Cliffs, NJ: Prentice Hall.
The Construction (Design and Management) Regulations 2007.
Working Time Regulations 1998.

30 Risk analysis and management

30.1 Introduction

Risk has been alluded to and discussed elsewhere in this book, but the subject is of such importance it is identified as a subject in its own right. Risk is associated with everything we do, as individuals, where even the air we breathe can potentially harm us because of pollutants and allergens, and in the workplace. In construction it is usually considered in terms of financial risk and the risks associated with safety (see 6.3 and 21.17), but many of the principles outlined here are relevant. Risk is defined by HM Treasury as 'uncertainty of outcome, whether positive opportunity or negative outcome'. However, others believe risk should not be confused with uncertainty, arguing the former is known about and an assessment of its probable impact made, whereas uncertainty is not known about and can have either a negative or positive effect. Clearly there is a conflict of views on this matter and the reader is well advised to seek clarification when reading others' discourses. There is a strong correlation between risk management and value engineering. The two subjects are linked together as any value management judgement can alter the risk.

Risk management is concerned with identifying relevant risks, assessing their likelihood and impact, and deciding how best to manage them. It is not about avoiding risk, for to do so would remove any entrepreneurial spirit in a team and life would become very boring. Risk taking is part of normal business practice; what needs to be done is to take calculated risks. There is a difference between accepting risk and ignoring it. There is a risk involved in crossing the road, but by taking simple steps such as looking in both directions first or, better still, using a pedestrian crossing, underpass or bridge, the risks are reduced. Risks occur at all stages of the project and within the company business as a whole. Both can be impacted upon by risks from outside such as government legislation and world events.

Is there a cost associated with managing risk? This is a difficult question to answer as it depends upon circumstances. In the case of the pedestrian crossing the road, then yes it costs to provide physical alternatives such as pedestrian crossings, but on the other hand what is the cost if a person is injured or killed if this provision is not available? If a client shifts the financial risk to the contractor, those that are vigilant will make allowance for this within their price for the works to counterbalance this disadvantage. Whether or not this makes the overall cost of the contract more expensive or cheaper is an interesting dilemma.

30.2 Dealing with risks

Having accepted that risks and risk taking are inherent to business, and construction is no exception, a procedure has to be adopted to deal with risk. The first stage is to identify the risks. Irrespective of the stage in the process it is necessary for the team to investigate the part of the process they are involved in and identify all potential risks. Further, they should look beyond their remit and consider all the activities prior to their area of responsibility that might affect them and follow activities such as maintenance issues and demolition when designing the new building – all of this in the context of the client's clearly stated priority of requirements. For example, these may be cost, programme, quality, prestigious design, impact on neighbours, environmental sensitivity, or today even, safety during the construction works.

30.3 Assessment of risks

Having established what the likely risks are, it is then possible to analyse what the implications are. It is difficult to quantify in a few words how to do this because the types of risks are so variable. For example, if it is a safety risk, it is possible to assess the frequency of an event occurring and the severity of the harm likely to result in the event of an accident occurring. This enables a risk assessment to be made (see 30.5). Assessing the probability of an unforeseen circumstance in the ground is a singular scenario and is more of a function of the quantity and quality of site investigation research carried out. Financial risks associated with property development in obtaining a return on one's investment are notoriously difficult to predict as has been identified (see 6.3) due to the many variables involved and the timescale of the investment. In any of these cases it is important to assess which risks are likely to have the greatest impact so that sufficient energies can be invested in establishing what they are and how to protect against them. Risk assessment is the systematic identification of risks and their relative risk compared against the others so management has a full and clear understanding of the risks faced.

30.4 Responding to risks

Having established the risk there are several ways to react to it. The most common probably in the development and contract process is to transfer a portion of the risk to others. For example, it would be unusual for a property developer of a significant project to take all the financial risk, and much of the risk will be transferred to other financial organisations that make a valued judgement as to the likely success or failure of the project. This is similar to the workings of Lloyds Insurance when the Names are offered the opportunity to make a proportion of the risk of insuring a specified situation, property or object. Laying off risk to others by necessity involves paying a premium to whoever takes on the risk. This is after all one of the ways they make their money. On assessing the risk, if they consider it to be small, their premium will be lower than if it is considered to be high. Their premium will also be calculated on their knowledge of the client. In the case of insurance, the greater the track record of no claims the lower the premium and in the case of property development, the proven successful developer would expect to pay less.

Transfer of risk between the client and the contractor is achieved by using clauses in the contract between the two parties. An example of this would be in the contract clauses on fluctuations. If the client requires a fixed price the contractor has to price with inflationary pressures in mind and if, through the pressures of competition, it underestimates, it will be taking the risk. On the other hand if the client uses a fluctuation clause the risk remains with the client rather than the contractor. In this case the risk is manageable as, unless there is some dramatic national or global economic incident, it is possible to make an informed decision about price. If the risk is more difficult to predict the contractor is more likely to add on a higher premium to cover for all eventualities. This clearly suggests in such cases the client would be better not trying to pass on the risk as it could cost more in the long term either because it will cost more or a contractor takes the risk, loses out and goes into liquidation. It is reasonable, therefore, to accept risk as part of one's business and responsibility rather than try to transfer it. In summary, when transferring risk it is essential the risk is understood, the most appropriate place is sought to manage it, and the cost of either keeping the risk or transferring it should be established.

An alternative to transfer of risk is to insure against risk. Homeowners will insure against risk for both contents and buildings cover in case of an accident. The greater the protection required the greater the premium will be. Likewise in contracting it is normal to have cover for indemnity against third party claims (liability insurance policy) and fire. Professional indemnity insurance will be sought for those involved in the design process, including architects, engineers and quantity and building surveyors, to cover for mistakes they might make. Contractors offering design and build services will also have this cover.

Reduction and elimination is another way to deal with risk. This is sometimes referred to as mitigation and we practise this in everyday use such as carrying an umbrella in case it rains or having a mobile phone in case the car breaks down. Examples in construction include value engineering studies producing an alternative design solution; improving the accuracy of information upon which estimates and construction programmes can be prepared, such as more detailed design or further site investigation; changing the method of construction so as to reduce safety risks; and selecting an alternative procurement strategy so as to allocate risk between the various parties in a more appropriate way.

A further approach to risk is to avoid taking it. Once the risks have been identified, it may be decided that the risks are too great and no further progress will be made. For example, a property developer may consider the probability of the project making an appropriate financial return is not good enough. Alternatively, a contractor decides there are too many unknowns about the project and excessive risk has been transferred in the conditions of contract making the overall risk too great, so declines to tender.

The final approach to risk is to ignore it altogether. In other words, bury one's head in the sand. Many in design and contracting still do not analyse the risk of the project they are embarked on, relying on the assumption the standard conditions of contract used will deal with risk, and then wonder why matters go wrong. This is because the balance of risk may not be pertinent to the specific project and also the standard forms do not cover for every eventuality.

Figure 30.1, a risk matrix adapted from the *Business Continuity Planning Guide* (1998), is an interesting way of assisting in deciding how to deal with the different ways of dealing with risk.

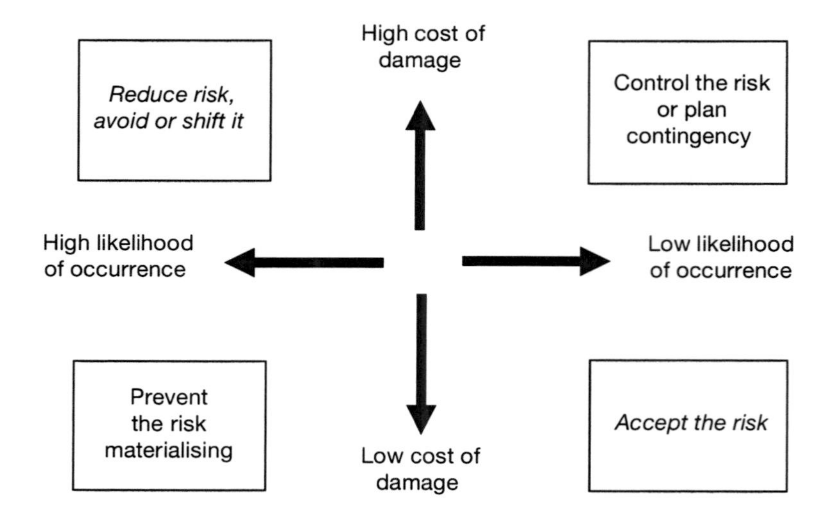

Figure 30.1 Risk matrix

Source: Adapted from *Business Continuity Planning Guide* (1998).

Risk should be reviewed at least once a year and more frequently if significant changes have occurred, be they internally or externally generated. A sudden rise in global oil prices might impact on a haulage business significantly. Countermeasures put in place may need to be checked, amended or removed altogether if the threat has disappeared. The Police force is continually doing this in the light of changing circumstances and threats.

30.5 Conducting a risk assessment

It would be difficult to assess risk across the whole of an organisation because of the scale of the task and its comprehension. It is therefore normal to break down the business into manageable and discrete activities; for example, production safety, computer security, development finance and suppliers.

Those involved in risk assessment need to be experienced and qualified to carry out this task and have sufficient knowledge of the area of work being assessed so they know what questions to ask and how to interpret the answers given. The kinds of questions asked could include:

- What currently are the hazards, failures, breakdowns or interruptions facing the organisation?
- If any of these occurred, what would the effects be on the organisation, business partners, customers, our staff, supply chain, local community, environment etc.?
- What is the likelihood of any of these hazards, breakdown, failure or interruptions occurring?
- Are we prepared for such an event?
- Is the risk acceptable to the organisation and others likely to be affected?
- Can the level of risk be controlled?

- Is there anything the organisation can do about it in terms of being prepared or taking steps for recovery?
- Are there any restraints that restrict preparation or recovery?

They will establish the facts by interviewing, physically inspecting the workplace and looking at data on previous incidents following a methodical sequence. As an example, an assessment looking into the risks associated with physical assets would normally include the following stages:

- listing all the department/company assets and grouping them into like categories such as, computers, printers, scanners and disc backup storage;
- establishing the replacement cost and categorising its value against a table of ranges of cost. This is because to replace an item costing £100,000 has a significantly greater impact on the business than replacing an item valued at £100;
- assessing its value to the organisation if it could not be replaced easily or quickly. It may be possible to replace an expensive item immediately, but a storage device costing only a few pounds but storing valuable data may be either irreplaceable or cause disruption to the business if damaged, destroyed, lost or stolen;
- considering the likely threats each asset is subjected to such as fire, theft or breakdown and for each of these threats, assessing the likelihood of this occurring on a rating scale;
- assessing, if the threat materialises, how widespread an impact it will have on the business;
- assessing the overall risk, either by a valued judgement or using calculations;
- drawing comparisons of all the risks to the assets.

30.6 Risk assessment calculations

The following is a typical example of one of the simpler approaches, to give the reader a flavour of the issues. To obtain a further insight into more sophisticated methods see the *Risk Handbook* by Chicken (1996). A matrix can be produced as shown in Table 30.1. The left-hand column shows the issues that might be assessed for a particular asset, the type and likelihood of a threat and its vulnerability in the event of the threat happening, sometimes referred to as the frequency of likely occurrence. The top row is different assets of similar groupings. A risk rating appropriate to each is entered into the boxes in the matrix. These range from very low significance to high significance and are given an appropriate rating scale number of 1 to 5. The example shown is fictitious, but is used to demonstrate how the assessment of the risk is calculated.

Table 30.1 Assessment of risk

	Asset A	*Asset B*	*Asset C*	*Asset D*	*Asset E*
Asset value (£)	2	4	1	5	2
Likelihood of threat	1	1	3	3	2
Vulnerability	3	4	1	2	5
Assessment of the risk	2.00	3.00	1.67	4.00	3.00

The bottom row showing the assessment of risk is calculated by adding together the three values (asset value, likelihood of threat and vulnerability) and dividing by three to obtain an average. The greater the value of assessment of risk is, the higher the risk to the asset in the business. The weakness of this simple system is it assumes the values attributed to these three issues are of equal importance relatively, which of course may not be the case. In this example the asset has been identified as a financial cost, but it could equally be a measure of the safety of the staff. From this data and calculations it can be demonstrated where the greater risks are in the organisation so that risks can be prioritised and appropriate action taken. Safety risk assessments can be seen in Chapter 21.17.

30.7 Monitoring and feedback

Throughout any project there should be an ongoing process of risk management reviews. It is suggested in the *Achieving Excellence in Construction Procurement Guide* (2003) there are several key times when this should occur:

- When the objectives and priorities of the main parties of a development project have been identified, a risk assessment of the potential project options should be made.
- Having decided upon a development project there are various options available that could meet the user's requirements. Once these alternatives have been considered it is necessary to carry out a detailed analysis looking at the balance cost and the risk of each solution before making the final decision.
- The type of procurement system best suited to the project should be selected, by applying a risk assessment of each of those available and considering how best to manage the risk in each case as outlined above (transfer, avoidance, reduction, acceptance or sharing).
- In the outline design process, the risk is associated with whole-life costing and life-cycle analysis and buildability aspects as errors here can have a significant long-term impact on the financial performance of both the building during construction and later on during its life in terms of running and maintenance costs, as well as the success or otherwise of its functional use.
- During the detailed design process, the risks are concerned with the selection of materials, especially finishes. These affect frequency of replacement or planned maintenance and ease or otherwise in cleaning. This process will be on-going during the construction of the building.
- The final stage of risk management is concerned with facilities management. The decision on how this will be accomplished may have already been looked at during the procurement process such as the service contract management phase of a PFI project. Alternatively, the management of this process may be done in-house, in which case the risk assessment needs to be carried out to decide how much is carried out by the company's employees and how much is out-sourced and how these subcontractors are controlled.

The final stage is feedback which involves analysing how well risks were managed and what lessons can be learned to improve risk management in the future. It is often overlooked as a result of new tasks and challenges, but should be encouraged.

30.8 When do risks occur throughout the process?

It is important to realise that risks associated with the development and construction of a building run through until the building is finally demolished and disposed of – it is a continuous process.

30.8.1 Development viability and feasibility stages

When developers see an opportunity for a possible development, they look at the expected return on their investment for a particular use of the project on completion, be that from rent or sales, and look at different sources of funding. The risks involved in this process include:

- deciding upon the use of the site;
- obtaining planning permission;
- estimating the costs of the construction works when the design is not complete so it is difficult to assess accurately and relies on ballpark figures – there are uncertainties about the ground conditions, decontamination costs, archaeological finds;
- obtaining the completed development on time;
- estimating the costs of facilities management;
- selecting the team to provide the information to enact the above;
- deciding on whether to sell or lease;
- estimating the costs involved in selling or leasing and if the latter, estimating the length of time to lease all the property;
- obtaining the returns estimated;
- obtaining the finance at the right interest rates and the implications if fluctuations have to be accounted for;
- accuracy of calculations, e.g. comparative, residual; and discounted cash flow methods.

Factors affecting the calculations are primarily economic and include the problems associated with the many variables in predicting interest rate changes, wage inflation, increase in the cost of materials and the state of the stock market. Also clients would not wish to commit finance to a project until all the risks have been identified and assessed and it has been agreed how they are to be managed.

30.8.2 Design stage

Much of the risk at this stage revolves around the strength of the relationship between the client and the design team and within the design group itself. The likely risks include:

- the lack of understanding of the client's needs by the design team;
- whether or not the client really understands what is needed. Often the client has a view of what is needed, but it is different from that of the people who have to live and work in the building;
- poor briefings by the client to the designers which of course the designers have a responsibility in rectifying;

- designing solutions that are complex, difficult, extend the contract duration and hence are costly to produce, i.e. buildability issues;
- designing the risk of crime out of buildings and/or the complex if more than one building such as a housing estate;
- breakdowns in communications between the design team partners;
- inadequate research into ground conditions;
- the use of inappropriate technologies;
- designing buildings difficult and/or expensive to maintain
- inaccurate cost planning;
- overshooting the budget;
- satisfying their own needs in producing an 'architectural masterpiece'.

It can be seen how value engineering at this stage can have a significant effect on improving the overall quality of the building performance and reducing risk (see Chapter 10).

30.8.3 Procurement stage

It is important the method of procurement is selected that best suits the type and needs of the project (see Chapter 11). Much of the risk depends upon the priority the client gives to the following:

- their experience and how much they wish to be involved in the construction process;
- separation of the design process from the management of the construction processes;
- the need to alter the specification and design as the project develops, building in a certain amount of flexibility into the briefing process;
- the ability to seek for remedies if dissatisfied with the design and construction processes;
- the complexity of the project;
- the speed at which the contract needs to be completed from inception to handover;
- how much flexibility there is from the budget price to the final price.

There is also risk to the contractor depending upon the procurement rate, which is in essence often reciprocal to that of the client.

30.8.4 Tendering stage

The following risks are inherent in the tendering process itself:

- While taking account of the amount of time predicted to be lost as a result of inclement weather, it is notoriously difficult to assess.
- In the euphoria of wanting to successfully obtain the contract and meet the client's needs there is a danger of making optimistic assumptions when programming.
- The time available for carrying out the estimate is limited so there is always the risk that not enough time is allocated to thinking through all the implications of the construction processes and the wrong method of works is selected as the basis for the tender.

- Deciding upon the levels of overheads and profit to be added is the greatest risk of all, which is why senior management takes the decision.
- Predicting the increased costs of plant, materials and labour throughout the duration of the project, especially if a long contract, is difficult – the risk is reduced in the event of having a fluctuation price contract.
- If during the course of the project shortages of labour, plant and materials occur, the price of obtaining them will rise.
- In the end, other than when insufficient overheads and profits have been added, the contract's success or failure is usually a reflection of the quality of the management.

It is interesting that many of the risks can be reduced significantly if all parties work with a spirit of cooperation rather than confrontation.

30.8.5 Construction stage

The first thought concerning risk in the construction process is naturally that of health and safety. However, there are other risks associated with this stage, some of which will have been assessed at the tendering stage. These include:

- sensitive projects that can attract objectors who may take positive action to prevent construction work from occurring – notable examples in recent years have been motorways and bypasses and airport runways;
- theft by the general public, the contractor's employees and organised criminals;
- terrorist activity on high-profile building works. This is not just about attacks on the building works, but implanting devices to set off later;
- vandalism;
- trespassers and possible injury incurred by them;
- fire, which may be started by accident or deliberately by vandals;
- inclement weather, which will almost certainly have been taken account of in the estimate, but is by its very nature unpredictable;
- industrial action – less likely than in the 1960s but still a possibility, not necessarily always on the site but also to a supplier;
- shortages of labour and materials;
- unforeseen construction problems;
- design deficiencies;
- delays in information being provided;
- complaints from neighbours about the nuisance value of the construction work;
- spillages of pollutants either into the ground or entering water courses and sewage systems;
- disposal of hazardous materials in general.

30.8.6 Commissioning and snagging stage

The amount of risk in the commissioning stage is a function of the complexity of the building. Clearly the greater the percentage of services in the project, the more it is likely to go wrong and this needs to be taken account of in the planning of the commissioning and installing processes. It is important to build in adequate time for testing and rectifying any faults found. How much snagging is involved is directly related to the methods in which quality is controlled and assured during the construction process.

30.8.7 Occupation of the building

The main question here is what is the impact on the business of the building users if something happens. This is developed further under business continuity development (see 30.9). Typical examples include:

- computer failure caused by damage of some kind such as fire, theft of the machine or information and computer viruses;
- fire destroying part or all of the premises;
- burglary, theft by personnel and vandalism;
- disclosure of information by company staff to others outside of the organisation, sometimes for financial gain and in other cases for political or mischievous purposes often referred to as 'leaks';
- assault on personnel in or near to the building;
- threats from terrorists;
- radiation contamination, which could occur in hospitals and science laboratories;
- the accuracy of the planned maintenance programme;
- the speed of response to maintenance and replacement requests, e.g. light bulbs in an operating theatre during an operation;
- sufficient retention monies withheld from the contractor to cover remedial work;
- whether changes of use can be accommodated;
- whether the budgets for facilities management are accurate.

30.8.8 Demolition stage

- Safety issues are of paramount importance and potentially are the greatest risk. These include not just the damages of collapse during the process, but also dusts, fibres and fumes resulting from previous use of the building and of course the discovery of asbestos.
- Disposal of the materials comes with its own risk. Materials need to be identified so either they can be safely disposed of or recycled.
- The financial risk is to do with having a methodology of demolition that can be carried out without short cuts having to be taken.
- Assumptions need to be made about the amount of material that can be recycled.
- The difficulties of dismantling need to be assessed.
- It is essential to take precautions against the dangers to third parties, especially if using explosives.
- The avoidance of pollution of the atmosphere and water courses during the process is of paramount importance.

30.9 Business continuity management (BCM)

Business continuity management is 'Planning which identifies an organisation's exposure to internal and external threats and synthesises hard and soft assets to provide effective prevention and recovery for an organisation, whilst maintaining competitive advantage and value system integrity' (Herbane *et al.* 1997).

There has been an increasing awareness of the sorts of issues raised above, in the occupation of the building, of the need to have a business plan in place that can be brought into action as a result of a serious disruption to a business. Notable examples reminding of the need, being the Saturday 15 June 1996 bomb blast in the centre of Manchester and the attack on the World Trade Centre twin towers in New York on 11 September 2001. One of the buildings severely damaged in the Manchester blast, later demolished, housed the offices of the Royal Sun Alliance. They had a recovery plan in place and had found alternative accommodation by the following Monday. They also found a warehouse to deal with recovering data from damaged computers and the documents taken from the damaged building, although in reality it was several days before access was permitted to recover the information while a structural assessment of the building was made. However, not all interruptions to business are as dramatic as these two examples, but can be equally damaging, as can be seen when the building housing the business is burnt to the ground. Many interruptions are less severe, but still cause many problems for all those concerned.

The concept of business continuity management has been a developing thought process much of which is based on the previous discussions in this chapter. Initially it was seen as dealing solely with the physical equipment, primarily computing, and was concerned with answering the question 'what would happen in the event of a technological failure?' This developed into, first, thinking about attacks against the system and the need to provide security to prevent the attacks being successful and, second, to implement procedures in the event of a successful attack, such as backup systems and the development of a survival plan to ensure survival and recovery.

Companies realised that recovery is not just about the physical attributes of the business, but there are people involved in this process who are in reality the most important resource the company utilises. It is all very well, as in the case of the Royal Sun Alliance in Manchester, transferring part of the business to spare office accommodation in their Liverpool offices, as well as finding space elsewhere in Manchester, but not if the activity going to Liverpool needs staff who are unable to work there because of their domestic needs, for example, a single parent with children at school who only works from 9.30 am until 3.00 pm.

There are four stages in the continuity management process as shown Figure 30.2.

30.9.1 Initiation

Initiation involves changing the behaviour of the managers and employees in the organisation. As in all significant changes within an organisation, whether it is in behavioural, procedural or process, the drive must come from the very top. They have to identify and determine what parts of the business are likely to need to be covered by the continuity plan, whether it can be developed internally or with outside assistance, and then set the policy. The likely impact on the business is part of the

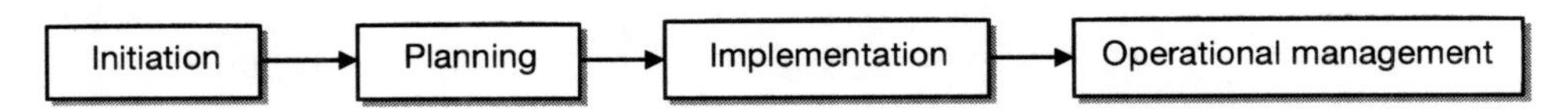

Figure 30.2 The continuity process

continuity planning process. These policy decisions need to then be clearly communicated to all stakeholders in the business. As with any policy, structures within the organisation need to be created and key staff selected and trained so the policy can be implemented. In other words it is necessary to have enthusiasts at each appropriate level. Sufficient finance also needs to be set aside, both for the initial setting up of the plan and its introduction and maintenance. How much is allocated depends upon what issues have been identified. In carrying out this work it is imperative the key staff are able to convince others in the organisation of the need, and one of the ways of doing this is to create scenarios that might happen and ask the question, 'What would the impact be on them and the company if this event took place?'

30.9.2 Continuity planning

The plan is a guide for action in the event of an interruption of part or all of the organisation's business. It is necessary to identify a list of crises that might affect the business. Some of the specific issues related to the development and construction processes have already been identified, but Table 30.2 modified from Elliot *et al.* (2002) gives a more general view of likely hazards.

Clearly not all will be relevant to the business of the organisation. The objective of the exercise is to establish, primarily, the likelihood and the impact such an event would have on the business. The timing of an event should be considered, as the work in some businesses will not be repetitive throughout the year. For example, what would happen if the system for communicating national exam results to school children broke down?

On identifying likely causes an analysis of the impact of such an event has to be established. So, for example, if there was an electrical power failure or fault, a whole range of effects could be expected besides the loss of power, such as power surges and loss of voltage that may have a knock-on effect on sensitive electrically driven equipment and machinery. This is referred to as business impact analysis and while not within the remit of this text to demonstrate methods of assessing the impact (see Elliot *et al.* 2002), it is important to realise who may be affected by an event.

Businesses do not work in isolation, as there are many stakeholders involved, both within the organisation and externally. Typically in a construction company they are as demonstrated in Figure 30.3.

Impact caused by interruption is a complex process. A breakdown caused by say one supplier does not just impact on the contractor. If the contract is delayed, besides possible penalties resulting from a late handover, the client's business is affected, as

Table 30.2 Likely hazards

Adverse weather	Floods	Negative publicity
Computer breakdown	Hostile takeover	Plant failure (e.g. tower crane collapse)
Computer viruses	Illegal activities	Product tampering
Computer failures	Industrial action	Sabotage by outsiders
Currency fluctuations	Kidnapping	Sabotage by staff
Design failure	Loss of important staff	Supplier/subcontractor goes bankrupt
Disease/epidemic	Major industrial accidents	Telecommunications failure
Fire	Natural disasters	Terrorism

Source: Adapted from Elliot *et al.* 2002.

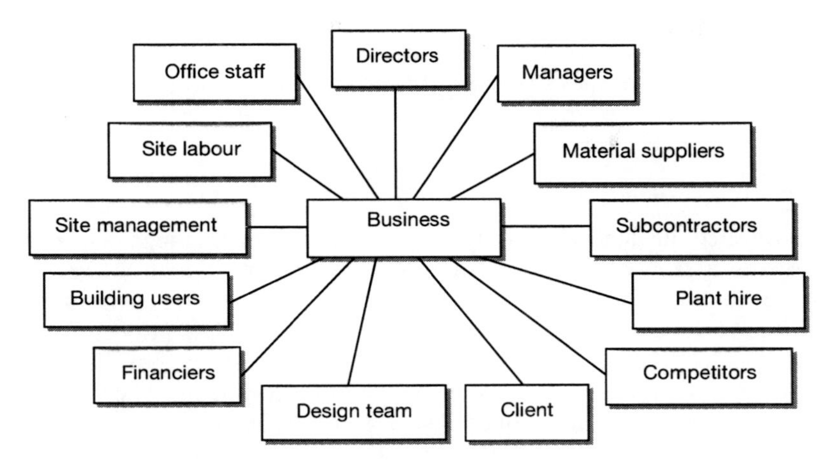

Figure 30.3 Main stakeholders in a construction company

are some or all of the other subcontractors and suppliers. Reputation may be damaged and that has a knock-on effect in obtaining work in the future, thereby introducing the possibility of redundancies within the company: a possible overstated bleak scenario, but is it? As supply chain management develops, it will be in the contractor's interest to assist those in the supply chain with their own business continuity plans as many of them will not have the necessary resources to carry it out themselves.

There are others who may be involved in some way as a result of an incident. The general public is affected by noise and pollution and in some cases is injured. The media report incidents that can in some cases be damaging to the reputation of the business and hence the flow of information to the media needs to be well controlled. Others who can impact on the business include protesters, trade unions and their members, regulatory bodies and governments.

After the analysis of the likely causes of interruptions, the next stage is to conduct a business impact evaluation. This involves first the setting of objectives in recovering from the interruption. For example, if the school is burnt down, the activity within it must be up and running elsewhere within one week, if the tower crane collapses, the site must be operational within one working week (subject to HSE approval) and if the telephones fail, alternative communication must be provided within 30 minutes.

Structured approach risk assessments have to be carried out to identify the likelihood of interruptions occurring and the consequences if they do, from which the priorities for a business recovery plan can be established. The final stage is to develop business interruption scenarios, i.e. what happens, and how do we deal with it if it occurs and do we decide how often these should be tested? This works well if there is wide participation within the organisation, bringing together all the inherent skill and expertise. This acts as the basis of producing the business continuity plan.

30.9.3 Implementation

In many senses producing the continuity plan is the easy part. Implementation of anything new requires skill and competent management. It comprises four stages, that of communicating the reasons for the plan and the implications to the individuals and the organisation as a whole; developing the culture and business ethic within the organisation to cope with the changes; introducing new control systems; and training the personnel to carry out new tasks.

30.9.4 Operational management

Having established the likely risks and introduced systems of management and control into the organisation, will they work in practice when an incident occurs? There are several stages to consider. How much time and effort is spent on each stage depends upon the likelihood and the magnitude of the effect if an incident occurs.

The first stage is to put together teams of staff who will be responsible for dealing with the incident, often referred to as the crisis management team. Their responsibility is for organising the stages mentioned below and dealing with incidents in general. How well they operate is often a function of the type of culture that exists in the organisation as a whole and the ability of the business to create teams and foster team building. This involves full participation of and contribution from all members, brainstorming sessions, complete openness and employing outside experts where necessary.

The second stage is to put in place a command and control structure that will be able to cope with incidents. This is important so that consistency of approach can be adopted in each case. This does not mean to say all levels of the command structure will be used for each incident as this depends upon its severity and the impact on the business. The more serious the problem, the higher up the chain of command the control will be. Specific personnel need to have clearly defined roles such as who will deal with the media and the emergency services. For example, if a fire occurs in a large sprawling industrial complex, it is no good a member of staff contacting the fire brigade saying there is a fire at the works. The fire services will need clear instructions as to where the fire is and how to get there as quickly as possible. This may mean advising security of the fire who will then liaise with the brigade.

The third stage is testing and auditing. A clear example of this is the rehearsals the emergency services conduct from time to time on the more dramatic incidents such as dealing with a 'weapons of mass destruction' incident or an aeroplane crash. The purpose of testing is to find out whether the systems in place will work, staff have the right attitude and have not become complacent, and to maintain awareness throughout the business. The frequency of tests will have been determined when planning. The types of test used are as follows:

- *Desk checks tests.* These are carried out regularly and frequently and involve checking that all the information contained in the plan are correct and up to date such as named persons and telephone numbers, also that the plan still matches the department/organisation's current work as this may have changed from when the last plan was introduced.
- *Walk-though tests.* All the named persons are brought together and asked to role-play their procedures. This highlights any communication difficulties and whether the activities are being completed within the prescribed times and sequence.
- *Simulation exercises.* Advanced notice is given to all those concerned. A scenario is then created that may restrict the participants in their use of standard com-

munication methods or parts of the building. Under observation they then have to set in motion the recovery plan and execute it to completion. The time allowed may be accelerated from that allowed for in a real incident. The observers look for breakdowns in the participants' performance either caused by their errors or the plan itself. A revised plan can then be introduced after the exercise and subsequent debrief.

- *Operational tests.* Here a complete department will be closed down and the employees have to relocate to another building either owned by the organisation or supplied by others and restart the business activity within a prescribed time. This tests the ability of the staff to carry out the exercise and whether there is enough backup resource to permit it to work.
- *Live exercises.* The difference between this type of test and the rest is that no prior notice is given. The simplest type would be evacuating the building in a fire drill, but could range up to dealing with a significant interruption of the work. These tests can raise the awareness of the impact on all of the organisation's employees. They can of course be expensive to enact.

The debrief is imperative in these tests. The objective is to establish whether the continuity plans work successfully in practice, and if not, why not, and to see how they should be modified for the future. The observers should also question their role to see whether they could improve their performance.

It should also be remembered that in the case of simulation, operational and live tests, the participants can suffer high levels of stress as they become more actively engaged in the process and attempt to put plans into operation which rarely go strictly to plan. These stress levels should be noted, as it may be necessary as a result of the findings to introduce training for staff and, more importantly, to educate management to manage stress in others better.

Bibliography

Achieving Excellence in Construction Procurement Guide 04: Risk and Value Management (2007) Office of Government Commerce.

Chicken, J. (1996) *Risk Handbook*. London: International Thomson Business Press.

Elliot, D., Swartz, E. and Herbane B. (2002) *Business Continuity Management: A Crisis Management Approach*. London: Routledge.

Herbane, B., Elliot, D. and Swartz, E. (1997) 'Contingency and Continua: Achieving Excellence through Business Continuity Planning'. *Business Horizons*, 40(6), 19–25.

Hughes, W., Champion, R. and Murdoch, J. (2015) *Construction Contracts: Law and Management*, 5th edn. London: Routledge.

Kelly, J., Morledge, R. and Wilkinson, S. (2002) *Best Value in Construction*. Oxford: Blackwell Science.

Property Advisors to the Civil Estate (1998) *Business Continuity Planning Guide*. London: PACE.

Part VI

Postscript

Environment and sustainability

31 Sustainability and the environment

31.1 Introduction

History will probably judge our era on how we have or have not dealt with the issues of sustainability and environment as it is one of if not the greatest threat to our survival on the planet. So what can those in the construction industry do?

The author argues that a good manager automatically considers safety, quality, the advancement of subordinates and their welfare as part of their everyday work. All of these can be given a section in their own right as they have been in the text. However, while sustainability and the environment should also be part of the manager's consideration, it is different. While many of the issues have been addressed throughout this text in the appropriate chapters (e.g. life cycle assessment and waste), the subject spans much more than the construction industry. It is hoped that this chapter will give the reader food for thought. The purpose of this chapter is to put it all into context.

31.2 Definitions

It is difficult to define the word sustainability. It is a very complex subject and would take several PhDs to do this and after that there would still be arguments. Suffice to say, it is a very inclusive term meaning different things to different people. However, the author believes the quote from the Brundtland Commission (1987) still holds good and is a very good starting point:

> Development that meets the needs of the present without compromising the ability of future generations to meet their own needs.

A more detailed way of looking at the subject was put forward by The Natural Step (TNS) and adopted by the UK Forum for the future which suggested four sustainability principles:

1. We cannot dig stuff up from the earth at a rate faster than it naturally returns and replenishes.
2. We cannot make chemical stuff at a rate faster than it takes nature to break it down.
3. We cannot cause destruction to the planet at a rate faster than it takes to renew.
4. We cannot do things that cause others to not be able to fulfil their basic needs.

They also produced the idea of The Natural Step Resource Funnel, adapted from TNS for this text, which is an effective way of comprehending what is quite difficult and is shown in Figure 31.1.

The upper wall of the funnel is the availability of resources and the ability of the ecosystem to continue to provide services and resources. The lower wall is the demand by societies for resources that are converted into basic goods and services, such as clothes, food, shelter and transportation, and other purchases providing a quality of life beyond that of survival.

The mechanisms that provide essential life-supporting goods and services for society's continued existence on the planet, such as food, fibre, clean air and water, productive subsoil and climate control are also in decline. However, the demand for these resources is increasing rapidly due to the population explosion, and both the developed and developing world's clamour to consume more.

As society's demand increases and the capacity to meet this demand declines, society moves into the narrower part of the funnel. As the funnel narrows there is less room for manoeuvre and fewer options are available as a result. If the availability of resources continues to decline and consumption continues to increase, the upper and lower walls of the funnel will collide. If solutions are found through innovation, creativity and change to achieve a balance, the walls of the funnel can then be made to open and we can produce a sustainable society. Since we all live in the funnel as individuals and families, working in education, business, industry and government etc., we all have the opportunity to change the impacts we are having and to be more strategic when making choices and long-term plans.

By becoming sustainable, critics sometimes ague that society will have to give up its quality of life and resort to relinquishing much of the comforts and benefits of today's way of life. Maybe if no action is taken, at some point in the future, that could become the truth. However, it is not suggested that this needs to happen, if society and its leaders accept both that there is a problem and its extent, and take steps to address the issues encouraging the development of appropriate technologies; then, other than making adjustments to the way we live, procure, manufacture and farm, and the decisions we make as the human race as a whole, in all probability we will survive without significant loss of quality of life and at the same time permit the developing world to improve its conditions also.

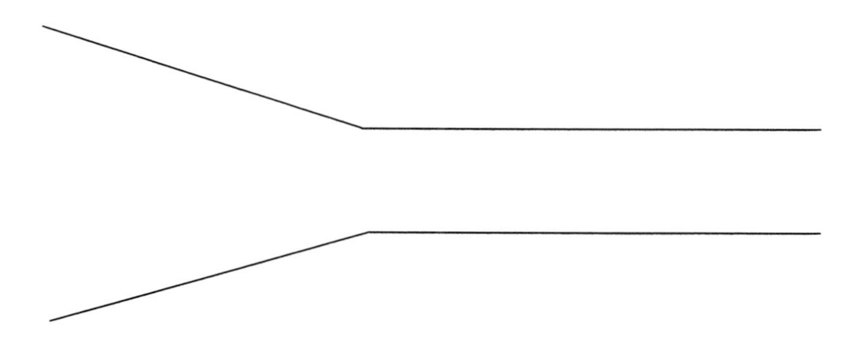

Figure 31.1 TNS Funnel
Source: Adapted from TNS.

It is also not the intention to debate the arguments of the effects of global warming in this text. The general consensus of the scientific community, including those who do not anticipate any significant effect on the climate, is that global warming is upon us. Irrespective of this, the author argues that in line with the statement of Brundtland and TNS we cannot continue to use resources to the extent we currently are; so alternative methods of construction have to be developed and implemented so that less is used in the construction of the building, and building materials are selected and used in such a way as to significantly reduce the amount consumed during the life of the building. While the energy for heating, cooling and lighting is often highlighted as the key issue, the materials used in the construction of the building are of equal importance as they require energy for extraction, manufacture and transport and can come from dwindling or non-renewable sources.

31.3 Technical or social-economic?

There is a danger that sustainability issues are looked at as a technical problem to be solved and engineered. This is a dangerous position to take even though this book is based upon exploring technical solutions. So before reading further it is important to place this aspect in context. Some examples:

- There was a mudslide in Columbia killing many people as a result. The authorities not wishing this to occur again, planted trees on the hillside to consolidate the soil. This worked successfully for many years until there was a further mudslide killing even more than the first. The local inhabitants had cut down the trees to obtain fuel for cooking and heating.
- It is not unusual in mountainous regions for the inhabitants to excavate into the ground and with the earth removed construct low walls, spanned with timber from the forests upon which further earth and grass is placed to insulate the occupiers and their animals from the winter cold. If these are in earthquake-prone areas and the quake occurs during the night, the loss of life can be extensive. There have been suggestions about providing lightweight insulated roofing, but this is beyond

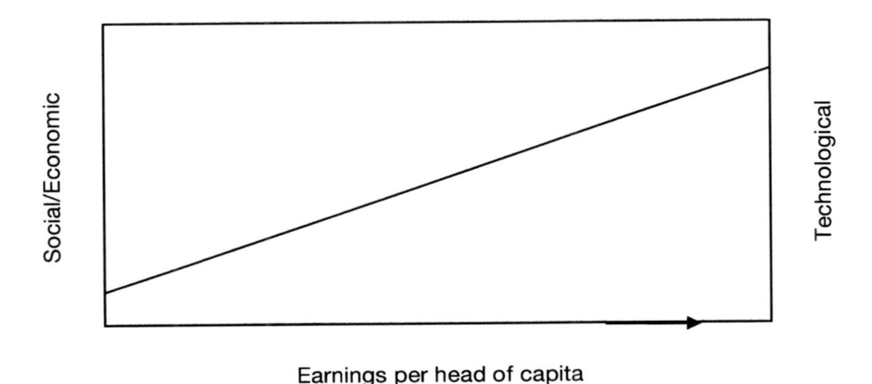

Figure 31.2 Social, economic and technological model

the wealth of the inhabitants. An alternative solution is to construct a structurally sound 'four-poster-bed' supporting the roof above where the occupants sleep. The roof would collapse elsewhere, but the people would survive.

There is a relationship between the amount of money available and the type of solution selected as demonstrated in Figure 31.2. The richer the nation, the more likely technical solutions will prevail than in the poorer society. However, even in the richer communities, technical solutions may not be the answer as there can be social impact as well. Many objections have been made to development because of the social or environmental impact, even though it might be argued that it is in the interest of the wider community to have the development take place.

To demonstrate the point, the Thames barrage, a technological solution, was constructed at great expense to protect London from flooding when certain high water conditions prevail. To attempt to do this in say Bangladesh, would be way beyond the financial capacity of the country. It is better therefore to have impending disaster warnings in place and sufficient accommodation on higher ground or on raised platforms and accept that the flooding will occur.

Solutions should also take account of social responsibility. Do they provide local sustainable employment as against importing labour? If the skills are not available locally can labour be trained? Technical solutions can be such that local labour is put out of work. For example, much of the road repair work in rural India is carried out by people who live and work by the roadside, breaking down rock and river stones to make suitably sized aggregate. From a western standpoint it is clearly very inefficient. Yet by mechanising the processes many hundreds of thousands would be made completely destitute. This does not mean to say that the standard of living of these people should be accepted, but only underlines how complex these issues are.

31.4 Integration in the construction industry

There is a significant lack of integration between the various interested parties associated with the construction processes. To demonstrate this, the question is posed, 'who determines sustainability decisions in construction?'

The obvious response might be the designer, but this is not the case. The designer is responsible for providing the client with what they want. They may try to influence the client, but in the end the paymaster will make the final decisions; but do they? The design has to be approved by the planners. The planners have to ensure that the designs satisfy planning legislation determined by central government (so they have an involvement) and are responsible to their political masters, in the case of local government, the elected councillors. They in turn may be anxious to attract industry to the locality to bolster local employment and increase the rate revenue. This is sustainable providing the labour is employed locally, but even if it is, if the company has to transport materials for manufacture and transport completed products long distances, then it can become unsustainable. Equally with house building, attracting more people to live in the area can improve the local amenities long term to satisfy their needs, but if they are just using the town as a dormitory, travelling to the city many miles away, the conflict can be seen again. The building control officers will work to ensure the building complies with the regulations, but they have little jurisdiction to promote sustainable solutions and will often find it easier to promote well tried-and-tested existing solutions.

The building services engineer should be ideally placed to conserve energy in the running of the building, but if paid as a percentage of the final costs of the building where is the incentive to find sustainable solutions? In any case they may be hamstrung by the requirements of the architect and client that result in high energy running costs.

The quantity surveyor is briefed to ensure the most cost-effective solution. If this is just for capital costs then the design will probably be unsustainable, because of the costs of maintenance. However, if life cycle analysis is involved, the situation will be improved.

Once the project has been awarded, the most influential position in the construction company is the purchasing manager. However, their brief, because profit margins are so small, is to obtain the most competitive prices available. The fact that the material selected may have to be transported long distances, has a high embodied energy or will off-gas pollutants into the enclosed atmosphere of the building for many years to come is not always a major consideration. Legislation such as COSHH should assist in this process, but in reality, most construction companies obtain the health data sheets of the products being used, but rarely look for less hazardous and more environmentally friendly substitutes.

Architecture also reflects society. The creation of fenced-off residential enclosures to keep the 'undesirables' out is an example of this. Yet, surely, a sustainable society is one that not only recognises but also celebrates and values the diversity of its citizens.

Solutions should not only satisfy the client but also have a social return on the investment. Producing a nice pleasant environment will reduce the amount of vandalism and violence. Using for example the GMP architectural design liaison department means that the opportunity for crime can often be significantly reduced in the building, car park areas and accompanying estate.

31.5 Where do we start?

The whole concept of 'the environment' is very hard to manage in the thought process because it covers such a wide and diverse number of topics. To assist in this process two different approaches devised by Curwell and March are demonstrated. The first is shown in Table 31.1. It is similar to the way the first and subsequent Building Research Establishment Environmental Assessment Method (BREEAM) systems of monitoring and classifying buildings were developed. It divided the problem into three main sections – global, community and individual – and considers in the second column which construction-related operations have a direct effect on the main sections. The third column demonstrates the environmental concerns that these activities have. It is not a complete list but gives the reader an indication the problems.

Table 31.2 takes the same basic information but subdivides it into discrete environmental issues in column one. Column two demonstrates the construction processes as in Table 31.1 and column three, the forms of environmental impact. Note that there is no direct horizontal link between those issues raised in columns three with those of column two: they can apply to them all.

There can be a danger in adopting this approach. In making these judgements care has to be taken to firstly not create further problems for the building users affecting their quality of life. For example, in the 1970s, resulting from the oil embargo of the

Table 31.1 Environmental concerns

GLOBAL	Extraction	Energy Timber Water
	Materials manufacture	Energy
	Buildings in use	Energy Sewage
	Demolition	CFCs Burning
COMMUNITY	Impact of planning	All those below
	Extraction	Noise/dust Traffic Landscape Ecology Water
	Impact of design	Ecology Aesthetics
	Construction phase	Site activities (noise/dust) Waste disposal Traffic
	Building in use and management of built resource	Noise/dust Traffic Legionella Fumes/dust Sewage
	Demolition	Noise/dust Waste disposal
INDIVIDUAL	Construction operatives	Health hazards Safety in general
	Building user	Building Sickness Syndrome Legionella Allergies Fibres Dusts Pesticides Gases (radon/volatiles)

oil-producing states and subsequent formation of OPEC, there was a rush to conserve energy, primarily by sealing the building, thereby reducing the ventilation rates. This coincided with an explosion of new, untested materials, which were now confined in the less ventilated air spaces, creating potential health problems.

Many houses were built without chimneys. Insulation, notably roof and cavity wall, was pushed as another means of saving energy. Coincidently at the same time there was a massive explosion in the availability and use of new materials on the market, notably oil-based products such as adhesives, sealants, timber preservatives (poisons by definition) and formaldehyde, a suspected carcinogen. The latter was used in chipboard, strand board and plywood, the bonding agent for manmade mineral

Table 31.2 Environmental concerns

ENERGY	Extraction Manufacture Transportation Site activities Building in use Planning	Global pollution Excessive use of resources
POLLUTION	Extraction Manufacture Transportation Air conditioning Site activities Demolition Pesticides	Community pollution Individual health Food chain Reduced standard of living
HEALTH	Water Sewage Air conditioning Traffic Site activities Building design Demolition	Legionella Dust Fibre Fumes Allergies Pesticides Building sickness syndrome
VISUAL IMPACT	Planning Extraction Transport	Community pollution Individual Standards of living
BIOLOGY/FAUNA	Aesthetics Extraction Traffic Building design Planning	Loss of species Protection of species
CLIMATE	Timber CFCs/HCFCs Air conditioning Demolition/burning Planning Site activities/burning Traffic	Global pollution Community Individual Agriculture Way of life

fibres, foam insulation, as a preservative in carpets and some types of wallpaper. So on the one hand buildings were being made more airtight and at the same time potentially more pollutants were being added to the internal air with the possibility of affecting the occupants' health. Concern has subsequently been expressed about the use of super-fine fibres especially if not contained (note glass-wool fibres are drawn and are larger than respirable sizes), and VOCs are being steadily reduced in paints, as these are an environmental hazard.

A further example suggested by the author is the relationship between asthma and the way we have changed our lifestyle over the last 60 years. As stated by Curwell *et al.* (2002, p. 6):

In the immediate post war period, the majority of householders lived without central heating and with either a polished or varnished timber floor or with linoleum with the occasional rug. As society became more affluent linoleum gave way to carpet and by the late sixties, cheap foam-backed broadloom carpets were readily available. At this time central heating was becoming the norm in new housing and retrofit was also occurring on a substantial scale. During the seventies and onwards, the double-glazing market expanded dramatically to such an extent that it is now the norm to install it in all new buildings.

The technology and life style changes affected relative humidity in buildings. Prior to central heating humidity levels would generally be higher. The advent of central heating initially reduced these levels, which subsequently increased again with the installation of double-glazing and other airtight measures. Add to this the reduction in air changes rates and the question should be posed whether or not there is any correlation between the increased incidences of ill health in terms of allergy and asthma. The house mite infestation implicated in the epidemic of asthma in the UK may be a symptom of changes in life style and technology which have occurred over the last sixty years.

31.6 The construction manager and sustainability

The construction manager often, depending on the method of procurement, is involved from the conception of the project until the handover of the building. This section identifies areas for consideration. Rather than use the approach demonstrated in Tables 31.1 and 31.2, the construction manager needs to identify issues at each stage of the development process. Here each phase of this process is subdivided into environmental, economic, social and institutional issues in line with the approach taken in *Sustainable Urban Development* (Curwell *et al.* 2005, ch. 6) with permission from the author.

31.6.1 The pre-project phase

This is the stage of the process when the various needs of the client are established, defined and assessed in order to establish whether the project is needed or whether alternative solutions to a new build are available.

Environmental issues

The first question is to determine whether there is a need to build, otherwise resources will be consumed unnecessarily. If the client already has a building for their activity, can it be used more effectively and altered to meet the client's requirements and if so, how is it best refurbished? Refurbishment is about extending the life of the existing building so as not to add to the unused or under-utilised building stock. It also has the advantage of concentrating development within the existing area without using more land and adding to urban sprawl.

If new build is the adopted option then building on a brown field site rather than green is a preferred solution, if possible. There will be traffic implications in terms of the volume of noise and pollution that will need to mitigated. An audit of the flora and fauna, especially but not exclusively, on green field sites needs to be carried out so that it can be protected and possibly enhanced. If it is to be destroyed during construction then it needs to be replaced.

Economic issues

In any project the key economic issue is the role the project plays in the business development plans. This means identifying the key client satisfaction, and the design and construction expectations. The key economic sustainability targets include: improved design and construction quality, reduced design and construction cost, reduced construction time and reduced defects at handover. Note that the longer the design stage, the longer there is no economic gain in the area and it is only when the construction phase begins that the local economy begins to profit.

Social issues

With any construction project there is a social impact on the community especially with large projects. Some people will gain, others will lose out and an assessment needs to be made. This means establishing the socio-economic status of the local community, i.e., average family income, levels of employment, school attainment etc.; their health and safety, i.e. mortality rates, crime incidence etc.; and family status such as the proportion of children living with both parents, ratio of males to females in the workforce, single parent families and so on.

With this information a sustainable plan can be developed for the training and employment of the local labour force, both for the construction process and the business occupying the premises on completion.

The designers and contractors should be concerned with the issues of fair employment and trade:

- an equal opportunities plan (for women, ethnic minorities, disabled etc.);
- targets to reduce the rate of fatal and non-fatal accidents in both the manufacturing and construction;
- job creation in the project for the locally employed.

The client and design team should where practical involve the local community in the design making decisions.

Institutional issues

The client, aided by the construction manager if employed at this stage, should ensure that all companies involved in the project have in place or develop a corporate social policy for the project. The three most important are:

- to establish an environmental management process for the project based on ISO 14000 principles. Targets should be set for resource reduction and pollution minimisation so the team can audit and justify all the resources inputs and waste outputs to and from the proposed building and throughout its development process. These include energy, materials and waste management;

- to ensure all team members, including the client and suppliers, publish independently validated reports including environmental, sustainable redevelopment and social responsibility issues;
- to ensure all firms involved in the project have a staff consultative committee and team members who should be regularly surveyed regarding their satisfaction in the way they are treated by their company.

36.6.2 Pre-construction phase

Environmental issues

The conceptual design should seek to maximise the utilisation of available solar, wind and water energy resources from the site and to minimise fossil fuel consumption. Passive techniques to be considered to reduce energy are:

- relatively narrow forms that permit natural day lighting and ventilation;
- thermal mass in the structure to reduce temperature fluctuations;
- well-proportioned and oriented windows and shading that reduce the risk of summer overheating;
- a super-insulated fabric.

Using these techniques can reduce energy consumption substantially.

Efficient spatial planning can also reduce the physical footprint of the building which in turn reduces the materials required to construct and reduces the cost of running the building and means less to dispose of or recycle at the end of the building's useful life. Strategies that seek to extend the life of buildings and their components through reused and/or recycled materials and components will reduce consumption and reduce pressure on landfill life.

Consideration of the economic balance between providing a basement with that of an extra storey for the housing of plant such as heating and ventilation, especially on brown sites is important. Where the previous use was heavy industry and the ground severely polluted, cleaning it before excavation and removal or reuse can be a significant issue.

The selection of materials that coincide with manufacturer's sizes, e.g. plasterboard, and a general understanding of their production processes can make for reduction in waste during the construction process (see Chapter 22).

Additional planting, soft landscape and habitats for fauna in confined areas is important. It is known to be beneficial to human health and well-being and needs fuller attention in the design process.

Economic issues

Consideration should be given to sourcing locally available construction materials and components. This also has social implications because it helps sustain local employment.

The provision of secure cycle storage and changing room and shower facilities goes to reducing the dependency upon the car. Depending upon the size of development, liaising with public transport providers to investigate the possibility of an improved transport system to the site during and after construction can mean reducing the number of car parking spaces provided.

When demolishing old buildings, there can be high-value materials such as carved stone, scarce materials such as copper and slates, and architectural features such as railings, gates and fireplaces that can be reused. Demolition contractors need to produce inventories of what they salvage to make it easier for designers to select recycled materials.

Water harvesting, the provision of local waste and sewerage treatment, combined heat and power plants and local wind power generation should all be considered. While these are generally for larger projects, water harvesting is appropriate for all kinds of development.

Social issues

The main consideration is to promote health, safety and well-being issues both to the users of the building and the local community. The importance of green space has been already mentioned. Designing against crime by working with the local police force and, if they have one, their design team such as based at the Greater Manchester Police can have a significant impact on the reduction of crime. They advise on planting, proposed building layouts, illumination, surveillance and the elimination of escape routes for criminals.

Accessibility to all areas is vital, not just on entry/exit and within the building, but also to the open spaces in the development and to and from the transport facilities for all sectors of the community, be they able-bodied, children, disabled or elderly.

Avoiding sick building syndrome problems in buildings is as important as the selection of building materials and components of low toxicity with minimum hazard to health to reduce the incidence of allergic reactions.

Institutional issues

All design team members should receive training in the concepts of the delivery of sustainable urban developments. These should be audited from time to time and appropriate training and development put in place if found deficient.

Using suppliers and services who can demonstrate sound sustainable development policies and practices, especially in terms of life cycle analysis, can save an inordinate amount of time and research.

31.6.3 The construction phases

Environmental issues

The main consideration is the minimisation of waste in the manufacturing and construction process (see Chapter 22) during the site assembly and reducing energy consumption. While the designer often specifies the material, it is the contractor that procures unless there is a nominated supplier.

Contractors should aim to reduce solid waste well below current standards and as far as possible dispose of waste by recycling the inorganic and composting organic material. Component manufacturers can assist by reducing the amount of packaging consistent with adequate protection during transit and storage prior to installation in the building.

Many of the construction processes are water intensive. Some can be reduced using, for example, gypsum board linings rather than plaster.

The selection of the correct plant for the job can improve energy efficiency. For example, using a larger capacity vehicle than is required will consume more fuel. Careful thought should be given in consultation with the suppliers to have full-load deliveries rather than part loads. This is not always possible because of the quantities needed and problems of storage on site for future work, but it may be that the supplier has to deliver a part load to another customer in the near vicinity and a compromise between all parties found even though this might conflict with just-in-time deliveries.

Damage to flora and fauna during construction can be avoided by minimising the area for temp works and accommodation; minimising excavation and earth movement and destruction of landscape and natural habitats; reinstatement of habitats on completion; arranging temp support for flora and fauna during the construction process; and reintroduction of reinstated habitat after construction.

Economic issues

The main role of the contractor is to deliver the completed building within budget, on time and to the right level of quality. Ideally this should be carried out with as much of the materials, labour and subcontractors resourced locally as possible.

Unnecessary costs are incurred using landfill sites; paying pollution penalties arising from poor practice and environmental management, e.g. escape of fuel; and inadequate protection of trees, landscape and other buildings.

Impacts from transport can be reduced by providing incentives for car-sharing, use of public transport and cycling. Management can use their position to persuade their suppliers and subcontractors to do the same with their employees. On large projects, rail and water transport may be competitive for bulk materials as well as having less impact on the environment.

Social issues

The local community is important. Considerate contractors believe the establishment of good relations with the local community is essential, such as advising on timing and duration of potential inconveniences, the phasing of operations to minimise disruptions, having an effective communication system such as a free newsletter delivered to residents, schools and businesses. A single-point rapid-response procedure for handling complaints means action can be taken promptly.

Other inconvenience issues include the control of dust and fumes and elimination of burning of mats on site; the regular cleaning of streets and pathways; consideration of the impact on the elderly and disabled caused by temporary walkways; scaffolding; and disruption to public transport due to bus stops being moved. The provision of illumination around the site assists in making it safer for the public at night as well as improving security.

Besides looking after the safety of workers and visitors to the site (see Chapter 21), consideration needs to be given to the general public both outside the site and for those that trespass within the site perimeter. Outside issues include falling objects, traffic moving in and out of the site and extra traffic congestion and parking.

The provision of safe viewing areas to allow the public to view progress on site has the double advantage, providing the site is well maintained, of keeping the public advised of progress, and of good public relations.

Institutional issues

The contractor must have environmental processes and procedures in place, e.g. ISO14001 with clear sustainable objectives for all those involved in the project. This responsibility should embrace the social and economic development objectives as well as environmental resource minimisation and efficiency objectives in terms of energy, materials and water management. Partnering between trade contractors and their suppliers is an important element in achieving this as well as the targets identified at the conception stage.

Conclusion

To summarise, the underlying goals of sustainable architecture and construction in this section are:

- use fewer resources – seek to reuse and recycle, minimise fossil fuel energy;
- respect the physical and cultural context – place the re/development in the continuum of history;
- minimise damage to the environmental systems (air, land, water, plants and animals);
- ensure participation of stakeholders in the decision-making process; and
- seek to persuade the client that they care.

36.7 Sustainable return on investment (SROI)

Sustainable return on investment (SROI) is a relatively new concept that the author believes will increasingly play a part in sustainability thinking. The RICS report (Bichard 2015) demonstrates the industry's interest. However, there is a lot of scope for growth and development. SROI measures extra financial value to the resources normally considered in conventional accounts. These are the environmental, economic and social values. As a tool it seeks to reduce inequality and declining environments by improving well-being by involving these values. An example taken from the RICS report was that for every £1 invested in antisocial behaviour workshops for young people in Salford, the project yielded £11.51 in return.

31.6 Final thoughts

If the science is correct and the earth does continue to warm but is held in check at some stage in the future, there will still be climate change within this period. Should we be mitigating against this now and designing and constructing our buildings so they can readily be adapted to the new climatic conditions they will be subjected to in the future? And what if governments are unable to put in measures to halt global warming in time and the tipping point is reached, how will this affect the design of our buildings?

Bibliography

Bichard, E. (2015) *Developing an Approach to Sustainable Return on Investment in the UK, Brazil and the USA*. London: RICS.

Bichard, E. and Cooper, C.L. (2008) *Positively Responsible: How Business Can Save the Planet*. Oxford: Butterworth-Heinemann.

Curwell, S. (2005) *Sustainable Urban Development*, vol. 1, ch. 6. London: Routledge.

Curwell, S.R., Fox, R., Greenberg, M. and March, C.G. (2002) *Hazardous Building Materials*. London: E & FN Spon.

Environment Agency and The Natural Step (2002) *Towards the Sustainable Use of Materials Resources*, 2020 Vision Series No. 4. Environment Agency.

Goodhew, S. (2016) *Sustainable Construction Process: a resouce text*. J. Wiley and Sons.

International Standards of the ISO 14001:2015 Environmental Management.

The Natural Step, *The Four System Conditions*. Available at: thenaturalstep.org (accessed 8 January 2017).

Index